Simple Rotor Analysis through Tutorial Problems

This book discusses various rotor systems, rotor dynamics and dynamics of rotating machinery problems through tutorials. Most of the covered problems can be derived and solved using hand calculations for deeper understanding of the subject. It correlates the examples provided in this book with real machinery where it can be used, and readers can analyse their own simple rotor system based on the variety of examples presented. All problems are supplemented by independent MATLAB® codes for exploring the subject with more ease with graphical outputs.

Features:

- Rotordynamics terminology and phenomena are introduced with very simple rotor-bearing models.
- In-depth analytical dynamic analysis of rotors mounted in flexible bearings and the effect of gyroscopic effects in simple rotor systems are covered.
- Offers the possibility for the reader to reproduce the results and see how the equations are derived and solved in rotor dynamics.
- A few examples of simple rotor-bearing-coupling systems, rotor-bearing-foundation systems, and two-spool rotors are covered.
- Directions are provided to extend the present exercise problems and their solutions.
- Examples are supplemented by MATLAB® codes with detailed solution steps.
- Includes multiple-choice questions and their solutions.

This book is aimed at senior undergraduate/graduate students in mechanical engineering, as well as scientists and practice engineers from the field of rotordynamics, rotating machinery/turbomachinery and aerospace engineering.

Simple Rotor Analysis through Tutorial Problems

Rajiv Tiwari

CRC Press
Taylor & Francis Group
Boca Raton London New York

CRC Press is an imprint of the
Taylor & Francis Group, an **informa** business

Designed cover image: © Rajiv Tiwari

First edition published 2024
by CRC Press
2385 NW Executive Center Drive, Suite 320, Boca Raton FL 33431

and by CRC Press
4 Park Square, Milton Park, Abingdon, Oxon, OX14 4RN

CRC Press is an imprint of Taylor & Francis Group, LLC

ISBN: 978-1-032-55556-0 (hbk)
ISBN: 978-1-032-63820-1 (pbk)
ISBN: 978-1-032-63821-8 (ebk)

DOI: 10.1201/9781032638218

Typeset in Times
by codeMantra

Contents

Preface

This tutorial book is of its first kind in the field of rotor system, rotor dynamics, or the dynamics of rotating machinery. Hardly any tutorial book is available on this topic, so definitely students, researchers and practice engineers, who are new in this field, will welcome this introductory book very well. An attempt is made to correlate examples provided in this book with real machinery, where it has an application. Readers can analyse their own simple rotor system based on a variety of examples presented in this book. Also, every chapter is supplemented with a large amount of multiple-choice questions (MCQs) and their descriptions, so that basic concepts can also be clarified in a capsule form.

The basic idea of writing this book in the form of solution to various exercise problems is that certain problem-solving skills cannot be described in a textbook in a detailed manner. Especially in the rotor system, a lot of concepts can be understood when a variety of different cases are attempted numerically. In this, the approach is not only to reach the end of solution but to try to discuss and have the possibility of exploring more concepts and methods during the process of solving it. Moreover, almost all problems are supplemented by MATLAB codes for exploring the subject with more ease with graphical outputs whenever it is necessary (Additional resources can be found for this publication at: https://www.routledge.com/authors/i16997-rajiv-tiwari). As a cautionary note, these computer codes are not general codes. These help to avoid tedious calculations and to understand step-by-step calculations more easily. Especially, if readers want to try out different numerical inputs of rotor configuration, then it is quite a time-saving exercise without such codes.

In this book, detailed steps are provided so that concepts can be understood more easily. Exploring more on the exercise problems through MATLAB codes so that one should not get stuck in the calculation of numbers. Graphical as well as detailed output of each step help in checking and comprehending the overall procedure. Throughout this book, it is ensured to make a point to the reader to give some hints in which direction they can explore more on a given exercise problem, and it broadens the thinking of reader so that they themselves can formulate new exercise problems and explore them with an understanding of the exercises presented herein. Wherever needed, the problem has been attempted with a fundamental approach, and the necessary background is covered so that reader should not feel gaps in grasping the concepts. However, a detailed and comprehensive coverage of related theory is given in my previous book *Rotor System: Analysis and Identification* by CRC Press, Boca Raton, USA, 2017.

Rotor systems have a lot of practical applications in all fields of industries. Rotor system theories have evolved through solving practical problems using simple mathematical models of various kinds of rotor systems. This tutorial book covers simple aspects of rotor systems, especially for linear analysis. This book begins with historical background and basic terminologies in rotor-bearing systems in Chapter 1 and then analysis of very simple rotor systems in Chapter 2 for the transverse vibration. Chapter 3 covers rolling element bearings (with some advanced analysis) and hydrodynamic bearings to have a theoretical calculation of stiffness and damping for rotordynamic analyses. Chapter 4 covers simple rigid/flexible long rotor mounted on flexible bearings/foundations to obtain critical speeds. The most important aspect of the gyroscopic effect on simple rotor systems is analysed in Chapter 5 using both quasi-static and dynamic approaches. Overall, this book contains around 110 Exercise Problems (a few of them descriptive) and 150 MCQs.

The author would like to acknowledge the contribution of various undergraduate, graduate and research scholar at IIT Guwahati who undertook this particular course on rotor dynamics during the past two and half decades. My heartfelt thanks to the help offered by the graduate students (notably Mr. Aakash Dewangan, Mr Shashikant K. Verma, Mr Thashreef A., Ms. Twinkle Mandawat, Ms. Beni J. Doley, Mr Manpreet Singh and Mr. Aditya S. Gangan), research scholars (Dr Gyan Ranjan, Mr Pantha Pratim Das, Mr Atul K Gautam, Dr Siva SrinivasR, Dr Nilakshi

Sarmah, Dr Prabhat Kumar and Dr D. Gayen) and the faculty at IIT Guwahati. I thank my gurus (Prof. (Late) J. S. Rao, Prof. K. Gupta, Prof. N. S. Vyas, Prof. M. I. Friswell, Prof. A. W. Lees, Dr. Arun Kumar and Prof. R. Markert) who introduced me to this subject and aspire me to excel in this field). I also acknowledge faculties (Prof. Jyoti K Sinha, Prof. Hassan Ouakad, Prof. Athanasios Chasalevris, Prof. F. Dohnal and Prof. Raghu Echempati), researchers and practice engineers (Dr. Soumendu Jana and Mr San Rajendra) who approached me for their own issues related to this topic or interacted with them during conferences, and that gave me a broader perspective of the subject, which is reflected in this book during the illustration of various examples. The difficult period of lock-down, during COVID-19 pandemic, forced me to complete this book early. I thank the publishing house (Dr. Gagandeep Singh, Senior Publisher, and Ms. Aditi Mittal, Editorial Assistant, (she/her) of CRC Press, Taylor and Francis Books India Pvt. Ltd.; Aimée Crickmore (she/her), Production Editor from Taylor & Francis, UK and Sathya Devi (she/her), Production Manager from CodeMantra, India) and its very efficient members for bringing out this book in a very short time and improving the overall quality of this book through rigorous editing. Finally, I thank my family members – my wife Vibha, son Antariksh and daughter Rimjhim – for their patience and supportive nature.

R. Tiwari

About the Author

Dr. Rajiv Tiwari was born in 1967 at Raipur, Madhya Pradesh, India. He graduated with a BE in 1988 (Mechanical Engineering) from Government College Engineering and Technology, Raipur, under Pt. Ravishankar University, Raipur, an M. Tech. (Mechanical Engineering) in 1991 and a PhD (Mechanical Engineering) in 1997 from the Indian Institute of Technology (IIT) Kanpur, India.

He started his career as a lecturer in 1996 at Regional Engineering College, Hamirpur (Himachal Pradesh), India, and worked there for 1 year. In 1997, he joined the Indian Institute of Technology Guwahati, as an assistant professor in the Department of Mechanical Engineering. He worked as a research officer at the University of Wales, Swansea, UK, for 1 year in 2001 on deputation. He was elevated to associate professor in 2002 and to Professor in 2007 at IIT Guwahati. He was the head of the Center of Educational Technology and Institute Coordinator of the National Programme on Technology Enhanced Learning (NPTEL) during 2005–2009, and the National Coordinator of the Quality Improvement Programme (QIP) for engineering college teachers during 2003–2009. He also visited the University of Darmstadt, Germany, under DAAD fellowship during May–July 2011.

He has been deeply involved in various research areas of rotordynamics, especially identifying mechanical system parameters (e.g. bearings, seals, gears and rotor crack dynamic parameters), diagnosing the faults of machine components (e.g. bearings, couplings, gearbox, pumps and induction motors) and applying active magnetic bearings to monitoring the condition of rotating machinery. His research areas also include rolling element bearing design and analysis for high-speed applications. He has completed three projects for the Aeronautical Research & Development Board (ARDB), India, on these topics. He has been offering consulting services for the last several years to Indian industries, like the Indian Space Research Organisation (ISRO), Trivendrum; Combat Vehicle R&D Establishment (CVRDE), Chennai; Gas Turbine Research Establishment (GTRE) Bangalore, National Bearings Company Ltd. (NBC) Jaipur, and Tata Bearings, Kharagpur, as well as other local industries in the northeast of India. One of the European power industries, Skoda Power, Czech Republic, has also consulted him on seal dynamic parameter estimates for steam turbine applications. Dr. Tiwari has authored more than 250 journal and conference papers. He has guided 57 M. Tech. students, and 17 PhD students and 7 more are currently pursuing research projects.

He has successfully initiated and organised a national-level symposium on rotor dynamics (NSRD-2003), four short-term courses on rotor dynamics (2004, 2005, 2008 and 2015) and a national workshop on "Use and Deployment of Web and Video Courses for Enriching Engineering Education" (2007) at IIT Guwahati, India. He has jointly organised an International Conference on Vibration Problems ICOVP 2015 at IIT Guwahati, Vibration Engineering and Technology of Machinery (VETOMAC-2021) jointly with BMS College of Engineering, Bengaluru, VETOMAC-2022 with Tribhuvan University, Nepal and VETOMAC-2023 with Indian Institute of Technology Roorkee. He has developed two web- and video-based freely available online courses under NPTEL: (1) Mechanical Vibration and (2) Rotor Dynamics, and under MHRD-sponsored virtual lab on the "Mechanical Vibration Virtual Lab." He has authored a textbook on *Rotor Systems: Analysis and Identification* from Taylor & Francis Group, CRC Press, USA, 2017. For consecutively last 3 years

(2020 through 2023), Prof. Tiwari has been featured among World's Top 2% Scientists List created by Stanford University. He has recently joined as Associate Editor to the *Journal of Vibration and Control* (JVC), Sage Publications, and *Journal of Vibration Engineering and Technology* (JVET), Springer. Nature Publications. Also, now he is in Worldwide Technical Committee on Rotor Dynamics of International Federation of Machines and Mechanisms (IFToMM). Recently, he has been included in The Bearings Sectional Committee, PGD 13, of Bureau of Indian Standards (BIS), which is responsible for the development of the national and international standards on the 'Magnetic Bearings', 'Aerospace Bearings' and 'Railway Bearings'.

Introduction

Rotor systems are known for high-power density with minimum vibrational energy, which makes them quieter as compared to reciprocating machines. Hence, as compared to reciprocating machines, the rotating machines are preferred for high-speed applications. The application of rotating machinery can be found in a variety of places, such as engines, turbines, generators, pumps, compressors, machine tools, dredgers, reaction flywheels, energy storage flywheels, mobiles, etc.

The reciprocating engines are preferred since they have internal combustion as compared to turbines, which have external combustion. The reciprocating pumps have applications for high-pressure generation, while centrifugal pumps have applications where high volumes need to be pumped. Apart from these in transmission systems, the rotating machinery has advantages, which involve shafts, bearings, couplings, gears, seals, flywheels, etc.

Due to the dynamic nature of the rotor systems apart from the static design, the dynamic design is also very important, especially with respect to critical speeds and instability conditions. The vast majority of the analysis involves finding the natural frequencies of the rotor system in the transverse, axial and torsional vibrations. Sources of instabilities in rotor systems are several, and the most common source is fluid-film bearings. Apart from this, the instability comes from seals, shaft asymmetry, rotor asymmetry, shaft material damping, bearing cage instability, etc. To perform such analysis, the rotor systems need to be modelled. As compared to the structural vibration, the rotor system differs due to gyroscopic effects, which makes the rotor whirl in the same direction as the spin speed of the rotor (i.e., forward whirl) and in the opposite direction (i.e., the backward whirl), along with associated whirl natural frequencies and critical speeds. The scope of this book is limited to simple rotor analysis for its dynamics.

1 A Brief History of Rotor Systems and Basic Terminologies

Exercise 1.1 Who was the famous person in 1869 who first analysed the rotor dynamics problem but wrongly predicted that it is impossible to operate industrial rotors at very high speeds?

***Solution*:** He was a famous Scottish mechanical engineer William John Macquorn Rankine (5 July 1820–24 December 1872). He predicted critical speed correctly by equating elastic force kx to centrifugal force $mr\omega^2$ of a whirl mass, m, attached by a spring, k, at a radius of r. The condition that these two forces are equal gives the critical speed condition $\omega^2 = k/m$. Since centrifugal force is proportional to the square of spin speed so he concluded that response will grow (unbounded) after critical speed. However, considering centrifugal force or Coriolis force as real force, it gives erroneous results (Meriam and Kraige, 2013) and he predicted instability above critical speed (supercritical speeds). Detailed discussion can be found in Vance et al. (2010) and Tiwari (2017).

Exercise 1.2 Who was the first engineer who experimentally demonstrated and reported that it was possible to rotate the rotor safely at very high speeds?

***Solution*:** Gustaf de Laval, a Swedish engineer, ran a steam turbine to supercritical speeds in 1889 in a stable condition. However, this critical speed invension was contrary to Rankine's prediction, i.e. unstable regimes at supercritical speeds.

Exercise 1.3 Who clarified theoretically the confusion of whether it is possible to rotate the rotor safely above critical speeds?

***Solution*:** Jeffcott (1919) developed a simple rotor model to clarify that it is possible to rotate the rotor safely above critical speeds. In honour of his name, the rotor model is now called as Jeffcott rotor model. Herein, he first published a clear understanding of rotor behaviour due to unbalance in the presence of damping using a mathematical model. The model consists of a massless flexible shaft simply supported at ends, and a rigid disc is mounted at its mid-span (Figure 1.1). Variant of Jeffcott rotor model in various ways has been extensively used by many researchers to understand several rotordynamic phenomena, and it is still being used for newer investigations. Already, more than 100 years have passed since the publication of this classical paper.

Exercise 1.4 Who were the first to use the terms the *whirling motion* and the *critical speed*?

***Solution*:** Rankine first used the term whirling motion (refer Figure 1.2) and Dunkerley used the term critical speed. Dunkerley (1894) showed that the critical speed of a rotating simply supported shaft and its natural frequency of transverse vibration are same. Relations of two different motions

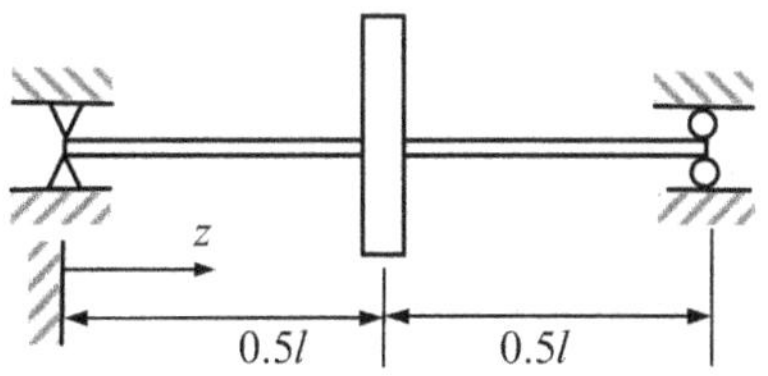

FIGURE 1.1 A simply-supported (Jeffcott) rotor system.

DOI: 10.1201/9781032638218-1

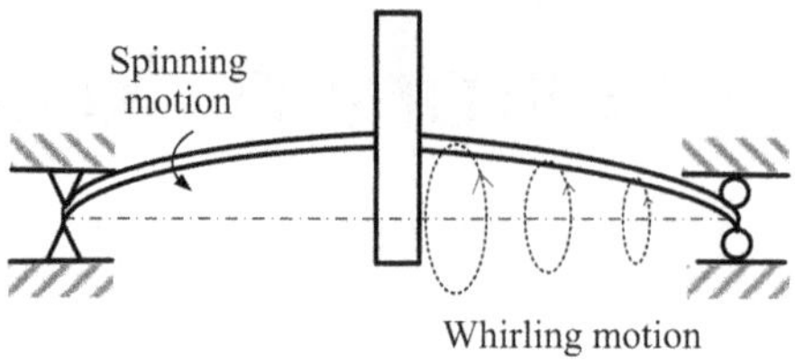

FIGURE 1.2 Whirling and spinning motion of a rotor system.

were provided, i.e. the former as orbital motion and the latter as oscillatory motion. He clarified that any unbalance would give force, which will excite natural frequency to give resonance condition of critical speeds, a synchronous whirl condition.

Exercise 1.5 Define the natural frequency and the critical speed of a rotor system.

***Solution*:** The natural frequency is a system property and depends upon the system stiffness and the mass distribution. When a dynamic system is given perturbation, it oscillates with a frequency equal to its natural frequency. Rotors can have the longitudinal (axial), transverse (bending) and/or torsional natural frequencies depending on the application in which they are used and their associated inertia and stiffness distribution.

In a rotor, due to unbalance, it gets a force called centrifugal force with an excitation frequency equal to its spin speed. When the spin speed coincides with the natural frequency of the system, it is called the critical speed (refer Figure 1.3). It is a resonance condition. In general, a rotor can have longitudinal (axial), transverse (bending) and torsional critical speeds.

Exercise 1.6 How many transverse critical speeds would there be for a two-disc (point masses), massless-flexible shaft rotor system?

***Solution*:** There will be two transverse critical speeds for a two-disc (point masses) massless-flexible shaft rotor system. It will have two degrees of freedom (DOFs) corresponding to transverse

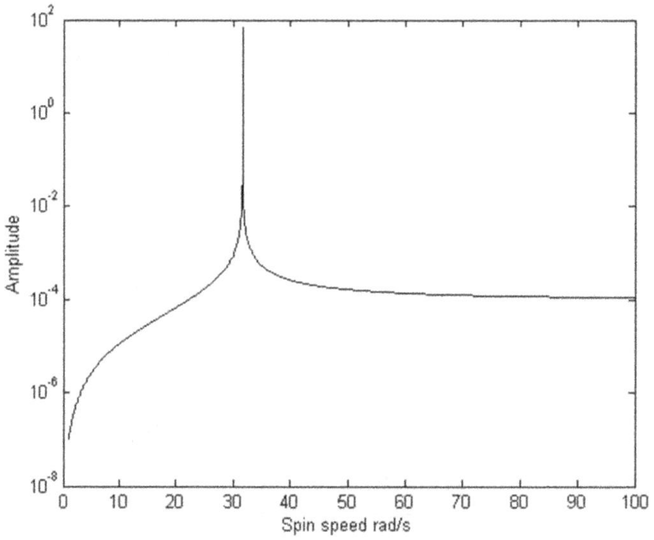

FIGURE 1.3 Plot of rotor amplitude with spin speed showing critical speed corresponding to peak amplitude.

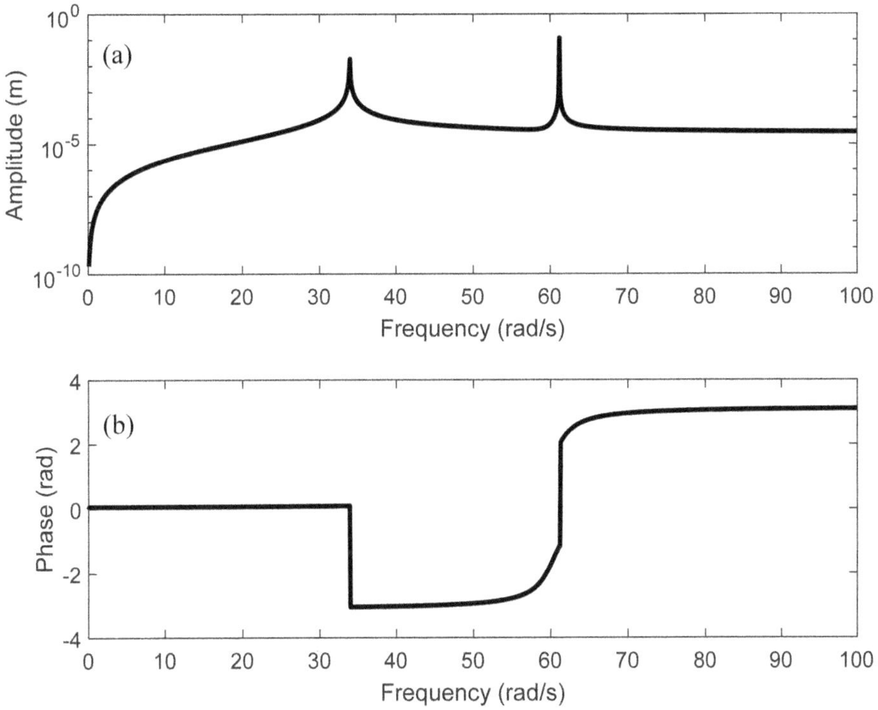

FIGURE 1.4 Variation of amplitude and phase with spin speed of the shaft showing (a) two critical speeds and (b) associated phase changes.

translational motion. If the discs are thin discs, then they will have four critical speeds since they will have diametral mass moments of inertia and will have four DOFs corresponding to transverse translational and rotational motions. A typical Bode plot showing two critical speeds is shown in Figure 1.4. At critical speeds, the phase change of order of π rad can be seen.

Exercise 1.7 Is natural frequency dependent on the spin speed of the rotor? If yes, then under what conditions?

***Solution*:** If we consider the gyroscopic effect, then the natural frequency of the rotor will depend upon the spin speed. Apart from this, often bearing dynamic parameters are speed-dependent (e.g. the hydrodynamic bearings) and that makes natural frequency dependent on the rotor spin speed.

Exercise 1.8 What is the most common cause of a synchronous motion in a rotor system?

***Solution*:** In the synchronous motion of the shaft, the orbital speed and its own spin speed are equal. The sense of rotation of the shaft spin and the whirling are also the same. The unbalanced force, in general, leads to synchronous whirl condition. Such force gives basically a forced response. Under this motion, the shaft will not have flexural vibration but whirls as a rigid body about the bearing axis.

Exercise 1.9 In a synchronous whirl of a rotor, what is the whirl frequency?

***Solution*:** The spin speed of the rotor will be the whirl frequency for an unbalanced rotor system, which leads to the synchronous whirl. For a balanced rotor, this condition will be achieved only at critical speed. In general, a rotor has asynchronous whirl (spin speed and whirl frequency are not the same), especially when the gyroscopic effect is present in the system.

Exercise 1.10 For a perfectly balanced rotor rotating at a speed, what is the frequency of whirl when it is perturbed from its equilibrium?

***Solution*:** Since no external force is present hence it will whirl (vibrate) at its natural frequency.

Exercise 1.11 In a general motion of a rotor, what is the whirl frequency?

***Solution*:** The asynchronous whirl motion occurs in the perfectly balanced rotor when it is perturbed. And due to this, it will have the whirl frequency as one of the natural frequencies of the rotor system and that may not be equal to the spin speed of the shaft.

Exercise 1.12 Do bearings and foundations have any effect on the critical speed of a rotor system?

***Solution*:** Bearings give stiffness and damping to the rotor system, and foundations give mass, stiffness and damping to the rotor system. So, yes, they affect the critical speed of a rotor system. We will see them in Chapter 4 through some simple examples.

Exercise 1.13 What is the internal and external damping in a rotor system?

***Solution*:** If the energy dissipation is due to the external interactions of the system like viscosity and from drag then it is called the external damping. The direction of external damping does not change with the rotor speed.

If the energy dissipation is due to the internal interactions of the rotor system, like internal friction between the inter-molecular layers (or between disc and shaft interface or shaft crack faces rubbing together) as a result of differential straining, then it is called the internal damping (or material (hysteretic) damping). The direction of internal damping does change with the rotor speed, and this may lead to instability in the system.

Exercise 1.14 How do you distinguish between rigid and flexible rotors?

***Solution*:** When the rotor rotates much below its first critical speed, it is considered a rigid rotor. But when it rotates near or above its first critical speed, it is considered as the flexible rotor. Near or above critical speed, flexible mode of vibration of rotor shafts are present and that has the effect of changing unbalanced force distribution in the rotor system. Due to this the balancing of the rigid and flexible rotors are quite different. In a rigid rotor case for dynamic balancing, only two balancing planes are sufficient, but in the case of a flexible rotor, it requires more than two balancing planes, depending on the speed of operation.

Exercise 1.15 Is there any difference between rigid and flexible rotor dynamic balancing?

***Solution*:** In general, two-plane balancing is used in the case of rigid rotors, and $(N+2)$ balancing planes are used in the case of flexible rotor system, where N is the number of natural modes that we want to balance. As mentioned in the answer to Exercise 1.14, the balancing force changes with vibration mode of rotor, so rigid and flexible rotor balancing have differences.

Exercise 1.16 What is a Campbell diagram?

***Solution*:** Due to various reasons, the whirl frequency of a rotor system changes with speed. For example, due to the gyroscopic effect or speed-dependent bearing dynamic parameters.

A Campbell diagram (refer Figure 1.5) represents the variation of the whirl frequency of a rotor system with its spin speed. It is used to locate the critical speed of the system. Also, often it is used to depict the instability regions in linear rotor systems due to negative damping or positive logarithmic decrement.

Exercise 1.17 Explain the inertia asymmetry and shaft asymmetry in a rotor system. What are the effects of these asymmetries on rotor behaviour?

***Solution*:** When inertias of rotor system in two principal radial directions are different (e.g., bladed disc), then it is called inertia asymmetry. Whereas, when the stiffness of the shaft system in two principal radial directions is different (e.g., non-circular shaft), it is called stiffness asymmetry. Both asymmetries give time-dependent system parameters and that leads to parametrically excited rotor systems. Such systems are prone to instability.

Exercise 1.18 What are the different active control mechanisms that can be applied in rotor systems?

***Solution*:** In general, for suppressing excessive vibrations, dampers are used in rotor systems. Electrorheological and magnetorheological fluids are often used for active control of rotor. Such fluid damping properties can be changed by supplying appropriate electric or magnetic field in very short time. Active magnetic bearings, which are non-contact type bearings, are being used to provide not only variable damping but stiffness based on feedback of rotor vibrations.

Exercise 1.19 Splitting of whirl natural frequencies occurs due to which factors in rotor systems?

***Solution*:** Commonly due to gyroscopic effect the rotors have the forward and backward whirls (refer Figure 1.5). Since the gyroscopic effect changes with the spin speed of the shaft, it is found that the whirl frequency of the rotor systems has splitting behaviour; one increases with speed and another decreases with speed. The one that increases is called the forward whirl, and the one that decreases is called the backward whirl. Such behaviour can be seen when bearings have anisotropy or rotor/shaft have asymmetry (either of inertia of rotor or stiffness of shaft), for example, a transverse crack on the shaft surface.

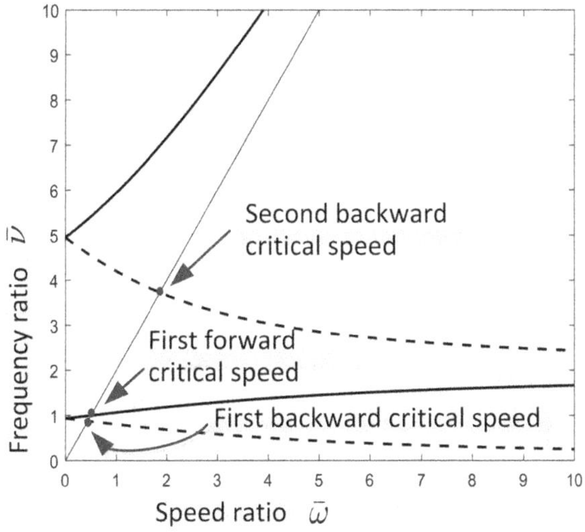

FIGURE 1.5 A typical Campbell diagram for the rotor system showing variation of whirl natural frequency with the spin speed of the shaft also showing critical speeds at intersection points (solid line: forward whirl, dashed line: backward whirl).

Exercise 1.20 Define the instability of a rotor system.

***Solution*:** It is a free vibration phenomenon. When the rotor system vibrates even in the absence of unbalanced effects (or any other external forces), it results in high levels of noise and component stress and a corresponding reduced fatigue life. This is called instability. The dynamic instability can be thought of as a negative damping in the system, wherein for every cycle of oscillations there is an accumulation of energy. So system vibration amplitude increases continuously, but it requires some source of energy from where the system will pump energy to the system, e.g. high-pressure steam or working fluid, transmission of high power, etc. In linear systems, the magnitude of these vibrations tends towards infinity, although in practice, shaft vibrations are often limited by the system's non-linearity. For more detailed treatment a book by Tiwari (2017) may be referred.

Exercise 1.21 What is Sommerfeld effect in transient rotor systems?

***Solution*:** Rotor dynamic systems are often analysed with ideal drive assumption. However, all drives are essentially non-ideal, i.e., they can only provide a limited amount of power. One basic fact often ignored in rotor dynamics studies is that the drive dynamics has complex coupling with the dynamics of the driven system. Increase in drive power input near resonance may contribute to increasing the transverse vibrations rather than increasing the rotor spin, which is referred to as the Sommerfeld effect.

Exercise 1.22 What is Morton's effect?

***Solution*:** Another phenomenon can occur in rotating machinery and give data similar to that resulting from a light rub. Inspection of the machinery affected by this condition will show that no rub is in fact occurring. Instead, it has been postulated that temperature gradients ("hot spots") occur on the shaft's journal surface as the result of high viscous shear stress in the bearing lubricant. Consequently, the shaft bows just as if a hot spot occurred from a rub, and a very similar vibration response is observed. This mechanism was first described in a 1994 paper by Keogh and Morton and has subsequently become known as the "Morton Effect." It generally occurs only under very specific conditions; in particular, on machines incorporating an overhung rotor design and which are heavily loaded. So Morton effect is due to journal thermal gradients (Keogh and Morton, 1994).

Exercise 1.23 What is the curve veering in the Campbell diagram? Why does it happen in rotor-bearing systems?

***Solution*:** Campbell diagram shows the whirl frequency variation with the spin speed of the rotor. In general, these whirl frequency curves do not intersect. However, in special cases, these do appear to intersect, but they swap the whirl frequency at the point of intersection (i.e., two modes form a coupled system and the curves repel each other, avoiding an intersection). Mathematically, it is the so-called curve veering in the eigenvalue problem.

Figure 1.6 shows a typical Campbell diagram with curve veering phenomenon, on which solid lines are for forward whirl and dashed lines for backward whirl. At intersection point of two frequency curves, swapping of the forward and backward whirl can be seen.

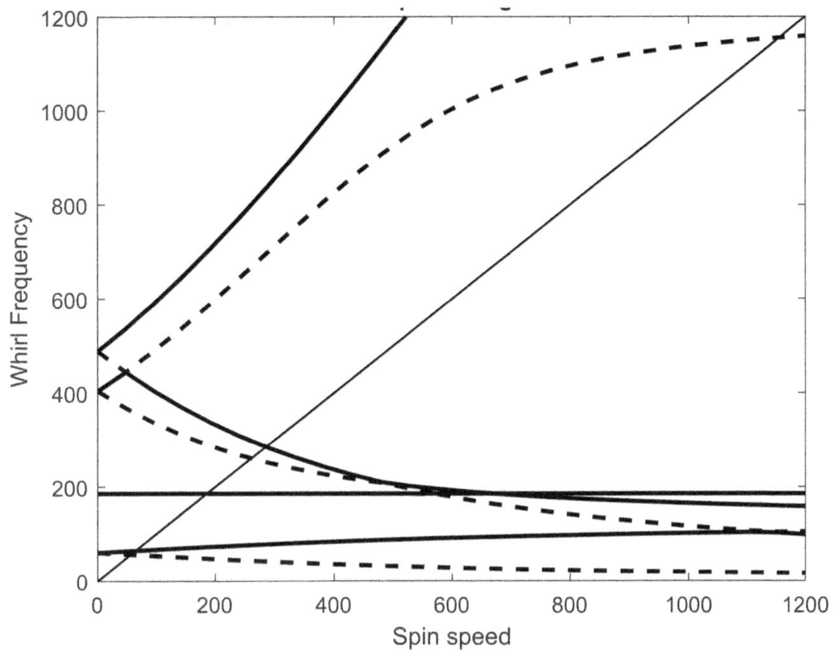

FIGURE 1.6 Campbell diagram showing curve veering phenomenon at intersection points of natural frequency curves.

Exercise 1.24 Multiple Choice Questions (MCQs)

i. In a rotor system, the whirl natural frequency depends on the spin speed of the rotor due to
   A. Gyroscopic couple only
   B. Fluid-film bearing only
   C. Both gyroscopic couple and fluid-film bearing
   D. Whirl natural frequency does not depend on the spin speed of the rotor

ii. In a rotor system, the backward whirl refers to
   A. sudden change of spin speed direction
   B. when rotor spin speed and whirling motion have opposite sense of rotation
   C. sudden reduction in spin speed value
   D. when both rotor spin speed and whirling motion have reversal in sense of rotation

iii. In a rotor system, instability may occur due to
   A. seals, bearings, shaft asymmetry, gyroscopic effect
   B. bearings, shaft asymmetry, seals, rubs
   C. rubs, bearings, gyroscopic effect, rotor asymmetry
   D. gyroscopic effect, rotor asymmetry, internal damping, seals

iv. Rotors can have following types of vibration
   A. Transverse and axial
   B. Torsional and transverse
   C. Axial and torsional
   D. Transverse, axial and torsional

v. Which of the following statement is true for a rotor system?
   A. rotational kinetic energy to be maximum
   B. vibration kinetic energy to be maximum
   C. both rotational and vibration kinetic energies to be maximum
   D. potential energy to be maximum

vi. Whirling motion in a rotor system refers to
   A. the movement of the centre of the deflected disc (or discs) in a plane perpendicular to the bearing axis
   B. spinning motion of disc about its own axis
   C. tilting motion of a disc about its diameter
   D. torsion motion of disc relative to shaft

vii. The whirl natural frequency in a perfectly balanced rotor mainly depends on
   A. effective stiffness only
   B. effective stiffness and damping
   C. effective damping only
   D. rotor spin speed only

viii. The whirl frequency in an unbalanced rotor mainly depends on
   A. effective stiffness and damping
   B. effective stiffness only
   C. effective damping only
   D. rotor spin speed only

ix. In rotor system, the self-centering phenomenon is
   A. rotation of rotor about its centre of gravity
   B. rotation of rotor about its bearing axis
   C. rotation of rotor about its geometrical centre
   D. rotation of rotor about its centre of percussion

x. Campbell diagram is
   A. a plot between whirl natural frequency and rotor spin speed
   B. a plot of vibration amplitude with rotor spin speed
   C. a plot of vibration phase with rotor spin speed
   D. a plot of rotor orbit

xi. In Campbell diagram, intersection of curves with $\nu=\omega$ line gives (ν whirl natural frequency and ω spin rotor speed)
   A. critical speeds
   B. whirl natural frequencies
   C. the phase of vibration response
   D. the amplitude of vibration response

xii. The critical speed is a
   A. free vibration phenomenon
   B. forced vibration phenomenon
   C. transient vibration phenomenon
   D. unstable vibration phenomenon

xiii. Synchronous whirl condition prevails when
   A. the rotor whirl frequency is same as rotor spin speed
   B. the rotor system is not operated at the critical speed
   C. the whirl frequency is not same as spin speed
   D. the whirl frequency is very slow as compared to rotor spin speed

xiv. A Jeffcott rotor is
   A. a flexible massless shaft with rigid disc at mid-span mounted on simply supports
   B. a flexible massless shaft with offset rigid disc mounted on simply supports
   C. a rigid shaft with rigid disc at mid-span mounted on flexible bearings
   D. a rigid shaft with offset rigid disc mounted on flexible bearings

xv. Condition monitoring of rotating machinery is related with
   A. predicting faults in machinery based on the measurement of physical parameters from machines
   B. preventive periodic maintenance of machinery
   C. investigating the cause of faults in machinery once it has occurred
   D. maintenance of machinery after failure

xvi. Unbalance is defined as
   A. offset of disc centre of gravity from its centre of rotation
   B. offset of disc centre of gravity from bearing centre
   C. product of disc mass and eccentricity
   D. product of eccentricity and square of spin speed of rotor

xvii. For a perfectly balanced Jeffcott rotor system, which is rotating at certain speed, ω, if the disc is given a small initial disturbance in transverse direction, the frequency of whirl will be
   A. same as the spin speed, ω
   B. whirl natural frequency
   C. zero
   D. depending on the initial disturbance

REFERENCES

Dunkerley, S., 1894, On the whirling and vibrations of shafts, *Philosophical Transactions of the Royal Society A*, **185**(1), 279–360. **https://doi.org/10.1098/rsta.1894.0008**

Jeffcott, H.H., 1919, The lateral vibration of loaded shafts in neighbourhood of a whirling speed: The effect of want of balance, *Philosophical Magazine*, Series 6, **37**, 304–314.

Keogh, P.S., and Morton, P.G. 1994, The dynamic nature of rotor thermal bending due to unsteady lubricant shearing within bearing, *Proceedings of the Royal Society of London Series A,* **445**, 273–290.

Meriam, J.L., and Kraige, L.G., 2013, *Engineering Mechanics Vol. 2: Dynamics*, Seventh Edition. Chichester: John Wiley and Sons.

Tiwari, R., 2017, *Rotor Systems: Analysis and Identification*. Boca Raton, FL: CRC Press.

Vance, J.M., Zeidan, F.Y., and Brian G. Murphy, 2010, *Machinery Vibration and Rotordynamics*. Hoboken, NJ: John Wiley and Sons.

FINAL REMARKS

This chapter basically introduces few questions and answers to history of rotor dynamics and related phenomena before analysing them in more detail in subsequent chapters through exercise problems. Hopefully, with the content of this chapter the readers will have curiosity and inquisitiveness about the subject to learn more about them. Few descriptive and MCQs helps readers to get introduce with the subject in very elementary level. MCQs are very simple and so more description of them is not given in this chapter but in subsequent chapters the MCQs are also discussed to get overall understanding of related concepts.

ANSWER TO MCQs

1.24.	i. C	ii. B	iii. B	iv. D	v. D	vi. A	vii. A
	viii. D	ix. A	x. A	xi. A	xii. B	xiii. A	xiv. A
	xv. A	xvi. C	xvii. B				

2 Transverse Vibrations of Simple Rotor Systems

Exercise 2.1 For a single-degree-of-freedom damped rotor system, obtain an expression for the frequency ratio ($\bar{\omega} = \omega/\omega_{nf}$) for which the damped response amplitude reaches maximum (i.e. location of the critical speed). Show that it occurs always at frequency ratio of more than one.

Solution: An expression for the unbalanced damped rotor response is given as (refer Tiwari (2017))

$$\bar{Y} = \frac{Y}{e} = \frac{\bar{\omega}^2}{\sqrt{\left(1-\bar{\omega}^2\right)^2 + (2\zeta\bar{\omega})^2}} \tag{2.1}$$

where e is eccentricity, the response becomes maximum, when the expression (2.1) becomes maximum. For undamped case (with damping ratio $\zeta = 0$), it occurs at $\bar{\omega} = 1$, in which case the denominator becomes equal to zero. But with damping, the denominator cannot be zero for a real value of $\bar{\omega}$, and it is zero for only complex value of $\bar{\omega}$, which is not a feasible solution. So, we need to differentiate the expression (2.1) and equate it to zero to find the value of $\bar{\omega}$ for which the slope of the $\bar{Y}$ versus $\bar{\omega}$ curve is zero.

$$\frac{d}{d\bar{\omega}}\left\{\bar{Y}\right\} = 0 \tag{2.2}$$

$$\text{or} \quad \frac{d}{d\bar{\omega}}\left\{\frac{\bar{\omega}^2}{\sqrt{(1-\bar{\omega}^2)^2 + (2\zeta\bar{\omega})^2}}\right\} = 0 \tag{2.3}$$

$$\text{or} \quad \frac{\left\{\sqrt{(1-\bar{\omega}^2)^2 + (2\zeta\bar{\omega})^2}\,(2\bar{\omega})\right\} - \left[\bar{\omega}^2 \frac{d}{d\bar{\omega}}\left\{\sqrt{(1-\bar{\omega}^2)^2 + (2\zeta\bar{\omega})^2}\right\}\right]}{\left\{\sqrt{(1-\bar{\omega}^2)^2 + (2\zeta\bar{\omega})^2}\right\}^2} = 0$$

$$\text{or} \quad \left\{\sqrt{(1-\bar{\omega}^2)^2 + (2\zeta\bar{\omega})^2}\,(2\bar{\omega})\right\} - \left[\left\{\frac{\bar{\omega}^2}{2}\left\{(1-\bar{\omega}^2)^2 + (2\zeta\bar{\omega})^2\right\}^{-\frac{1}{2}}\right\}\left\{\begin{matrix} 2(1-\bar{\omega}^2)(-2\bar{\omega}) + \\ 4\zeta^2 2\bar{\omega} \end{matrix}\right\}\right] = 0$$

$$\text{or} \quad 2\bar{\omega}\left\{(1-\bar{\omega}^2)^2 + (2\zeta\bar{\omega})^2\right\} - 2\bar{\omega}^2\left\{-(1-\bar{\omega}^2)\bar{\omega} + 2\zeta^2\bar{\omega}\right\} = 0$$

$$\text{or} \quad 1 + \bar{\omega}^4 - 2\bar{\omega}^2 + 4\zeta^2\bar{\omega}^2 + \bar{\omega}^2 - \bar{\omega}^4 - 2\zeta^2\bar{\omega}^2 = 0$$

$$\text{or} \quad -\bar{\omega}^2 + 2\zeta^2\bar{\omega}^2 + 1 = 0 \Rightarrow \bar{\omega}^2 = \frac{1}{1-2\zeta^2} \tag{2.4}$$

which gives

$$\bar{\omega} = \sqrt{1/(1-2\zeta^2)} \tag{2.5}$$

DOI: 10.1201/9781032638218-2

This is the expression for the frequency ratio at which damped response becomes maximum. Hence, it is always more than 1 for the underdamped system, i.e. $0<\zeta<1$. For undamped case, it gives maximum amplitude at $\bar{\omega}=1$. Now, Eq. (2.5) can be written as

$$\omega=\sqrt{1/(1-2\zeta^2)}\omega_{nf} \tag{2.6}$$

For the underdamped system $\zeta>0$, hence ω is always more than ω_{nf} at resonance, however, difference will be small. Figure 2.1 shows variation of amplitude ratio with respect to frequency ratio for various values of damping ratio. It can be seen that for $\zeta \geq 1/\sqrt{2}$ the amplitude ratio asymptotically approaches equal to value of 1. Figure 2.2 shows variation of amplitude ratio and phase with respect to frequency ratio for various values of damping ratio. It can be seen that the phase is $\pi/2$ at resonance (at frequency ratio = 1) for all damping values.

Exercise 2.2 Let us define a new frequency ratio in terms of the damped natural frequency, that is, $\bar{\omega}_d=\omega/\omega_{nf_d}$ with $\omega_{nf_d}=\omega_{nf}\sqrt{1-\zeta^2}$. Obtain an expression for the amplitude ratio (Y/e) and the

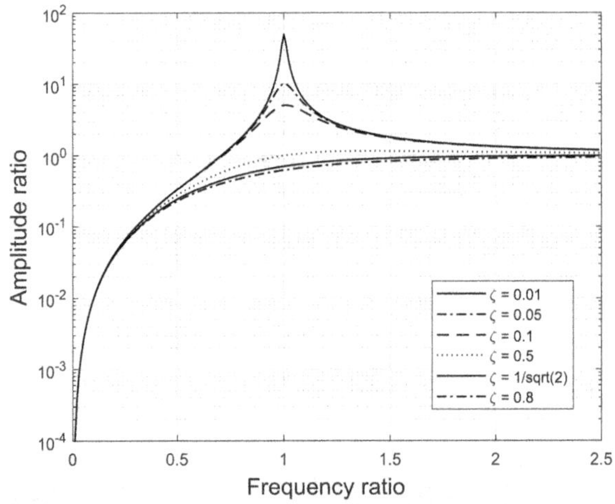

FIGURE 2.1 Variation of amplitude ratio with respect to frequency ratio for various values of damping ratio.

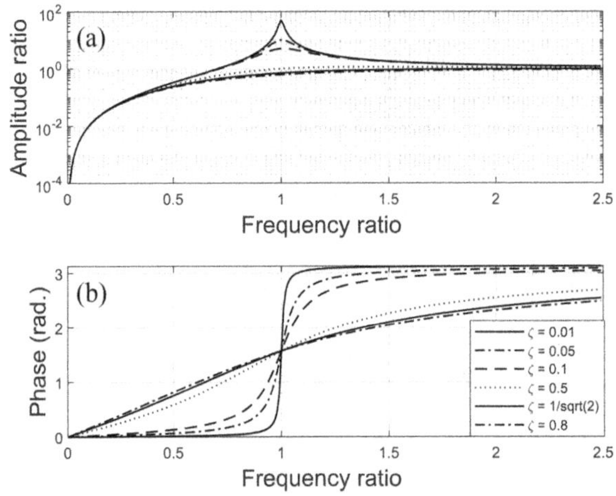

FIGURE 2.2 Variation of amplitude ratio and phase with respect to frequency ratio for various values of damping.

phase, ϕ, in terms of the new frequency ratio defined. Plot the amplitude ratio and the phase versus the new frequency ratio and discuss the results. Obtain an expression for the frequency ratio ($\bar{\omega}_d = \omega / \omega_{nf_d}$) for which the damped response amplitude reaches maximum. In the expression so obtained of frequency ratio, what is the maximum value of damping ratio that is feasible for the under-damped system?

Solution: A new frequency ratio is defined as

$$\bar{\omega}_d = \frac{\omega}{\omega_{nfd}} = \frac{\omega}{\omega_{nf}\sqrt{1-\zeta^2}} = \frac{\bar{\omega}}{\sqrt{1-\zeta^2}} \tag{2.7}$$

Rearranging the aforementioned equation, we get

$$\bar{\omega} = \bar{\omega}_d \sqrt{1-\zeta^2} \tag{2.8}$$

The expression for the amplitude ratio and the phase in terms of the conventional frequency ratio, $\bar{\omega}$, is given by

$$\frac{Y}{e} = \frac{\bar{\omega}^2}{\sqrt{\left(1-\bar{\omega}^2\right)^2 + (2\zeta\bar{\omega})^2}} \quad \text{and} \quad \phi = \tan^{-1}\left(\frac{2\zeta\bar{\omega}}{1-\bar{\omega}^2}\right) \tag{2.9}$$

On substituting for $\bar{\omega}$ from (2.8) into (2.9), we get

$$\frac{y}{e} = \frac{\bar{\omega}_d^2\left(1-\zeta^2\right)}{\sqrt{\left\{1-\bar{\omega}_d^2\left(1-\zeta^2\right)\right\}^2 + 4\zeta^2\bar{\omega}_d^2\left(1-\zeta^2\right)}} \quad \text{and} \quad \phi = \tan^{-1}\left\{\frac{2\zeta\bar{\omega}_d\left(1-\zeta^2\right)^{0.5}}{1-\bar{\omega}_d^2\left(1-\zeta^2\right)}\right\} \tag{2.10}$$

The frequency ratio at which the response becomes maximum is given by (on differentiating Eq. (2.10) and equating to zero as per Exercise 2.1),

$$\bar{\omega} = \sqrt{1/(1-2\zeta^2)} \tag{2.11}$$

From (2.8), Eq. (2.11) becomes,

$$\begin{aligned} \bar{\omega} &\equiv \bar{\omega}_d\sqrt{1-\zeta^2} = \sqrt{1/(1-2\zeta^2)} \\ \text{or} \quad \bar{\omega}_d &= \sqrt{\frac{1}{\left(1-\zeta^2\right)\left(1-2\zeta^2\right)}} \end{aligned} \tag{2.12}$$

From the above, in order to have real value of frequency ratio for the underdamped system, we have the following condition

$$\frac{1}{\left(1-\zeta^2\right)\left(1-2\zeta^2\right)} > 0$$

For underdamped systems, we have $0 < \zeta < 1$ so that $\left(1-\zeta^2\right)$ is always positive. So for above condition to satisfy, we have

$$\text{or} \quad 1-2\zeta^2 > 0 \quad \text{or} \quad \zeta < \frac{1}{\sqrt{2}}$$

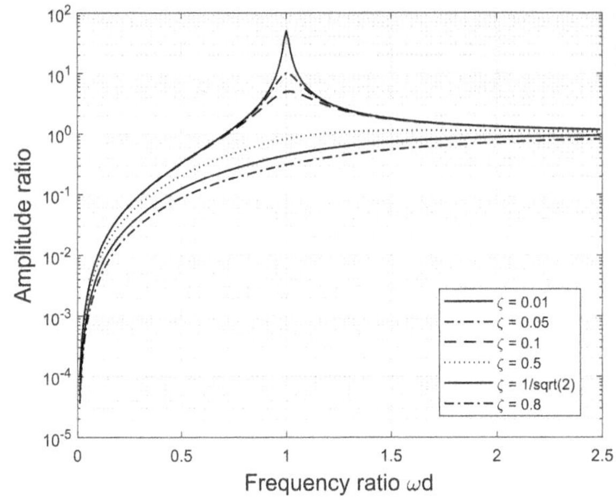

FIGURE 2.3 Variation of amplitude ratio with respect to damped frequency ratio for various values of damping ratio.

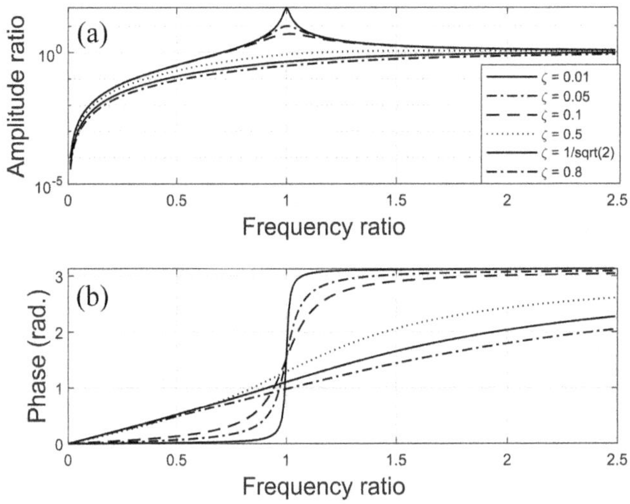

FIGURE 2.4 Variation of amplitude ratio and phase with respect to damped frequency ratio for various values of damping ratio.

which is same as in Exercise 2.1. Figure 2.3 shows variation of amplitude ratio with respect to damped frequency ratio for various values of damping ratio. It can be seen that for $\zeta \geq 1/\sqrt{2}$ the amplitude ratio asymptotically approaches equal to value of 1. Figure 2.4 shows variation of amplitude ratio and phase with respect to damped frequency ratio for various values of damping ratio.

Exercise 2.3 Obtain transverse critical speeds of a cantilever rotor system as shown in Figure 2.5. Take the mass of the disc as $m = 10$ kg and the diametral mass moment of inertia as $I_d = 0.02$ kg-m^2. The shaft diameter, d, is 10 mm, and the total length of the shaft span, l, is 0.2 m. The shaft is assumed to be massless, and its Young's modulus $E = 2.1 \times 10^{11}$ N/m^2. Ignore the gyroscopic effect and take one plane motion only. Influence coefficients are given as $\alpha_{yf} = l^3/3EI; \alpha_{yM} = \alpha_{\varphi f} = l^2/2EI$; $\alpha_{\varphi M} = l/EI$, where I is the second moment of area of the shaft cross-section. Subscripts f and M represent force and moment, respectively; y and ϕ represent the translational and rotational displacements, respectively.

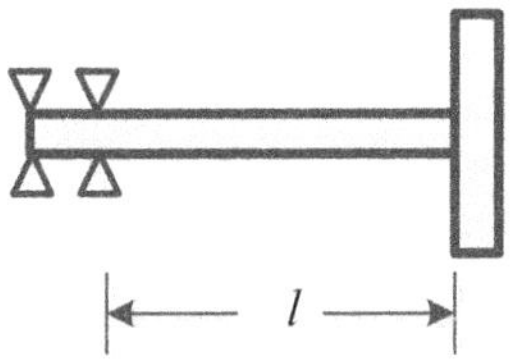

FIGURE 2.5 A cantilever shaft with a disc at the free end.

***Solution*:** We will attempt this problem in three ways: (a) coupled rotational and translational motions; (b) For a point mass and (c) uncoupled rotational and translational motions.

a. Coupled rotational and translational motion: Influence coefficients for a cantilever beam are given as follows (refer Appendix 2.1)

$$\alpha_{xf} = l^3/3EI = 2.5869 \times 10^{-5} \text{ m/N}; \quad \alpha_{\varphi M} = l/EI = 1.9402 \times 10^{-3} \text{ 1/N-m};$$

and

$$\alpha_{xM} = \alpha_{\varphi f} = l^2/2EI = 1.9402 \times 10^{-4} \text{ 1/N}.$$

For the present problem, we have to neglect the gyroscopic effect. Hence, motions in orthogonal transverse planes are uncoupled. Equations of motion in a single plane can be written as (refer Tiwari (2017)):

$$\begin{bmatrix} m & 0 \\ 0 & I_d \end{bmatrix} \begin{Bmatrix} \ddot{x} \\ \ddot{\varphi}_y \end{Bmatrix} + \begin{bmatrix} \alpha_{xf} & \alpha_{xM} \\ \alpha_{\varphi f} & \alpha_{\varphi M} \end{bmatrix}^{-1} \begin{Bmatrix} x \\ \varphi_y \end{Bmatrix} = \begin{Bmatrix} 0 \\ 0 \end{Bmatrix} \tag{2.13}$$

where the inverse of influence coefficient matrix represents the stiffness matrix, and x and φ_y are the transverse translational and rotational displacements. In the free vibration during the simple harmonic motion, we have

$$\left(-\omega_{nf}^2 \begin{bmatrix} m & 0 \\ 0 & I_d \end{bmatrix} + \begin{bmatrix} \alpha_{xf} & \alpha_{xM} \\ \alpha_{\varphi f} & \alpha_{\varphi M} \end{bmatrix}^{-1} \right) \begin{Bmatrix} x \\ \varphi_y \end{Bmatrix} = \begin{Bmatrix} 0 \\ 0 \end{Bmatrix} \tag{2.14}$$

where ω_{nf} is the natural frequency of the system. For the non-trivial solution, we have

$$\left| -\omega_{nf}^2 \begin{bmatrix} m & 0 \\ 0 & I_d \end{bmatrix} + \begin{bmatrix} \alpha_{yf} & \alpha_{yM} \\ \alpha_{\varphi f} & \alpha_{\varphi M} \end{bmatrix}^{-1} \right| = 0 \tag{2.15}$$

which gives the frequency equation, as follows:

$$mI_d\omega_{nf}^4(\alpha_{xf}\alpha_{\varphi M} - \alpha_{\varphi f}^2) - \omega_{nf}^2(m\alpha_{xf} + I_d\alpha_{\varphi M}) + 1 = 0 \tag{2.16}$$

On substituting numerical values for the present problem, it gives

$$2.509325 \times 10^{-9}\omega_{nf}^4 - 2.97482 \times 10^{-4}\omega_{nf}^2 + 1 = 0 \tag{2.17}$$

It can be solved to give two natural frequencies, as

$$\omega_{nf1} = 58.84 \text{ rad/s} \quad \text{and} \quad \omega_{nf2} = 339.24 \text{ rad/s}$$

b. For a point mass ($I_d = 0$) Eq. (2.16) becomes

$$1 - \omega_{nf}^2 m\alpha_{xf} = 0 \tag{2.18}$$

$$\text{or} \quad \omega_{nf}^2 = \frac{1}{m\alpha_{xf}} = \frac{1}{10 \times 2.5869 \times 10^{-5}} = 3{,}865.71 \Rightarrow \omega_{nf} = 62.17 \text{ rad/s.}$$

which is close to the fundamental frequency obtained in the previous case.

c. For the uncoupled translational and rotational motions ($\alpha_{yM} = \alpha_{\varphi f} = 0$), Eq. (2.16) becomes

$$mI_d\omega_{nf}^4\alpha_{xf}\alpha_{\varphi M} - \omega_{nf}^2(m\alpha_{xf} + I_d\alpha_{\varphi M}) + 1 = 0 \tag{2.19}$$

On putting values of the present rotor system, we get

$$1.00373 \times 10^{-8}\omega_{nf}^4 - 2.97482 \times 10^{-4}\omega_{nf}^2 + 1 = 0 \tag{2.20}$$

which gives, $\omega_{nf1} = 62.17$ rad/s and $\omega_{nf2} = 160.53$ rad/s.

Uncoupled motion in the cantilever beam is not realistic but to have some approximate solution assessment this case has been considered. The fundamental natural frequency is same as in the case (b), however, second natural frequency is quite different as for the case (a).

Exercise 2.4 Obtain the transverse critical speed of a rotor system as shown in Figure 2.6. Assume the mass of the disc $m = 5$ kg, the diametral mass moment of inertia $I_d = 0.02$ kg-m^2, and Young's modulus of the shaft $E = 2.1 \times 10^{11}$ N/m^2. Take the shaft segment lengths as $a = 0.3$ m and $b = 0.7$ m. The diameter of the shaft is 10 mm. Ignore the gyroscopic effect.

For the present case, influence coefficients are given as $\alpha_{yf} = \dfrac{a^2(a+b)}{3EI}$, $\alpha_{\varphi M} = \dfrac{(3a+b)}{3EI}$, and $\alpha_{\varphi f} = \alpha_{yM} = \dfrac{a(3a+b)}{6EI}$, where I is the second moment of area of the shaft cross-section. Subscripts f and M represent force and moment, respectively, and y and φ represent the translational and rotational displacements.

Solution: The given data for the present problem are

$$m = 5 \text{ kg}; \quad I_d = 0.02 \text{ kg-m}^2; \quad a = 0.3 \text{ m}; \quad b = 0.7 \text{ m}; \quad d = 0.01 \text{ m};$$

$$E = 2.1 \times 10^{11} \text{ N/m}^2; \quad I = \frac{\pi}{64}d^4 = 4.9087 \times 10^{-10} \text{ m}^4.$$

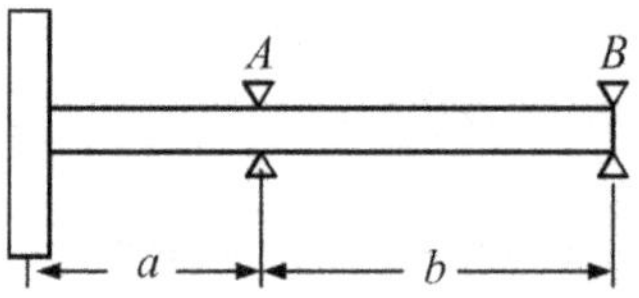

FIGURE 2.6 An overhung rotor system.

Influence coefficients for the rotor system of the present problem are as follows (refer Appendix 2.1):

$$\alpha_{xf} = \frac{a^2(a+b)}{3EI} = 2.9103\times10^{-4}\ \text{m/N};\quad \alpha_{\varphi M} = \frac{(3a+b)}{3EI} = 5.1738\times10^{-3}\ \text{1/N-m};$$

$$\alpha_{\varphi f} = \alpha_{xM} = \frac{a(3a+2b)}{6EI} = 1.1156\times10^{-3}\ \text{1/N}.$$

For the present problem, we have to neglect gyroscopic effect. Hence, motions in both orthogonal transverse planes are uncoupled. Equations of motion in a single plane can be written as (refer Tiwari (2017))

$$\begin{bmatrix} m & 0 \\ 0 & I_d \end{bmatrix}\begin{Bmatrix} \ddot{x} \\ \ddot{\varphi}_y \end{Bmatrix} + \begin{bmatrix} \alpha_{xf} & \alpha_{xM} \\ \alpha_{\varphi f} & \alpha_{\varphi M} \end{bmatrix}^{-1}\begin{Bmatrix} x \\ \varphi_y \end{Bmatrix} = \begin{Bmatrix} 0 \\ 0 \end{Bmatrix} \tag{2.21}$$

For the free vibration with simple harmonic motion, we have

$$\left(-\omega_{nf}^2\begin{bmatrix} m & 0 \\ 0 & I_d \end{bmatrix} + \begin{bmatrix} \alpha_{xf} & \alpha_{xM} \\ \alpha_{\varphi f} & \alpha_{\varphi M} \end{bmatrix}^{-1}\right)\begin{Bmatrix} x \\ \varphi_y \end{Bmatrix} = \begin{Bmatrix} 0 \\ 0 \end{Bmatrix} \tag{2.22}$$

where ω_{nf} is the natural frequency of the system. For the non-trivial solution, we have

$$\left| -\omega_{nf}^2\begin{bmatrix} m & 0 \\ 0 & I_d \end{bmatrix} + \begin{bmatrix} \alpha_{xf} & \alpha_{xM} \\ \alpha_{\varphi f} & \alpha_{\varphi M} \end{bmatrix}^{-1} \right| = 0 \tag{2.23}$$

which gives the frequency equation, as

$$mI_d\omega_{nf}^4(\alpha_{xf}\alpha_{\varphi M} - \alpha_{xM}^2) - \omega_{nf}^2(m\alpha_{xf} + I_d\alpha_{\varphi M}) + 1 = 0 \tag{2.24}$$

On substituting numerical values for the present problem, it gives

$$2.61086\times10^{-8}\omega_{nf}^4 - 1.558574\times10^{-3}\omega_{nf}^2 + 1 = 0 \tag{2.25}$$

It can be solved for the two natural frequencies as

$$\omega_{nf1} = 25.47\ \text{rad/s} \quad \text{and} \quad \omega_{nf2} = 242.97\ \text{rad/s}$$

For the point mass ($I_d = 0$): Eq. (2.24) becomes

$$1 - \omega_{nf}^2 m\alpha_{yf} = 0 \quad \text{or} \quad \omega_{nf}^2 = \frac{1}{m\alpha_{yf}} = \frac{1}{5\times2.9103\times10^{-4}},\quad \Rightarrow \omega_{nf} = 26.22\ \text{rad/s}.$$

which is close to the fundamental natural frequency obtained from coupled transverse translational and rotational motions.

Exercise 2.5 Obtain the bearing reaction forces and moments of a cantilever rotor at rotor speeds of (i) 0.5 ω_{nf_1}, (ii) 0.5(ω_{nf_2}+ω_{nf_1}) and (iii) 1.5 ω_{nf_2}; where ω_{nf_1} and ω_{nf_2} are the first and second transverse natural frequencies, respectively. Take the mass of the disc to be m=10 kg and the diametral mass moment of inertia to be I_d=0.02 kg-m². The disc has a residual unbalance of 25 g-cm. The shaft diameter is 10 mm, and the total shaft span is 0.5 m. The shaft is assumed to be massless and its Young's modulus E=2.1 × 10^{11} N/m². Take one plane motion only. Influence coefficients are given as $\alpha_{xf} = l^3/3EI$, $\alpha_{xM} = \alpha_{\varphi f} = l^2/2EI$, $\alpha_{\varphi M} = l/EI$, where I is the second moment of area of the shaft cross-section. Subscripts f and M represent force and moment, respectively, and y and φ represent the translational and rotational displacements.

***Solution*:** The given data for the present problem are

$$m = 10\,\text{kg}; \quad I_d = 0.02\ \text{kg-m}^2. \quad U = me = 25\ \text{g-cm};$$
$$l = 0.5\ \text{m}; \quad d = 0.01\ \text{m}; \quad E = 2.1\times 10^{11}\ \text{N/m}^2; \quad I = \frac{\pi}{64}d^4 = 4.9087\times 10^{-10}\ \text{m}^4.$$

Influence coefficients for the present rotor system are as follows (refer Appendix 2.1)

$$\alpha_{xf} = \frac{l^3}{3EI} = 4.0420\times 10^{-4}\ \text{m/N}; \quad \alpha_{\varphi M} = \frac{l}{EI} = 4.8504\times 10^{-3}\ \text{1/N-m};$$

$$\alpha_{xM} = \alpha_{\varphi f} = \frac{l^2}{2EI} = 1.2126\times 10^{-3}\ \text{1/N}.$$

For the present problem, we have to the neglect gyroscopic effect. Hence, the motion in both orthogonal transverse planes is uncoupled. Equations of motion in a single plane can be written as,

$$\begin{bmatrix} m & 0 \\ 0 & I_d \end{bmatrix} \begin{Bmatrix} \ddot{x} \\ \ddot{\varphi}_y \end{Bmatrix} + \begin{bmatrix} \alpha_{xf} & \alpha_{xM} \\ \alpha_{\varphi f} & \alpha_{\varphi M} \end{bmatrix}^{-1} \begin{Bmatrix} x \\ \varphi_y \end{Bmatrix} = \begin{Bmatrix} 0 \\ 0 \end{Bmatrix} \tag{2.26}$$

For the free vibration, we have

$$\left(-\omega_{nf}^2 \begin{bmatrix} m & 0 \\ 0 & I_d \end{bmatrix} + \begin{bmatrix} \alpha_{xf} & \alpha_{xM} \\ \alpha_{\varphi f} & \alpha_{\varphi M} \end{bmatrix}^{-1} \right) \begin{Bmatrix} x \\ \varphi_y \end{Bmatrix} = \begin{Bmatrix} 0 \\ 0 \end{Bmatrix} \tag{2.27}$$

where ω_{nf} is the natural frequency of the system. For non-trivial solution, we have

$$\left| -\omega_{nf}^2 \begin{bmatrix} m & 0 \\ 0 & I_d \end{bmatrix} + \begin{bmatrix} \alpha_{xf} & \alpha_{xM} \\ \alpha_{\varphi f} & \alpha_{\varphi M} \end{bmatrix}^{-1} \right| = 0 \tag{2.28}$$

which gives frequency equation as,

$$mI_d\omega_{nf}^4(\alpha_{xf}\alpha_{\varphi M} - \alpha_{\varphi f}^2) - \omega_{nf}^2(m\alpha_{xf} + I_d\alpha_{\varphi M}) + 1 = 0 \tag{2.29}$$

On substituting values for present problem, it gives

$$9.82852\times 10^{-8}\omega_{nf}^4 - 4.139\times 10^{-3}\omega_{nf}^2 + 1 = 0 \tag{2.30}$$

It can be solved to give two natural frequencies, as

$$\omega_{nf1} = 15.59\ \text{rad/s} \quad \text{and} \quad \omega_{nf2} = 204.89\ \text{rad/s}.$$

Now the bearing forces are given as (refer Tiwari (2017)),

$$\left\{ \begin{array}{c} R_b \\ M_b \end{array} \right\} = \mathbf{A} \left\{ \begin{array}{c} f_x \\ M \end{array} \right\} \quad \text{with} \quad \mathbf{A} = \left[\begin{array}{cc} 1 & 0 \\ l & 1 \end{array} \right] \tag{2.31}$$

where R_b and M_b are reaction force and reaction moment at bearing, respectively. However, f_x and M are force and moment developed at the disc and these are given as

$$\left\{ \begin{array}{c} f_x \\ M \end{array} \right\} = \mathbf{K} \left\{ \begin{array}{c} x \\ \varphi_y \end{array} \right\} \quad \text{with} \quad \mathbf{K} = \left[\begin{array}{cc} \alpha_{xf} & \alpha_{xM} \\ \alpha_{\varphi f} & \alpha_{\varphi M} \end{array} \right]^{-1} = \left[\begin{array}{cc} 9.896 \times 10^3 & -2.474 \times 10^3 \\ -2.474 \times 10^3 & 8.246681 \times 10^2 \end{array} \right] \tag{2.32}$$

From EOMs, displacement vectors are related with the unbalance force, as

$$\left(\mathbf{K} - \omega^2 \mathbf{M}\right) \left\{ \begin{array}{c} x \\ \varphi_y \end{array} \right\} = \left\{ \begin{array}{c} me\omega^2 \\ 0 \end{array} \right\} \quad \text{or} \quad \left\{ \begin{array}{c} x \\ \varphi_y \end{array} \right\} = \mathbf{Z}^{-1} \left\{ \begin{array}{c} me\omega^2 \\ 0 \end{array} \right\} \tag{2.33}$$

$$\text{with,} \quad \mathbf{Z}^{-1} = \left(\mathbf{K} - \omega^2 \mathbf{M}\right)^{-1} = \left[\begin{array}{cc} k_{11} - m\omega^2 & k_{12} \\ k_{21} & k_{22} - I_d\omega^2 \end{array} \right]^{-1}$$

Hence, on substituting Eqs. (2.32) and (2.33) into Eq. (2.31), the bearing reaction force vector is

$$\left\{ \begin{array}{c} R_b \\ M_b \end{array} \right\} = \mathbf{AKZ}^{-1} \left\{ \begin{array}{c} me\omega^2 \\ 0 \end{array} \right\} = \mathbf{AKZ}^{-1} \left\{ \begin{array}{c} 2.5 \times 10^{-4} \omega^2 \\ 0 \end{array} \right\} \tag{2.34}$$

By using present rotor system data in various matrices, we get

$$\text{At } \omega = 0.5\omega_{nf1}, \text{ we have } \left\{ \begin{array}{c} R_b \\ M_b \end{array} \right\} = \left\{ \begin{array}{c} 0.0202 \text{ N} \\ 0.0101 \text{ Nm} \end{array} \right\}$$

$$\text{At } \omega = 0.5\left(\omega_{nf1} + \omega_{nf2}\right), \text{ we have } \left\{ \begin{array}{c} R_b \\ M_b \end{array} \right\} = \left\{ \begin{array}{c} 0.0156 \text{ N} \\ -0.0179 \text{ Nm} \end{array} \right\}$$

$$\text{At } \omega = 1.5\omega_{nf2}, \text{ we have } \left\{ \begin{array}{c} R_b \\ M_b \end{array} \right\} = \left\{ \begin{array}{c} -0.3977 \text{ N} \\ -0.0873 \text{ Nm} \end{array} \right\}$$

The bearing force and moment changes with speed and have much higher values post critical range (i.e. Case (c)).

Exercise 2.6 Find transverse natural frequencies (or critical speeds) of a cantilever rotor system as shown in Figure 2.7. Consider the shaft as massless and is made of steel with a Young's modulus of 2.1×10^{11} N/m^2. A disc is mounted at the free end of the shaft with a mass of 10 kg and a diametral mass moment of inertia of 0.04 kg-m^2. Do not consider the gyroscopic effect (please note that for this condition only, transverse natural frequencies and critical speeds will be the same). In the diagram, all dimensions are in cm.

***Solution*:** First, we need to calculate influence coefficients since the present rotor system is not having standard uniform cross-section shaft even though the end condition is standard, i.e. the cantilever. Two cases are considered (i) the diametral mass moment of inertia of the disc is ignored and only transverse translational displacement is considered and (ii) only the diametral mass moment of

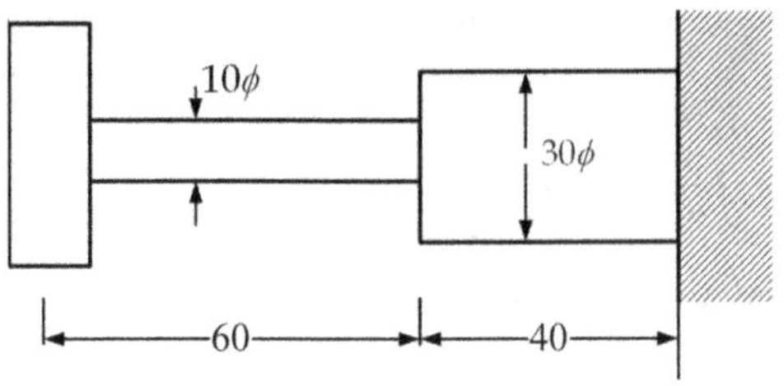

FIGURE 2.7 A stepped shaft with cantilever end conditions.

inertia of the disc is considered, i.e. only rotational displacement is considered. These two cases will result in single natural frequency of the rotor system. Both transverse translational and rotational displacements can be considered (refer chapter 8 of Tiwari (2017)), and there we will see that the system has two natural frequencies. In Chapter 5, we will see that while considering the gyroscopic effect also we will have four natural frequencies, two each for the forward and backward whirls.

Case I: Let us apply a vertical force F, at the free end of the cantilever beam as shown in Figure 2.8.

Due to this force F, the bending moment $M = Fz$ is acting on the beam at an axial location of, z and it has the same expression at either shaft section. The strain energy stored due to the bending is given as

$$U = \int_0^{0.6} \frac{M^2}{2EI_1} dz + \int_{0.6}^{1} \frac{M^2}{2EI_2} dz \tag{2.35}$$

The translational displacement is expressed as,

$$\delta = \frac{\partial U}{\partial F} = \int_0^{0.6} \frac{M\left(\frac{\partial M}{\partial F}\right)}{EI_1} dz + \int_{0.6}^{1} \frac{M\left(\frac{\partial M}{\partial F}\right)}{EI_2} dz = \int_0^{0.6} \frac{Fx^2}{EI_1} dz + \int_{0.6}^{1} \frac{Fx^2}{EI_2} dz = \frac{F}{EI_1} \frac{z^3}{3}\bigg|_0^{0.6} + \frac{F}{EI_2} \frac{z^3}{3}\bigg|_{0.6}^{1} \tag{2.36}$$

Therefore, we get

$$\delta = 6.9846 \times 10^{-8} F + 0.3130 \times 10^{-8} F = 7.2976 \times 10^{-8} F \tag{2.37}$$

Hence, the stiffness for the translational motion is given as

$$k = \frac{F}{\delta} = 1.3703 \times 10^7 \text{ N/m}$$

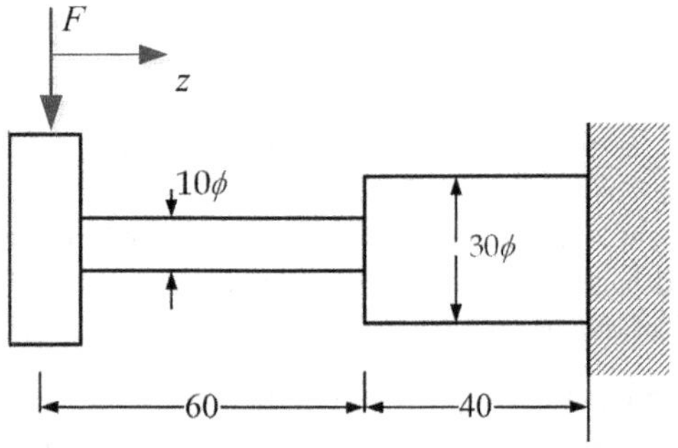

FIGURE 2.8 The rotor with a force at the free end.

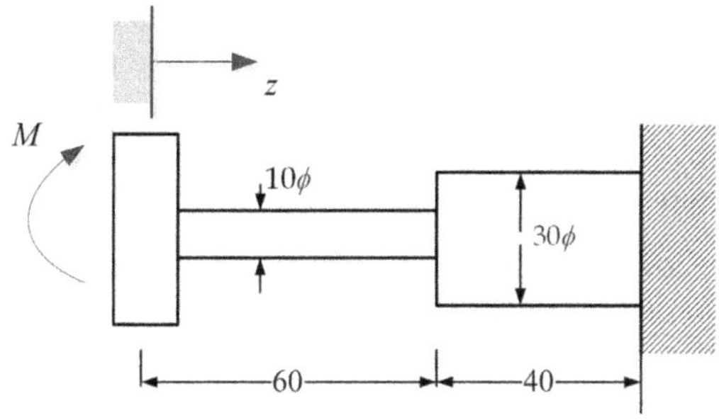

FIGURE 2.9 The rotor with a moment at free end.

The natural frequency related to the pure translational motion is then given as

$$\omega_{nf} = \sqrt{\frac{k}{m}} = 1170.60 \text{ rad/s}$$

Case II: Now let us apply only a moment M at the free end (refer Figure 2.9).

The shaft will have bending moment due to this external moment and the strain energy stored is given as

$$U = \int_0^{0.6} \frac{M^2}{2EI_1} dz + \int_{0.6}^{1} \frac{M^2}{2EI_2} dz \tag{2.38}$$

The rotational displacement due to this moment is,

$$\varphi = \frac{\partial U}{\partial M} = \int_0^{0.6} \frac{M}{EI_1} dz + \int_{0.6}^{1} \frac{M}{EI_2} dz = \frac{M}{EI_1} z\Big|_0^{0.6} + \frac{M}{EI_2} z\Big|_{0.6}^{1} \tag{2.39}$$

Therefore, we get

$$\varphi = 5.8205 \times 10^{-7} M + 0.0479 \times 10^{-7} M = 5.8684 \times 10^{-7} M \tag{2.40}$$

The bending rotational stiffness for the transverse rotational motion is given as

$$k = \frac{M}{\varphi} = 1.7040 \times 10^6 \text{ Nm/rad}$$

Hence, the natural frequency related to the pure rotational motion is,

$$\omega_{nf} = \sqrt{\frac{k}{I_d}} = \sqrt{\frac{1{,}704{,}033.52}{0.04}} = 6{,}526.93 \text{ rad/s}$$

In the aforementioned analyses, the translational and rotational motions have been considered uncoupled. To consider the coupling, we need to obtain strain energy for application of force as well as moment, simultaneously. Then corresponding translational and rotational displacements can be obtained in terms of force as well moment applied. This will give coupling terms (between translational and rotational displacements) also in terms of influence coefficients. For more details, chapter 8 of Tiwari (2017) can be referred.

Exercise 2.7 (a) While the Jeffcott rotor is whirling, with the help of the centre of gravity, the centre of spinning of the disc, and the bearing axis, draw their relative positions in an axial plane when the rotor is (i) below the critical speed, (ii) at the critical speed and (iii) above the critical speed. (b) Define the following terms: the natural frequency and the critical speed of a rotor, and the synchronous and asynchronous whirls.

***Solution*:** (a) Figure 2.10 shows an orbital motion of an unbalanced shaft about the bearing centre. The orbital and spin motion is assumed in the counter clock direction. It can be seen that below the critical speed, the centre of gravity of the disc (i.e.., the unbalance force, F) is always outside the orbit. Due to damping the displacement (which is in the radial direction) is lagging behind the unbalance force direction. When the rotor is at the critical speed, the phase becomes 90° and when the rotor is beyond the critical speed then it becomes around 180° (slightly less than 180° with damping). This makes the unbalance force directed towards the bearing centre, which makes the rotor exhibit self-centring behaviour.

(b) Definition of various terms are as follows:

Natural Frequency: The natural frequency is the frequency at which a rotor system tends to oscillate (usually transversely or sometimes torsionally or axially) in the absence of any driving external force (or moment) when the rotor is given as momentary disturbance. Free vibrations of a rotor with an elastic shaft are called natural vibrations and occur at a frequency called the natural frequency (or more specifically whirl natural frequency). Unlike structures, the rotor natural frequency can be rotor speed-dependent.

Critical Speed: If the forcing frequency is equal to the natural frequency, the amplitude of vibration increases rapidly. This phenomenon is known as the resonance. Usually forcing in rotor is due to the unbalance and its frequency is equal to the spin speed of the shaft. So the speed of the rotor at the resonance is called the critical speed. With the gyroscopic effect in the rotor system, we can have the concept of forward and backward critical speeds.

If the excitation frequency of the force is equal to the rotor speed (both in magnitude and sense of rotation) the force is said to be synchronous one, which is usually caused by the unbalance force. The synchronous motion is represented corresponding to the intersection of the frequency curve with a straight line of positive 45° slope in the Campbell diagram (i.e., a plot of natural whirl frequency with the spin speed of the shaft). The intersection of this synchronous line with the forward and backward natural whirling frequency curves (which occurs due to gyroscopic effect, anisotropic bearings, asymmetric shaft and rotor or any other asymmetry due to defects in rotor, like the cracks,

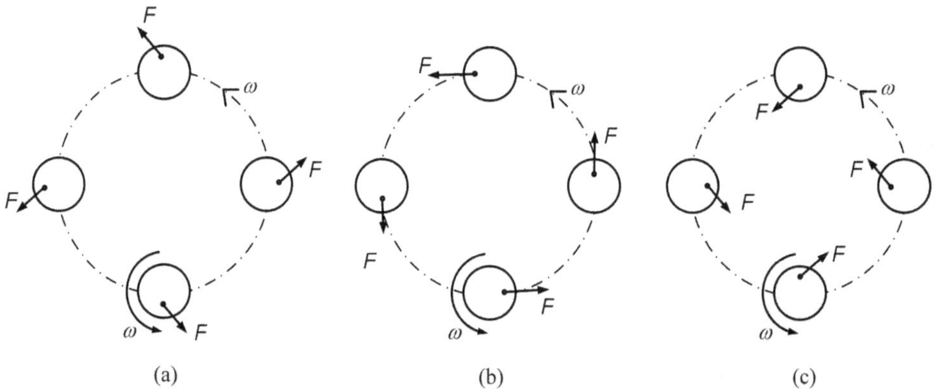

FIGURE 2.10 Whirling of the shaft at different speed conditions (a) below critical speed (b) at critical speed (c) above critical speed.

gear transmission errors, misalignment, rub, etc.) is analogous to the resonance in the non-rotating system. These intersection points are termed as the critical speeds (forward or backward depending upon the corresponding whirl natural frequencies). Like the resonances, the critical speeds are associated with a local peak in amplitude accompanied by a shift in phase of the order of 90° (after crossing the critical speed the change in the phase of the order of 180°).

Synchronous whirl: When the whirling frequency of a rotor is equal to the rotating speed of the rotor both in magnitude and sense of rotation, then is called as synchronous whirl. Usually, the whirling caused by the unbalance is the synchronous one. The shaft whirls in bend configuration about bearing axis without any reversal of flexural stress (i.e., as a rigid body).

Asynchronous whirl: When the whirling frequency of a rotor is not equal to the rotating speed of the rotor (both in magnitude and sense of rotation), it is called as asynchronous whirl. Asynchronous whirl can have both forward (i.e., the sense of whirl direction is the same as the spin speed) and backward (the sense of whirl direction is opposite to the spin speed) whirls. Due to the gyroscopic effect such whirls are observed in rotor systems. For such motion a reversal of flexural stresses take place. As a special case when the magnitude of the whirl frequency is equal to the spin speed of the shaft but the sense of rotation is opposite then it is called anti-synchronous whirl. When rotor-stator interaction (e.g., rubs) takes place due to friction, such whirl occurs, however, it is a very rare phenomenon.

Exercise 2.8 In a design stage of a rotor-bearing system it has been found that one of the critical speeds is very close to the fixed operating speed of the rotor. List the design modifications a designer can make to overcome this problem.

Solution: We can change the critical speed of system by changing the rotor mass and its distribution. It can be also done by changing effective stiffness of system. For that, the diameter and the length can be altered. By adding auxiliary support, we can increase the stiffness of system thereby increasing the critical speed of system. By changing the position of supports and types of supports (bearings and its pedestal), the effective stiffness of system can be modified. If the above methods are not feasible or effective then we can add external dampers to the system to minimise amplitudes at resonance condition.

Exercise 2.9 A cantilever shaft of 1 m length (l) and 30 mm diameter (d) has a disc of 5 kg mass (m) attached at its free end, with a negligibly small diametral mass moment of inertia. The shaft has a through hole parallel to the shaft axis of diameter 3 mm (d_i), which is vertically below the shaft centre, with the distance between the centers of the shaft and the hole as 6 mm (e). Consider no cross-coupling in two orthogonal directions as well as between the translational and rotational displacements, and obtain the transverse natural frequencies of the shaft system in two principal planes. Consider the shaft to be massless and Young's modulus $E = 2.1 \times 10^{11}\,\text{N/m}^2$.

Solution: The rotational and translational displacements are uncoupled. The disc has negligible diametral mass moment of inertia. So only translational displacement is considered.

Given data for the rotor system are,

$$l = 1\text{ m},\ d = 0.03\text{ m},\ m = 5\text{ kg},\ I_d = 0,\ e = 6\text{ mm},\ d_i = 3\text{ mm},\ E = 2.1\times10^{11}\text{ N/m}^2.$$

The equivalent stiffness of shaft in two principal directions is obtained as (refer Figure 2.11)

$$I_1 = \frac{\pi\left(d^4 - d_i^4 - 16d_i^2e^2\right)}{64} = 3.9502\times10^{-8}\text{ m}^4;\quad I_2 = \frac{\pi\left(d^4 - d_i^4\right)}{64} = 3.9757\times10^{-8}\text{ m}^4;$$

$$k_{eq1} = \frac{3EI_1}{l^3} = 24{,}886.47\text{ N/m};\qquad k_{eq2} = \frac{3EI_2}{l^3} = 25{,}046.79\text{ N/m}.$$

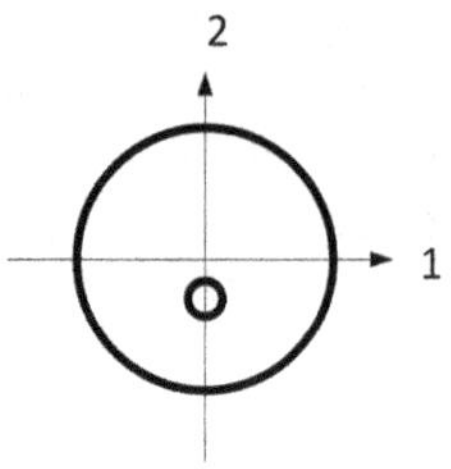

FIGURE 2.11 Shaft cross section with a through hole.

So, in two orthogonal transverse directions, natural frequencies are

$$\omega_{nf1} = \sqrt{\frac{k_{eq1}}{m}} = \sqrt{\frac{24{,}886.47}{5}} = 70.55 \text{ rad/s}, \quad \text{and} \quad \omega_{nf2} = \sqrt{\frac{k_{eq2}}{m}} = \sqrt{\frac{25{,}046.79}{5}} = 70.78 \text{ rad/s}$$

Now try obtaining the natural frequencies (four numbers), when the thin disc has the diametral mass moment of inertia of 0.01 kg-m^2 (refer Chapter 8 of Tiwari (2017)).

Exercise 2.10 For the Jeffcott rotor with an axially offset disc consider pure rotational displacement (tilting) of the disc (without translational displacement) and obtain the transverse natural frequency for the tilting motion. Let EI is the flexural rigidity of the shaft, a and b are the axial location of the disc from the left and right bearings with $l=a+b$, m is the mass of the disc and r is the radius of the disc.

***Solution*:** Let the moment applied at disc location C be M, a and b are the axial location of the disc from the left and right bearings (A and B, respectively) with $l=a+b$. Also let R_A and R_B be the reactions at the left and right bearings due to moment applied at the disc.

A Jeffcott rotor model with an axially offset disc is shown in Figure 2.12. The free body diagram of the rotor system is shown in Figure 2.13. From the free body diagram on taking force and moment equilibrium, we have

$$R_A + R_B = 0 \tag{2.41}$$

$$\text{and} \quad \sum M_A = 0 \quad \text{or} \quad M - R_B l = 0 \tag{2.42}$$

Therefore, reactions are bearings are

$$R_B = M/l; \quad R_A = -M/l \tag{2.43}$$

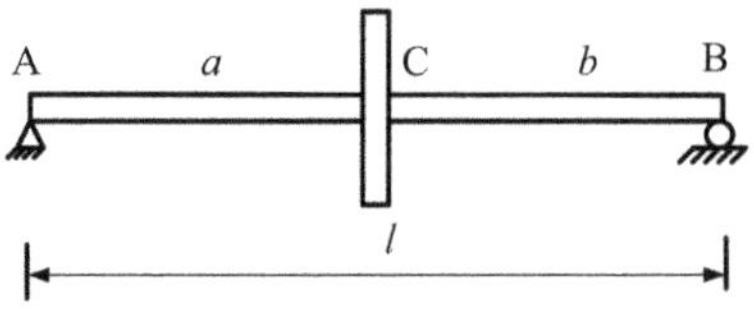

FIGURE 2.12 A Jeffcott rotor with an axially offset disc.

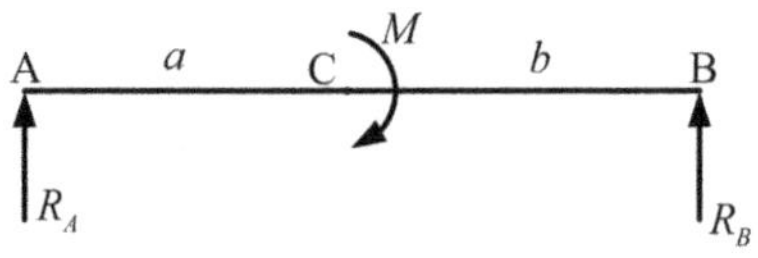

FIGURE 2.13 A free body diagram of the rotor system.

The strain energy in the shaft section AC (the bending moment at z distance from bearing A is $R_A z$)

$$U_1 = \int_0^a \frac{(R_A z)^2}{2EI} dz \tag{2.44}$$

The strain energy in the shaft section CB (the bending moment at z distance from bearing B is $R_B z$)

$$U_2 = \int_0^b \frac{(R_B z)^2}{2EI} dz \tag{2.45}$$

Hence, the total strain energy will be

$$U = U_1 + U_2 = \frac{1}{2EI}\left[R_A^2 \frac{z^3}{3}\bigg|_0^a + R_B^2 \frac{z^3}{3}\bigg|_0^b \right] = \frac{1}{2EI}\left[\frac{M^2}{l^2}\frac{a^3}{3} + \frac{M^2}{l^2}\frac{b^3}{3} \right] = \frac{1}{6EI}\frac{M^2}{l^2}\left[a^3 + b^3\right] \tag{2.46}$$

The angular deflection at C due to tilting motion is obtained as

$$\varphi = \frac{\partial U}{\partial M} = \frac{1}{3EI}\frac{M}{l^2}\left[a^3 + b^3\right] \tag{2.47}$$

Therefore, the influence coefficient is

$$\alpha_{\varphi M} = \frac{\varphi}{M} = \frac{1}{3EI}\frac{\left(a^3 + b^3\right)}{l^2} \tag{2.48}$$

As, $b = l - a$, on putting this in Eq. (2.46), we get

$$\alpha_{\varphi M} = \frac{1}{3EI}\frac{\left(l^2 - 3la + 3a^2\right)}{l} \tag{2.49}$$

Equivalent moment stiffness of the shaft would be

$$k_{M\varphi} = \frac{1}{\alpha_{\varphi M}} = \frac{3EIl}{\left(l^2 - 3la + 3a^2\right)} \tag{2.50}$$

and the diametral mass moment of inertia for thin disc is given as

$$I_d = \tfrac{1}{4} m r^2 \tag{2.51}$$

where m is the mass of the disc and r is the radius of the disc. Hence, the natural frequency would be (pure rotational motion will take place when $a=b$)

$$\omega_{nf} = \sqrt{k_{M\varphi}/I_d} \quad \text{or} \quad \omega_{nf} = \sqrt{\frac{12EIl}{mr^2\left(l^2 - 3la + 3a^2\right)}} \equiv \sqrt{\frac{12EIl^2}{mr^2\left(a^3 + b^3\right)}} \quad \text{with} \quad a = b \tag{2.52}$$

It should be noted that the stiffness and natural frequency for the pure translation motion (pure translation motion will take place when $a = b$), respectively, are given as

$$k_{fy} = \frac{1}{\alpha_{yf}} = \frac{48EI}{l^3} \quad \text{and} \quad \omega_{nf} = \sqrt{\frac{48EI}{ml^3}} \tag{2.53}$$

Exercise 2.11 What are the length and the diameter of a cantilever shaft if the transverse critical speed has to be fixed at 100 Hz (by considering pure translational motion only) and it has 2 kg of mass at its free end? Because of space limitations, the length of the shaft should not exceed 30 cm. $E = 2.1 \times 10^{11}$ N/m^2.

Solution: For pure translation motion of the cantilever shaft with a disc at free end (refer **Appendix 2.1**), we have

$$\omega_{nf} = \sqrt{\frac{k_{eq}}{m}} \quad \text{with} \quad k_{eq} = \frac{F}{\delta} = \frac{3EI}{l^3} \tag{2.54}$$

So that Eq. (2.54) can be written as

$$\omega_{nf} = \sqrt{\frac{3EI}{ml^3}} \tag{2.55}$$

Now, according to the problem, the critical speed is specific as

$$\omega_{cr} = 100 \times 2\pi \text{ rad/s}$$

Since we have $\omega_{cr} = \omega_{nf}$, so from Eq. (2.55), we get the following condition

$$(100 \times 2\pi)^2 = \frac{3 \times 2.1 \times 10^{11} \times \pi d^4}{64 \times 2 \times .03^3} \Rightarrow d^4 = \frac{(100 \times 2\pi)^2 \times 64 \times 2 \times .03^3}{3 \times 2.1 \times 10^{11} \times \pi} \tag{2.56}$$

Solving the aforementioned equation, we get $d = 0.0288$ m.

Exercise 2.12 The transverse critical speed of a rotor system, as shown in Figure 2.14, is fixed at 5.98 rad/s. Take the disc as a point mass with $m = 5$ kg. What will be the overhung shaft length, a? Take the shaft length as $b = 0.7$ m. The diameter of the shaft is 10 mm. Neglect the gyroscopic effect. $E = 2.1 \times 10^{11}$ N/m^2.

Solution: For the present problem, we must ignore the gyroscopic effect. Hence, the transverse motions in both orthogonal transverse planes are uncoupled. Equations of motion in a single plane can be written, as

$$\begin{bmatrix} m & 0 \\ 0 & I_d \end{bmatrix} \begin{Bmatrix} \ddot{x} \\ \ddot{\varphi}_y \end{Bmatrix} + \begin{bmatrix} \alpha_{xf} & \alpha_{xM} \\ \alpha_{\varphi f} & \alpha_{\varphi M} \end{bmatrix}^{-1} \begin{Bmatrix} x \\ \varphi_y \end{Bmatrix} = \begin{Bmatrix} 0 \\ 0 \end{Bmatrix} \tag{2.57}$$

where m is the mass of disc, I_d is the diamatral mass moment of inertia of the disc, α is the influence coefficient, x is the transverse translational displacement and φ_y is the transverse rotational displacement. In subscripts, f and M represent force and moment, respectively. For the free vibration, we have

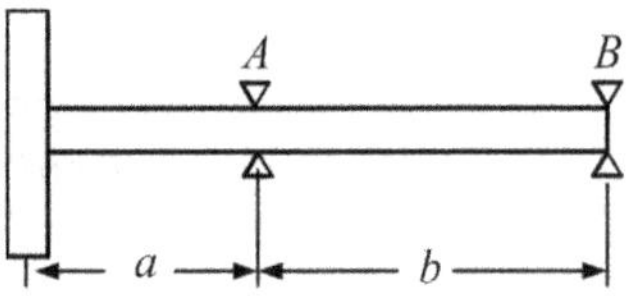

FIGURE 2.14 An overhung rotor system.

$$\left(-\omega_{nf}^2\begin{bmatrix} m & 0 \\ 0 & I_d \end{bmatrix}+\begin{bmatrix} \alpha_{xf} & \alpha_{xM} \\ \alpha_{\varphi f} & \alpha_{\varphi M} \end{bmatrix}^{-1}\right)\begin{Bmatrix} x \\ \varphi_y \end{Bmatrix}=\begin{Bmatrix} 0 \\ 0 \end{Bmatrix} \tag{2.58}$$

where ω_{nf} is the natural frequency of the system. For the non-trivial solution, we have

$$\left|-\omega_{nf}^2\begin{bmatrix} m & 0 \\ 0 & I_d \end{bmatrix}+\begin{bmatrix} \alpha_{xf} & \alpha_{xM} \\ \alpha_{\varphi f} & \alpha_{\varphi M} \end{bmatrix}^{-1}\right|=0 \tag{2.59}$$

which gives the frequency equation, as

$$mI_d\omega_{nf}^4(\alpha_{xf}\alpha_{\varphi M}-\alpha_{\varphi f}^2)-\omega_{nf}^2(m\alpha_{xf}+I_d\alpha_{\varphi M})+1=0 \tag{2.60}$$

For the point mass ($I_d = 0$), the aforementioned equation becomes,

$$1-\omega_{nf}^2 m\alpha_{xf}=0;\quad \omega_{nf}^2=\frac{1}{m\alpha_{xf}} \Rightarrow \quad \omega_{nf}^2=\frac{3EI}{ma^2(a+b)} \tag{2.61}$$

We have the following rotor parameters

$$m=5\text{ kg};\quad b=0.7\text{ m};\quad d=0.01\text{ m};\quad E=\ 2.1\times10^{11}\text{ N/m}^2;\quad I=4.9087\times10^{-10}\text{ m}^4$$

Putting values of parameters, we get

$$5.98^2=\frac{3\times2.1\times10^{11}\times4.9087\times10^{-10}}{5\times a^2(a+0.7)} \Rightarrow \quad a^3+0.7a^2-1.729569=0 \tag{2.62}$$

Solving for a, we get $a = 1.0067$ m (other two values are not feasible). Obtain now the diameter of the shaft from the data available, whether it gives d=10 mm.

Exercise 2.13 The transverse critical speed of a rotor system as shown in Figure 2.15 is fixed at 5.98 rad/s. Take the disc as a point mass with m=5 kg. What will be the diameter of the uniform shaft, d? Take shaft length a=0.3 m, b=0.7 m. Ignore the gyroscopic effect. E=2.1 × 10^{11} N/m^2.

***Solution*:** For the present problem, we have to ignore the gyroscopic effect. Hence, motions in both orthogonal transverse planes are uncoupled. Equations of motion in a single plane can be written as,

$$\begin{bmatrix} m & 0 \\ 0 & I_d \end{bmatrix}\begin{Bmatrix} \ddot{x} \\ \ddot{\varphi}_y \end{Bmatrix}+\begin{bmatrix} \alpha_{xf} & \alpha_{xM} \\ \alpha_{\varphi f} & \alpha_{\varphi M} \end{bmatrix}^{-1}\begin{Bmatrix} x \\ \varphi_y \end{Bmatrix}=\begin{Bmatrix} 0 \\ 0 \end{Bmatrix} \tag{2.63}$$

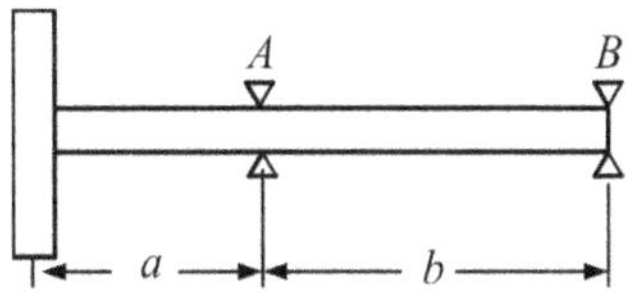

FIGURE 2.15 An overhung rotor system.

For the free vibration, we have

$$\left(-\omega_{nf}^2\begin{bmatrix} m & 0 \\ 0 & I_d \end{bmatrix}+\begin{bmatrix} \alpha_{xf} & \alpha_{xM} \\ \alpha_{\varphi f} & \alpha_{\varphi M} \end{bmatrix}^{-1}\right)\begin{Bmatrix} x \\ \varphi_y \end{Bmatrix}=\begin{Bmatrix} 0 \\ 0 \end{Bmatrix} \tag{2.64}$$

where ω_{nf} is the natural frequency of the system. For the non-trivial solution, we have

$$\left|-\omega_{nf}^2\begin{bmatrix} m & 0 \\ 0 & I_d \end{bmatrix}+\begin{bmatrix} \alpha_{xf} & \alpha_{xM} \\ \alpha_{\varphi f} & \alpha_{\varphi M} \end{bmatrix}^{-1}\right|=0 \tag{2.65}$$

which gives the frequency equation, as

$$mI_d\omega_{nf}^4(\alpha_{xf}\alpha_{\varphi M}-\alpha_{xM}^2)-\omega_{nf}^2(m\alpha_{xf}+I_d\alpha_{\varphi M})+1=0 \tag{2.66}$$

For the point mass ($I_d=0$), the aforementioned equation becomes (refer Appendix 2.1)

$$1-\omega_{nf}^2 m\alpha_{xf}=0 \Rightarrow \omega_{nf}^2=\frac{1}{m\alpha_{xf}} \Rightarrow \omega_{nf}^2=\frac{3EI}{ma^2(a+b)} \tag{2.67}$$

We have following rotor parameters

$$m=5\text{ kg};\quad a=0.3\text{ m};\quad b=0.7\text{ m};\quad E=2.1\times10^{11}\text{ N/m}^2;\quad I=\frac{\pi}{64}\times d^4\text{ m}^4$$

On putting the values of the parameters, we get

$$5.98^2=\frac{3\times2.1\times10^{11}\times I}{5\times0.3^2\times(0.3+0.7)} \Rightarrow I=2.554314\times10^{-11}\text{m}^4 \Rightarrow \frac{\pi}{64}d^4=2.554314\times10^{-11}$$

which gives

$$d=4.7761\times10^{-3}\text{ m}\quad\text{or}\quad d=4.776\text{ mm}.$$

Exercise 2.14 The transverse critical speed of a rotor system as shown in Figure 2.15 is fixed at 5.98 rad/s. Take disc as a point mass with $m=5$ kg. What will be the diameter of the uniform shaft, d? Take shaft length $2a=b=0.7$ m. $E=2.1\times10^{11}$ N/m^2. Neglect the gyroscopic effect.

***Solution*:** For the present problem, we have to ignore the gyroscopic effect. Hence, motions in both orthogonal transverse planes are uncoupled. Equations of motion in a single plane can be written (refer Tiwari, 2017), as

$$\begin{bmatrix} m & 0 \\ 0 & I_d \end{bmatrix}\begin{Bmatrix} \ddot{x} \\ \ddot{\varphi}_y \end{Bmatrix}+\begin{bmatrix} \alpha_{xf} & \alpha_{xM} \\ \alpha_{\varphi f} & \alpha_{\varphi M} \end{bmatrix}^{-1}\begin{Bmatrix} x \\ \varphi_y \end{Bmatrix}=\begin{Bmatrix} 0 \\ 0 \end{Bmatrix} \tag{2.68}$$

For free vibration we have,

$$\left(-\omega_{nf}^2\begin{bmatrix} m & 0 \\ 0 & I_d \end{bmatrix}+\begin{bmatrix} \alpha_{xf} & \alpha_{xM} \\ \alpha_{\varphi f} & \alpha_{\varphi M} \end{bmatrix}^{-1}\right)\begin{Bmatrix} x \\ \varphi_y \end{Bmatrix}=\begin{Bmatrix} 0 \\ 0 \end{Bmatrix} \tag{2.69}$$

where ω_{nf} is the natural frequency of the system. For the non-trivial solution, we have

$$\left| -\omega_{nf}^2 \begin{bmatrix} m & 0 \\ 0 & I_d \end{bmatrix} + \begin{bmatrix} \alpha_{xf} & \alpha_{xM} \\ \alpha_{\varphi f} & \alpha_{\varphi M} \end{bmatrix}^{-1} \right| = 0 \tag{2.70}$$

which gives frequency equation as,

$$mI_d\omega_{nf}^4(\alpha_{xf}\alpha_{\varphi M} - \alpha_{\varphi f}^2) - \omega_{nf}^2(m\alpha_{xf} + I_d\alpha_{\varphi M}) + 1 = 0 \tag{2.71}$$

For the point mass ($I_d = 0$), the aforementioned equation becomes (refer Appendix 2.1)

$$1 - \omega_{nf}^2 m\alpha_{xf} = 0 \Rightarrow \quad \omega_{nf}^2 = \frac{1}{m\alpha_{xf}} \Rightarrow \quad \omega_{nf}^2 = \frac{3EI}{ma^2(a+b)} \Rightarrow \quad \omega_{nf}^2 = \frac{8EI}{mb^3} \tag{2.72}$$

We have following rotor parameters

$$m = 5 \text{ kg}; \quad a = 0.35 \text{ m}; \quad b = 0.7 \text{ m}; \quad E = 2.1 \times 10^{11} \text{ N/m}^2; \quad I = \frac{\pi}{64} \times d^4 \text{ m}^4.$$

On putting values of parameters, we get

$$5.98^2 = \frac{8 \times 2.1 \times 10^{11} \times I}{5 \times 0.7^3} \Rightarrow \quad I = 3.65054 \times 10^{-11} \text{ m}^4;$$

$$\text{and} \quad \frac{\pi}{64} d^4 = 3.65054 \times 10^{-11} \Rightarrow \quad d = 5.222 \times 10^{-3} \text{ m} \Rightarrow \quad d = 5.222 \text{ mm}.$$

Exercise 2.15 For a Jeffcott rotor with a disc at the midspan, influence coefficients are given as $\alpha_{yf} = l^3/(48EI)$, $\alpha_{yM} = \alpha_{\varphi f} = 0$, $\alpha_{\varphi M} = l/(12EI)$, where l is the span length and EI is the modulus of rigidity of the shaft. Let m and I_d be the mass and the diametral mass moment of inertia, respectively, of the disc. Obtain the natural frequencies of the rotor system.

***Solution*:** Equations of motion in a single plane can be written as,

$$\begin{bmatrix} m & 0 \\ 0 & I_d \end{bmatrix} \begin{Bmatrix} \ddot{y} \\ \ddot{\varphi}_x \end{Bmatrix} + \begin{bmatrix} \alpha_{yf} & \alpha_{yM} \\ \alpha_{\varphi f} & \alpha_{\varphi M} \end{bmatrix}^{-1} \begin{Bmatrix} y \\ \varphi_x \end{Bmatrix} = \begin{Bmatrix} 0 \\ 0 \end{Bmatrix} \tag{2.73}$$

For the free vibration, we have

$$\left(-\omega_{nf}^2 \begin{bmatrix} m & 0 \\ 0 & I_d \end{bmatrix} + \begin{bmatrix} \alpha_{yf} & \alpha_{yM} \\ \alpha_{\varphi f} & \alpha_{\varphi M} \end{bmatrix}^{-1} \right) \begin{Bmatrix} y \\ \varphi_x \end{Bmatrix} = \begin{Bmatrix} 0 \\ 0 \end{Bmatrix} \tag{2.74}$$

where ω_{nf} is the natural frequency of the system. For a non-trivial solution, we have

$$\left| -\omega_{nf}^2 \begin{bmatrix} m & 0 \\ 0 & I_d \end{bmatrix} + \begin{bmatrix} \alpha_{yf} & \alpha_{yM} \\ \alpha_{\varphi f} & \alpha_{\varphi M} \end{bmatrix}^{-1} \right| = 0 \tag{2.75}$$

which gives the frequency equation as

$$m I_d \omega_{nf}^4 (\alpha_{yf}\alpha_{\varphi M} - \alpha_{\varphi f}^2) - \omega_{nf}^2 (m\alpha_{yf} + I_d \alpha_{\varphi M}) + 1 = 0 \tag{2.76}$$

As given for present rotor system, for $\alpha_{\varphi f} = 0$, the aforementioned equation becomes

$$m I_d \omega_{nf}^4 \alpha_{yf}\alpha_{\varphi M} - \omega_{nf}^2 (m\alpha_{yf} + I_d \alpha_{\varphi M}) + 1 = 0 \tag{2.77}$$

which can be resolved into two factors, as

$$(m\alpha_{yf}\omega_{nf}^2 - 1)(I_d \alpha_{\varphi M}\omega_{nf}^2 - 1) = 0 \tag{2.78}$$

which gives (refer Appendix 2.1),

$$\omega_{nf1} = \sqrt{1/(\alpha_{yf} m)} = \sqrt{48EI/(ml^3)} \quad \text{and} \quad \omega_{nf2} = \sqrt{1/(\alpha_{\varphi M} I_d)} = \sqrt{12EI/(I_d l)}$$

Exercise 2.16 Obtain transverse natural frequencies of a rotor system as shown in Figure 2.16. The mass of the disc is m=5 kg, and the diametral mass moment of inertia is I_d=0.02 kg-m^2. Lengths of the shaft are a=0.3 m and b=0.7 m. The diameter of the shaft is 10 mm. Bearing *A* has the roller support, and Bearing *B* has the fixed support condition. Ignore the mass of the shaft and the gyroscopic effect of the disc. $E = 2.1 \times 10^{11}$ N/m^2.

Solution**:** ***Method 1***: For simplicity of the analysis, the shaft is considered massless. The first step would be to obtain the influence coefficients corresponding to the disc location acted on by the concentrated force and moment. We need to derive translational and rotational displacements at the disc location due to the force f_y and the moment M acting at this location as shown in Figure 2.17. Using the *energy method* these influence coefficients are obtained as follows.

The intermediate support has single vertical reaction R_A and fixed support has three reactions R_{Bx}, R_{By} and M_{Bx} (in fact the horizontal component is zero since no net external force in the horizontal direction).

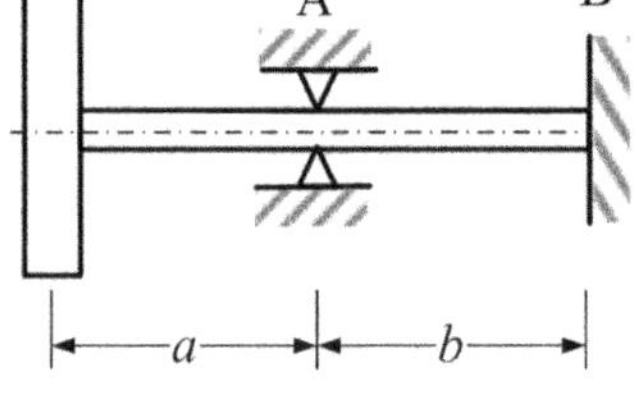

FIGURE 2.16 An overhung rotor system with one end fixed condition.

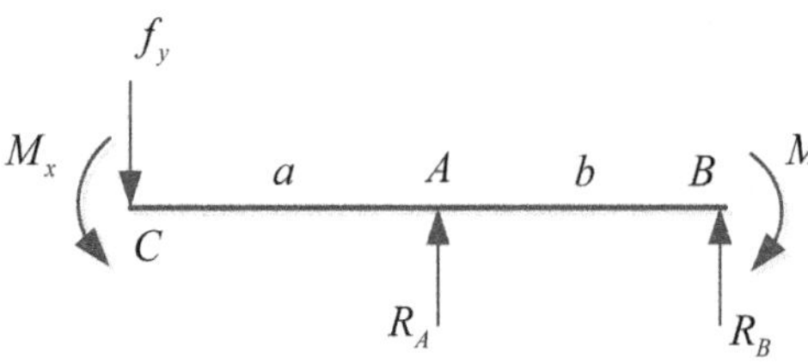

FIGURE 2.17 Free body diagram of the shaft system.

Here there are a total of four unknowns (i.e., R_A, R_{Bx}, R_{By} and M_{Bx}) and we have only three equilibrium equations for two-dimensional problem. The fourth unknown will be determined by using the deflection at A equal to zero. As at point A, there is a roller support, therefore at this point the deflection will be zero, i.e., $y_A = 0$.

For a cantilever beam with a point force (refer Figure 2.18) and moment (refer Figure 2.19) at free end (or any other point on the shaft), the translational deflection between the load point and fixed support is given as (refer Appendix 2.1)

$$y = \frac{Pz^2}{6EI}(3l - z) \quad \text{and} \quad y = \frac{Mz^2}{2EI} \tag{2.79}$$

Now we will use Eq. (2.79) to obtain translational deflection at point A in Figure 2.17. For force $P = f_y$ at $l = (a+b)$ distance, the translational deflection at $z=b$ will be

$$y_1 = \frac{Pz^2}{6EI}(3l - z) = \frac{f_y b^2}{6EI}(3l - b) \tag{2.80}$$

For moment M_x at $l = (a+b)$ distance, the translational deflection at $z=b$ will be

$$y_2 = \frac{Mz^2}{2EI} = \frac{M_x b^2}{2EI} \tag{2.81}$$

For force $P = -R_{Ay}$ at $l = b$ distance, the translational deflection at $z = b$ will be

$$y_3 = \frac{Pz^2}{6EI}(3l - z) = \frac{-R_{Ay}b^2}{6EI}(3b - b) = -\frac{R_{Ay}b^2}{6EI}(2b) = -\frac{R_{Ay}b^3}{3EI} \tag{2.82}$$

So total translational deflection at support A can be obtained by summing deflection components from Eqs. (2.80) through (2.82), which is in fact equal to zero. Now, applying the *superposition principle* to get the deflection at point A and equating it to zero. So, we have

$$y_A = y_{A1} + y_{A2} + y_{A3} = \frac{f_y b^2}{6EI}(3l - b) + \frac{M_x b^2}{2EI} - \frac{R_{Ay}b^3}{3EI} = 0 \tag{2.83}$$

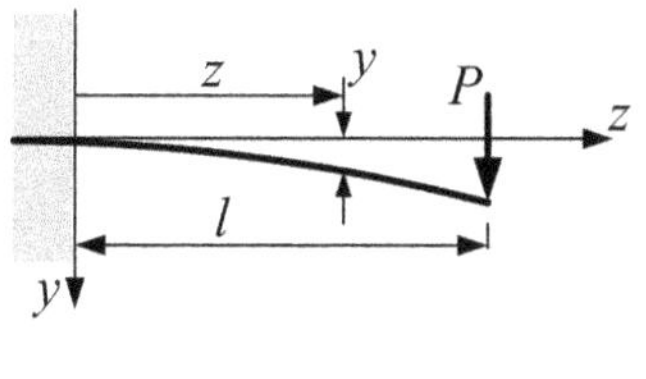

FIGURE 2.18 A cantilever shaft with a point force at free end.

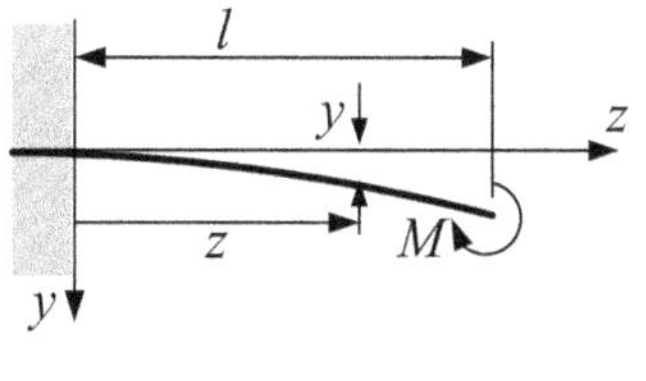

FIGURE 2.19 A cantilever shaft with a point moment at free end.

This can be used to solve for reaction at support A, as

$$\frac{R_{Ay}b^3}{3EI} = \frac{f_y b^2}{6EI}(3l-b) + \frac{M_x b^2}{2EI} \quad \text{or} \quad R_{Ay} = \frac{(3l-b)f_y}{2b} + \frac{3M_x}{2b} \equiv \frac{(3a+2b)f_y}{2b} + \frac{3M_x}{2b} \tag{2.84}$$

On taking the force balance in the free body diagram (refer Figure 2.17) in the vertical direction, we get

$$f_y = R_{Ay} + R_{By} \Rightarrow R_{By} = f_y - R_{Ay}$$

$$\text{or} \quad R_{By} = f_y - \left\{ \frac{(3a+2b)f_y}{2b} + \frac{3M_x}{2b} \right\} = -\frac{3af_y}{2b} - \frac{3M_x}{2b} \tag{2.85}$$

On taking the moment balance in the free body diagram (Figure 2.17), we get

$$\sum M_A = 0 \quad \Rightarrow R_{By}b - M_{Bx} + f_y a + M_x = 0; \quad \Rightarrow M_{Bx} = R_{By}b + f_y a + M_x$$

$$\text{or} \quad M_{Bx} = \left(-\frac{3af_y}{2b} - \frac{3M_x}{2b} \right) b + f_y a + M_x \Rightarrow M_{Bx} = -\frac{1}{2}\left(f_y a + M_x \right) \tag{2.86}$$

The strain energy in shaft section AC is (with bending moment at z with origin at C is given by $(f_y z + M)$)

$$U_1 = \int_0^a \frac{\left(f_y z + M_x \right)^2}{2EI} dz \tag{2.87}$$

The strain energy in shaft section BA is (with the bending moment at z with origin at C is given by $\left\{ f_y z + M_x - R_{Ay}(z-a) \right\}$)

$$U_2 = \int_a^l \frac{\left\{ f_y z + M_x - R_{Ay}(z-a) \right\}^2}{2EI} dz \quad \text{with} \quad R_{Ay} = \frac{(3a+2b)f_y}{2b} + \frac{3M_x}{2b}$$

$$\text{or} \quad U_2 = \frac{1}{2EI} \int_a^l \left\{ f_y z + M_x - R_{Ay}(z-a) \right\}^2 dz$$

$$= \frac{1}{2EI} \int_a^l \left(f_y z + M_x - \left\{ \frac{(3a+2b)f_y}{2b} + \frac{3M_x}{2b} \right\} (z-a) \right)^2 dz$$

$$\text{or} \quad U_2 = \frac{1}{2EI} \int_a^l \left[\left\{ \frac{-3az + (3a^2 + 2ab)}{2b} \right\} f_y + \left\{ \frac{-3z + (3a+2b)}{2b} \right\} M_x \right]^2 dz \tag{2.88}$$

Therefore, total strain energy is

$$U = U_1 + U_2 = \int_0^a \frac{\left(f_y z + M_x \right)^2}{2EI} dz + \frac{1}{2EI} \int_a^l \left[\left\{ \frac{-3az + (3a^2 + 2ab)}{2b} \right\} f_y + \left\{ \frac{-3z + (3a+2b)}{2b} \right\} M_x \right]^2 dz \tag{2.89}$$

Hence, the transverse deflection at C is given as

$$\delta_C = \frac{\partial U}{\partial f_y} = \int_0^a \frac{(f_y z + M_x)z}{EI} dz + \frac{1}{EI}\int_a^l \left[\left\{ \frac{-3az + (3a^2 + 2ab)}{2b} \right\} f_y + \left\{ \frac{-3z + (3a + 2b)}{2b} \right\} M_x \right] \left\{ \frac{-3az + (3a^2 + 2ab)}{2b} \right\} dz$$

$$\text{or} \quad \delta_C = \left(\frac{a^3}{3EI} + \frac{a^2 b}{4EI} \right) f_y + \left(\frac{a^2}{2EI} + \frac{ab}{4EI} \right) M_x \equiv \alpha_{yf} f_y + \alpha_{yM} M_x \tag{2.90}$$

Therefore, influence coefficients are

$$\alpha_{yf} = \frac{1}{2EI}\left(\frac{ba^2}{2} + \frac{2a^3}{3} \right) = \frac{1}{EI}\left(\frac{ba^2}{4} + \frac{a^3}{3} \right) \quad \text{and} \quad \alpha_{yM} = \frac{1}{EI}\left(\frac{ab}{4} + \frac{a^2}{2} \right) \tag{2.91, 2.92}$$

The angular deflection at C will be

$$\varphi_C = \frac{\partial U}{\partial M_x} = \int_0^a \frac{(f_y z + M_x)}{EI} dz + \frac{1}{EI}\int_a^l \left[\left\{ \frac{-3az + (3a^2 + 2ab)}{2b} \right\} f_y + \left\{ \frac{-3z + (3a + 2b)}{2b} \right\} M_x \right] \left\{ \frac{-3z + (3a + 2b)}{2b} \right\} dz$$

$$\text{or} \quad \varphi_C = \frac{b}{8EI}(2M_x + 2f_y a) + \frac{1}{2EI}(2M_x a + f_y a^2) = \frac{1}{EI}\left(\frac{ab}{4} + \frac{a^2}{2} \right) f_y + \frac{1}{EI}\left(\frac{b}{4} + a \right) M_x \tag{2.93}$$

Therefore, corresponding influence coefficients are given as

$$\alpha_{\varphi f} = \frac{1}{EI}\left(\frac{ab}{4} + \frac{a^2}{2} \right) \quad \text{and} \quad \alpha_{\varphi M} = \frac{1}{EI}\left(\frac{b}{4} + a \right) \tag{2.94, 2.95}$$

On putting the values of a and b in the aforementioned equations, we get

$$\alpha_{yf} = 2.4010 \times 10^{-4} \text{ m/N}; \quad \alpha_{yM} = \alpha_{\varphi f} = 9.4584 \times 10^{-4} \text{ 1/N} \quad \text{and}$$

$$\alpha_{\varphi M} = 4.6079 \times 10^{-3} \text{ 1/N-m}$$

Equations of motion in a single plane can be written as (refer Tiwari, 2017)

$$\begin{bmatrix} m & 0 \\ 0 & I_d \end{bmatrix} \begin{Bmatrix} \ddot{x} \\ \ddot{\varphi}_y \end{Bmatrix} + \begin{bmatrix} \alpha_{yf} & \alpha_{yM} \\ \alpha_{\varphi f} & \alpha_{\varphi M} \end{bmatrix}^{-1} \begin{Bmatrix} x \\ \varphi_y \end{Bmatrix} = \begin{Bmatrix} 0 \\ 0 \end{Bmatrix} \tag{2.96}$$

For the free vibration, we have

$$\left(-\omega_{nf}^2 \begin{bmatrix} m & 0 \\ 0 & I_d \end{bmatrix} + \begin{bmatrix} \alpha_{yf} & \alpha_{yM} \\ \alpha_{\varphi f} & \alpha_{\varphi M} \end{bmatrix}^{-1} \right) \begin{Bmatrix} x \\ \varphi_y \end{Bmatrix} = \begin{Bmatrix} 0 \\ 0 \end{Bmatrix}$$

where ω_{nf} is the natural frequency of the system. For the non-trivial solution, we have

$$\left| -\omega_{nf}^2 \begin{bmatrix} m & 0 \\ 0 & I_d \end{bmatrix} + \begin{bmatrix} \alpha_{yf} & \alpha_{yM} \\ \alpha_{\varphi f} & \alpha_{\varphi M} \end{bmatrix}^{-1} \right| = 0$$

which gives the frequency equation, as

$$mI_d\omega_{nf}^4(\alpha_{yf}\alpha_{\varphi M} - \alpha_{\varphi f}^2) - \omega_{nf}^2(m\alpha_{yf} + I_d\alpha_{\varphi M}) + 1 = 0 \tag{2.97}$$

On substituting numerical values for the present problem, it gives

$$2.1129 \times 10^{-8}\omega_{nf}^4 - 1.2921 \times 10^{-3}\omega_{nf}^2 + 1 = 0$$

which gives,

$$\omega_{nf1} = 27.99 \text{ rad/s} \quad \text{and} \quad \omega_{nf2} = 245.49 \text{ rad/s}.$$

Exercise 2.17 Consider a rotor system as shown in Figure 2.20 for obtaining the transverse natural frequency. Two flexible *massless* shafts are connected by a coupling (i.e., a pin joint). A thin disc of mass 3 kg is attached to one of the shafts (the shaft on the left) and it is not interfering with the relative motion between the two shafts. The other ends of shafts have fixed conditions. Take the length of each of the shafts as 0.5 m and the diameter as 0.05 m. Young's modulus $E = 2.1\ (10)^{11}$ N/m^2.

Solution: First we need to calculate influence coefficients. We may consider two cases (a) uncoupled transverse translational and rotational motions and (b) coupled case.

(a) *Method 1 (Single DOF)*: Let us apply a vertical force F at the free end of one beam that is at the pin location as shown in Figure 2.21. The force F gets divided into two parts as shown in Figure 2.22 due to the symmetry (i.e. of equal dimensions) of two shafts. The reaction at the pin join will be equal and opposite irrespective of shaft dimensions but the net force will be different in case shafts have different dimensions.

Now since both cantilever beams are acting as parallel so total stiffness of the rotor system will be

$$k_{eq} = \frac{3EI}{(l/2)^3} + \frac{3EI}{(l/2)^3} = \frac{24EI}{l^3} + \frac{24EI}{l^3} = \frac{48EI}{l^3}$$

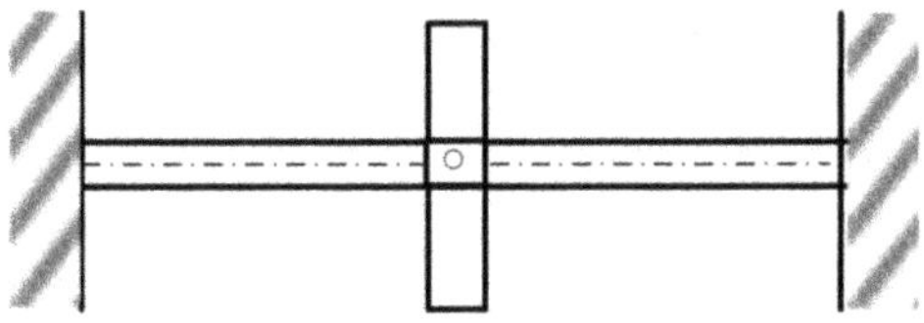

FIGURE 2.20 Two shafts connected by a coupling.

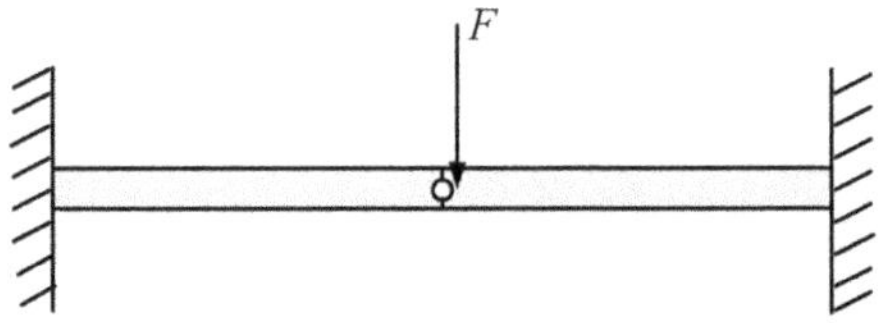

FIGURE 2.21 Two cantilever shafts connected by a pin joint with an external force on one of shaft.

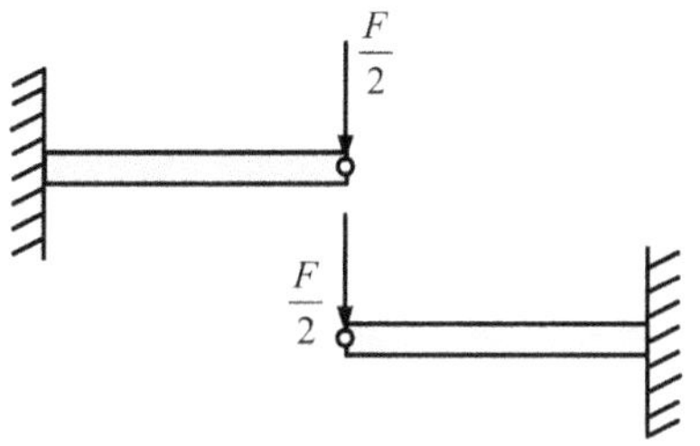

FIGURE 2.22 Free body diagram of two shafts.

Given data are: $m = 3$ kg, $l_1 = 0.5$ m, $l_2 = 0.5$ m, $d = 0.05$ m, $E = 210$ GPa, $I = \dfrac{\pi d^4}{64} = 3.0680 \times 10^{-7}\,\text{m}^4$
So, the stiffness will be

$$k_{eq} = \frac{48EI}{l^3} = \frac{48 \times 2.1 \times 10^{11} \times 3.0680 \times 10^{-7}}{1^3} = 3.0925 \times 10^6\,\text{N/m}$$

So the natural frequency of the rotor system is

$$\omega_{nf} = \sqrt{\frac{k_{eq}}{m}} = \sqrt{\frac{3.0925 \times 10^6}{3}} = 1015.30\ \text{rad/s.}$$

Method 2: The strain energy stored due to bending in one shaft is given as

$$U = \int_0^{0.5} \frac{M^2}{2EI} dz \quad \text{with} \quad M = Fz/2 \tag{2.98}$$

The translational displacement is expressed as

$$\delta = \frac{\partial U}{\partial F} = \int_0^{0.5} \frac{M\left(\dfrac{\partial M}{\partial F}\right)}{EI_1} dz = \int_0^{0.5} \frac{Fz^2}{4EI_1} dz = \frac{F}{4EI_1} \frac{z^3}{3}\Bigg]_0^{0.5} = \frac{F}{4EI_1} \frac{0.5^3}{3} \tag{2.99}$$

$$\text{or} \quad \delta = 1.6168 \times 10^{-7} F = \alpha_{yf} F \tag{2.100}$$

Hence, the stiffness for the transverse translational motion is given as

$$k = \frac{F/2}{\delta} = 1.5463 \times 10^6\ \text{N/m} \tag{2.101}$$

There are two such shafts. Their stiffness acts in parallel. Hence, they will get added. Hence, the natural frequency related to the pure translational motion is, (the same as in the previous method)

$$\omega_{nf} = \sqrt{\frac{2k}{m}} = 1{,}015.30\ \text{rad/s.}$$

(a) *Method 2* (*Two DOFs*): Let us take R as the reaction force at pin joint due to the load F in one of the shafts (as mentioned in the left shaft near pin location) (refer Figure 2.22). On taking free body diagram, the new force on the left shaft would be $(F - R)$ and on the right shaft will be R. If these two shafts are joined together the reaction force must cancel each other. The pin joint allows the rotation at the joint but the translational displacement of both shafts will be the same at the common point (i.e. at the pin joint). Now the translational deflection at the free end of the left cantilever beam is given by

$$y_a = \frac{(F-R)l^3}{3EI} \tag{2.102}$$

Similarly, the translational deflection at the free end of the right cantilever beam is given by

$$y_b = \frac{Rl^3}{3EI} \tag{2.103}$$

It should be noted that the length of both shafts is the same $l_1 = l_2 = l$. Also, we know that $y_a = y_b$, which gives

$$\frac{(F-R)l^3}{3EI} = \frac{Rl^3}{3EI} \tag{2.104}$$

On solving for the unknown reaction force R, we get

$$R = \frac{F}{2} \tag{2.105}$$

It should be noted that Eq. (2.104) can be used to obtain reactions force when the lengths of two shafts are different.

Now considering an external moment M is only applied near the pin location on the left shaft and due to this moment, the reaction at pin is, let us, say R_1. As mentioned earlier, the pin joint allows the rotation at the joint but the translational displacement of both shafts will be the same at the common point (i.e. at the pin joint). Thus, the deflections at the free end of the cantilever shafts are given by

$$y_a = \frac{Ml^2}{2EI} - \frac{R_1 l^3}{3EI}; \quad y_b = \frac{R_1 l^3}{3EI} \tag{2.106}$$

Again, it should be noted that the length of both shafts is same $l_1 = l_2 = l$. Since $y_a = y_b$, so we will get

$$y_a = y_b \quad \Rightarrow \frac{Ml^2}{2EI} - \frac{R_1 l^3}{3EI} = \frac{R_1 l^3}{3EI} \quad \Rightarrow \frac{Ml^2}{2EI} = \frac{2R_1 l^3}{3EI}; \quad R_1 = \frac{3M}{4l} \tag{2.107}$$

Now the bending moment is obtained, when both force F and moment M are acting near the pin, in the left shaft, as

$$M_x = \frac{F}{2}z - R_1 z + M = \left(\frac{F}{2} - \frac{3M}{4l}\right)z + M \tag{2.108}$$

Now the strain energy is obtained, in the left shaft, as

$$U = \int_0^l \frac{M_z^2 dz}{2EI} = \frac{1}{2EI}\left\{\left(\frac{F}{2} - \frac{3M}{4l}\right)z + M\right\}^2 dz = \frac{1}{2EI}\int_0^l \left\{\left(\frac{F}{2} - \frac{3M}{4l}\right)^2 z^2 + M^2 + 2\left(\frac{F}{2} - \frac{3M}{4l}\right)zM\right\}dz \tag{2.109}$$

After integration and putting the limits, we get

$$U = \frac{1}{2EI}\left[\left(\frac{F}{2} - \frac{3M}{4l}\right)^2 \frac{l^3}{3} + \left(\frac{F}{2} - \frac{3M}{4l}\right) l^2 M + M^2 l\right]$$

$$\text{or} \quad U = \frac{1}{2EI}\left[\frac{F^2 l^3}{12} + M^2 l\left(\frac{3}{16} + 1 - \frac{3}{4}\right) + \frac{MFl^2}{4}\right] \tag{2.110}$$

On solving, we get

$$U = \frac{1}{2EI}\left[\frac{7M^2 l}{16} + \frac{MFl^2}{4} + \frac{F^2 l^3}{12}\right] \tag{2.111}$$

The translational displacement is obtained as

$$y(l) = \frac{dU}{dF} = \frac{1}{2EI}\left(\frac{Ml^2}{4} + \frac{Fl^3}{6}\right) = \left(\frac{l^3}{12EI}\right)F + \left(\frac{l^2}{8EI}\right)M = \alpha_{yf}F + \alpha_{yM}M \tag{2.112}$$

Now, the angular displacement is obtained, as

$$\varphi_z(l) = \frac{dU}{dM} = \frac{1}{2EI}\left(\frac{7Ml}{8} + \frac{Fl^2}{4}\right) = \left(\frac{l^2}{8EI}\right)F + \left(\frac{7l}{16EI}\right)M = \alpha_{\varphi f}F + \alpha_{\varphi M}M \tag{2.113}$$

Hence, we have the influence coefficient matrix, as

$$\boldsymbol{\alpha} = \begin{bmatrix} \alpha_{yf} & \alpha_{yM} \\ \alpha_{\varphi f} & \alpha_{\varphi M} \end{bmatrix} = \begin{bmatrix} \dfrac{l^3}{12EI} & \dfrac{l^2}{8EI} \\ \dfrac{l^2}{8EI} & \dfrac{7l}{16EI} \end{bmatrix} \tag{2.114}$$

Given data are: $m = 3$ kg, $l_1 = 0.5$ m, $l_2 = 0.5$ m, $d = 0.05$ m, $E = 210$ GPa, $I = \dfrac{\pi d^4}{64} = 3.0680 \times 10^{-7}\ \text{m}^4$

On putting values in the influence coefficient matrix, we have

$$\boldsymbol{\alpha} = \begin{bmatrix} 1.2934 \times 10^{-6} & 1.9402 \times 10^{-6} \\ 1.9402 \times 10^{-6} & 6.7906 \times 10^{-6} \end{bmatrix} \tag{2.115}$$

Hence the stiffness matrix will be

$$\mathbf{K} = \boldsymbol{\alpha}^{-1} = 10^6 \begin{bmatrix} 1.3530 & -0.3866 \\ -0.3866 & 0.2577 \end{bmatrix} \tag{2.116}$$

We have the mass matrix, as $\mathbf{M} = \begin{bmatrix} 3 & 0 \\ 0 & 0.05 \end{bmatrix}$ so that $\mathbf{M}^{-1} = \begin{bmatrix} 0.333 & 0 \\ 0 & 20 \end{bmatrix}$.

From the eigenvalue problem, we have

$$\left(\mathbf{M}^{-1}\mathbf{K} - \omega_{nf}^2 \mathbf{I}\right)\boldsymbol{\eta} = \mathbf{0} \tag{2.117}$$

We have

$$\mathbf{M}^{-1}\mathbf{K} = 10^6 \begin{bmatrix} 0.4510 & -0.1289 \\ -7.7313 & 5.1542 \end{bmatrix} \tag{2.118}$$

Hence, the eigenvalue problem becomes

$$\left(\mathbf{M}^{-1}\mathbf{K} - \omega_{nf}^2 \mathbf{I}\right)\boldsymbol{\eta} = \begin{bmatrix} 0.4510 \times 10^6 - \omega_{nf}^2 & -0.1289 \times 10^6 \\ -7.7313 \times 10^6 & 5.1542 \times 10^6 - \omega_{nf}^2 \end{bmatrix} \begin{Bmatrix} y \\ \varphi_x \end{Bmatrix} = \begin{Bmatrix} 0 \\ 0 \end{Bmatrix} \tag{2.119}$$

For the non-trivial solution on putting the determinant of the matrix to zero, we get

$$0.1500\omega_{nf}^4 - 8.4077 \times 10^5 \omega_{nf}^2 + 1.9924 \times 10^{11} = 0 \tag{2.120}$$

On solving, we get natural frequencies, as

$$\omega_{nf1} = 497.94 \text{ rad/s} \quad \text{and} \quad \omega_{nf2} = 2314.57 \text{ rad/s}.$$

Exercise 2.18 For a Jeffcott rotor the following energy expression are given:

$$T = \tfrac{1}{2} m\dot{x}_G^2 + \tfrac{1}{2} m\dot{y}_G^2; \quad U = \tfrac{1}{2} kx^2 + \tfrac{1}{2} ky^2; \quad \delta W_{nc} = (-c\dot{x})\delta x + (-c\dot{y})\delta y$$

$$\text{with} \quad x_G = x + e\cos\omega t \quad \text{and} \quad y_G = y + e\sin\omega t.$$

where T is the kinetic energy, U is the potential energy, δW_{nc} is the non-conservative virtual work done, x and y are the coordinates of disc geometrical centre (i.e., generalised coordinates), x_G and y_G are the coordinates of disc centre of gravity, e is the eccentricity, m is the mass of the disc, k is the stiffness of the shaft, c is the viscous damping in the rotor system, and ω is the spin speed of the rotor. Using Lagrange's equation (refer to chapter 7 of Tiwari (2017)) obtain equations of motion of the rotor system.

***Solution*:** Given KE and PE energies, respectively, are

$$T = \tfrac{1}{2} m\dot{x}_G^2 + \tfrac{1}{2} m\dot{y}_G^2 = \tfrac{1}{2} m(\dot{x} - e\omega\sin\omega t)^2 + \tfrac{1}{2} m(\dot{y} + e\omega\cos\omega t)^2 \tag{2.121}$$

$$U = \tfrac{1}{2} kx^2 + \tfrac{1}{2} ky^2 \tag{2.122}$$

And non-conservative virtual work done is

$$\delta W_{nc} = (-c\dot{x})\delta x + (-c\dot{y})\delta y \tag{2.123}$$

Using Lagrange's equation, for coordinate y, we have

$$\frac{d}{dt}\left(\frac{\partial T}{\partial \dot{y}}\right) - \frac{\partial T}{\partial y} + \frac{\partial U}{\partial y} = \frac{\partial(\delta W_{nc})}{\partial y} \tag{2.124}$$

So that

$$\frac{d}{dt}\left(\frac{\partial T}{\partial \dot{y}}\right) = m(\ddot{y} - e\omega^2\sin\omega t); \quad \frac{\partial T}{\partial y} = 0; \quad \frac{\partial U}{\partial y} = ky; \quad \frac{\partial(\delta W_{nc})}{\partial y} = -c\dot{y} \tag{2.125}$$

Hence, equations of motion will be

$$m(\ddot{y} - e\omega^2 \sin\omega t) + ky = -c\dot{y} \quad \text{or} \quad m\ddot{y} + c\dot{y} + ky = me\omega^2 \sin\omega t \tag{2.126}$$

Similarly using Lagrange's equation for the coordinate, x, we have

$$\frac{d}{dt}\left(\frac{\partial T}{\partial \dot{x}}\right) - \frac{\partial T}{\partial x} + \frac{\partial U}{\partial x} = \frac{\partial(\delta W_{nc})}{\partial x} \tag{2.127}$$

So that

$$\frac{d}{dt}\left(\frac{\partial T}{\partial \dot{x}}\right) = m(\ddot{x} - e\omega^2 \cos\omega t); \quad \frac{\partial T}{\partial x} = 0; \quad \frac{\partial U}{\partial x} = kx; \quad \frac{\partial(\delta W_{nc})}{\partial x} = -c\dot{x} \tag{2.128}$$

Hence, the equation of motion becomes

$$m(\ddot{x} - e\omega^2 \cos\omega t) + kx = -c\dot{x} \quad \text{or} \quad m\ddot{x} + c\dot{x} + kx = me\omega^2 \cos\omega t \tag{2.129}$$

Exercise 2.19 For a Jeffcott rotor with an offset disc, the following energy expressions are given:

$$T = \tfrac{1}{2}m\dot{x}_G^2 + \tfrac{1}{2}m\dot{y}_G^2 + \tfrac{1}{2}I_d\dot{\varphi}_y^2 + \tfrac{1}{2}I_d\dot{\varphi}_x^2 \quad \text{with} \quad x_G = x + e\cos\omega t \quad \text{and} \quad y_G = y + e\sin\omega t$$

$$\text{and} \quad U = \tfrac{1}{2}k_{11}x^2 + \tfrac{1}{2}k_{11}y^2 + \tfrac{1}{2}k_{22}\varphi_x^2 + \tfrac{1}{2}k_{22}\varphi_y^2 + \tfrac{1}{2}k_{12}\varphi_x y + \tfrac{1}{2}k_{12}\varphi_y x; \quad \text{with} \quad k_{12} = k_{21}$$

where T is the kinetic energy, U is the potential energy, x and y are the translational coordinates of the disc geometrical center, x_G and y_G are the coordinates of the disc center of gravity, φ_x and φ_y are rotational coordinates of the disc, e is the eccentricity, m is the mass of the disc, k is the stiffness of the shaft, subscripts 1 and 2 represent, respectively, the translational and rotational displacements, and ω is the spin speed of the rotor. Using the Lagrange's equation (refer to chapter 7 of Tiwari (2017)) obtain equations of motion of the rotor system.

***Solution*:** Given KE and PE energies are

$$\begin{aligned} T &= \tfrac{1}{2}m\dot{x}_G^2 + \tfrac{1}{2}m\dot{y}_G^2 + \tfrac{1}{2}I_d\dot{\varphi}_x^2 + \tfrac{1}{2}I_d\dot{\varphi}_y^2 \\ &= \tfrac{1}{2}m(\dot{x} - e\omega\sin\omega t)^2 + \tfrac{1}{2}m(\dot{y} + e\omega\cos\omega t)^2 + \tfrac{1}{2}I_d\dot{\varphi}_x^2 + \tfrac{1}{2}I_d\dot{\varphi}_y^2 \end{aligned} \tag{2.130}$$

$$\text{and} \quad U = \frac{1}{2}k_{11}x^2 + \tfrac{1}{2}k_{11}y^2 + \tfrac{1}{2}k_{22}\varphi_x^2 + \tfrac{1}{2}k_{22}\varphi_y^2 + \tfrac{1}{2}k_{12}\varphi_x y + \tfrac{1}{2}k_{12}\varphi_y x \tag{2.131}$$

The Lagrange's equation for coordinate x, we have

$$\frac{d}{dt}\left(\frac{\partial T}{\partial \dot{x}}\right) - \frac{\partial T}{\partial x} + \frac{\partial U}{\partial x} = \frac{\partial(\delta W_{nc})}{\partial x} \tag{2.132}$$

So that

$$\frac{d}{dt}\left(\frac{\partial T}{\partial \dot{x}}\right) = m(\ddot{x} - e\omega^2 \cos\omega t); \quad \frac{\partial T}{\partial x} = 0; \quad \frac{\partial U}{\partial x} = k_{11}x + \tfrac{1}{2}k_{12}\varphi_y + \tfrac{1}{2}k_{21}\varphi_y \tag{2.133}$$

Hence, the equation of motion is

$$m(\ddot{x} - e\omega^2 \cos\omega t) + k_{11}x + \tfrac{1}{2}k_{12}\varphi_y + \tfrac{1}{2}k_{21}\varphi_y = 0 \tag{2.134}$$

$$\text{or} \quad m\ddot{x} + k_{11}x + k_{12}\varphi_y = me\omega^2 \cos\omega t \quad \text{with} \quad k_{12} = k_{21} \tag{2.135}$$

Similarly, the Lagrange's equation for coordinate y, we have,

$$\frac{d}{dt}\left(\frac{\partial T}{\partial \dot{y}}\right) - \frac{\partial T}{\partial y} + \frac{\partial U}{\partial y} = \frac{\partial(\delta W_{nc})}{\partial y} \tag{2.136}$$

$$\text{and} \quad \frac{d}{dt}\left(\frac{\partial T}{\partial \dot{y}}\right) = m(\ddot{y} - e\omega^2 \sin\omega t); \quad \frac{\partial T}{\partial y} = 0; \quad \frac{\partial U}{\partial y} = k_{11}y + \tfrac{1}{2}k_{12}\varphi_y + \tfrac{1}{2}k_{21}\varphi_y \tag{2.137}$$

Hence, the equation of motion, becomes

$$m(\ddot{y} - e\omega^2 \sin\omega t) + k_{11}y + \tfrac{1}{2}k_{12}\varphi_y + \tfrac{1}{2}k_{21}\varphi_y = 0 \tag{2.138}$$

$$\text{or} \quad m\ddot{y} + k_{11}y + k_{12}\varphi_y = me\omega^2 \sin\omega t \tag{2.139}$$

Similarly, the Lagrange's equation for the coordinate φ_x, we have

$$\frac{d}{dt}\left(\frac{\partial T}{\partial \dot{\varphi}_x}\right) - \frac{\partial T}{\partial \varphi_x} + \frac{\partial U}{\partial \varphi_x} = \frac{\partial(\delta W_{nc})}{\partial \varphi_x} \tag{2.140}$$

So that

$$\frac{d}{dt}\left(\frac{\partial T}{\partial \dot{\varphi}_x}\right) = I_d\ddot{\varphi}_x; \quad \frac{\partial T}{\partial \varphi_x} = 0; \quad \frac{\partial U}{\partial \varphi_x} = k_{22}\varphi_x + \tfrac{1}{2}k_{12}y + \tfrac{1}{2}k_{21}y \tag{2.141}$$

Hence, the equation of motion becomes,

$$I_d\ddot{\varphi}_x + k_{22}\varphi_x + \tfrac{1}{2}k_{12}y + \tfrac{1}{2}k_{21}y = 0 \tag{2.142}$$

$$\text{or} \quad I_d\ddot{\varphi}_x + k_{22}\varphi_x + k_{12}y = 0 \tag{2.143}$$

Similarly, the Lagrange's equation for coordinate φ_y, we get

$$\frac{d}{dt}\left(\frac{\partial T}{\partial \dot{\varphi}_y}\right) - \frac{\partial T}{\partial \varphi_y} + \frac{\partial U}{\partial \varphi_y} = \frac{\partial(\delta W_{nc})}{\partial \varphi_y} \tag{2.144}$$

So that

$$\frac{d}{dt}\left(\frac{\partial T}{\partial \dot{\varphi}_y}\right) = I_d\ddot{\varphi}_y; \quad \frac{\partial T}{\partial \varphi_y} = 0; \quad \frac{\partial U}{\partial \varphi_y} = k_{22}\varphi_y + \tfrac{1}{2}k_{12}x + \tfrac{1}{2}k_{21}x \tag{2.145}$$

Hence, the equation of motion becomes

$$I_d\ddot{\phi}_y + k_{22}\phi_y + k_{12}x = 0 \tag{2.146}$$

Interestingly, in the present system, the coupling of translational and angular displacements can be seen in a plane (either in y–z or z–x). Such a system is possible when we have an axially asymmetrical rigid rotor mounted on flexible bearing having centre of gravity offset from mid-span. However, the bearings do not have cross-coupled terms but it is anisotropic.

Exercise 2.20 For a Jeffcott rotor to alleviate the crossing of critical speed during run-up a switching technique by an auxiliary bearing is applied. The stiffness of the rotor without and with auxiliary bearing are k_1 and k_2, respectively. Let the amplitudes of the rotor without and with auxiliary bearing are y_1 and y_2, respectively. Let m be the mass of the rotor and ω be the spin speed of the shaft. Obtain the switching frequency, ω_c, at which the system changes from the auxiliary bearing to without auxiliary bearing. Also, obtain the maximum non-dimensional amplitude of vibration, $|\bar{y}_{\max}| = |\bar{y}/e|$. Give the plot of y_1 and y_2 with the spin speed and show the switching frequency location.

Solution: The non-dimensional unbalance response y/e is given as (refer eq. (2.16) of Tiwari (2017)),

$$\bar{y}_1 = y_1/e = \frac{m\omega^2}{k_1 - m\omega^2} \quad \text{(Without auxiliary bearing)} \tag{2.147}$$

$$\bar{y}_2 = y_2/e = \frac{m\omega^2}{k_2 - m\omega^2} \quad \text{(With auxiliary bearing)} \tag{2.148}$$

At the switching frequency (refer Figure 2.23, the switching frequency is after critical speed of the original system so corresponding displacement will be negative), we have

$$\bar{y}_2 = -\bar{y}_1 \Rightarrow \frac{m\omega_c^2}{k_2 - m\omega_c^2} = -\frac{m\omega_c^2}{k_1 - m\omega_c^2} \Rightarrow (k_1 - m\omega_c^2) = -(k_2 - m\omega_c^2)$$
$$\text{or} \quad 2m\omega_c^2 = k_1 + k_2 \Rightarrow \omega_c = \sqrt{\frac{k_1 + k_2}{2m}} \tag{2.149}$$

Substituting this in $\bar{y}_1$ (Eq. (2.147)), we get

$$\text{Maximum amplitude} = \frac{\frac{1}{2}(k_1 + k_2)}{k_1 - \frac{1}{2}(k_1 + k_2)} \Rightarrow \bar{y} = \left|\frac{y}{e}\right| = \frac{k_1 + k_2}{k_1 - k_2} \tag{2.150}$$

Figure 2.23 shows the variation of amplitude with spin speed of rotor without and with auxiliary bearing for a typical value of system parameters given in the figure title.

Exercise 2.21 Obtain the transverse natural frequency of the rotor-bearing system as shown in Figure 2.24. Consider the shaft to be rigid. The bearing on the left is simply supported, and the bearing on the right is having two springs, and each spring has the stiffness, k. Take L = 1 m, a = 0.3 m, k = 1 kN/m, m = 5 kg, and I_d = 0.02 kg-m^2. Consider a single-plane motion and neglect the gyroscopic couple effect.

Solution: The present problem will be solved by two methods (i) Method 1: Lagrange's equation (ii) Method 2: Newton's second law.

Method 1: Let us take y as the translational displacement of mass, m. The present system has only single DOF. Let φ_x be the tilt of the mass (or the shaft as shown in Figure 2.25) and is related with y, as

$$\varphi_x = y/l \tag{2.151}$$

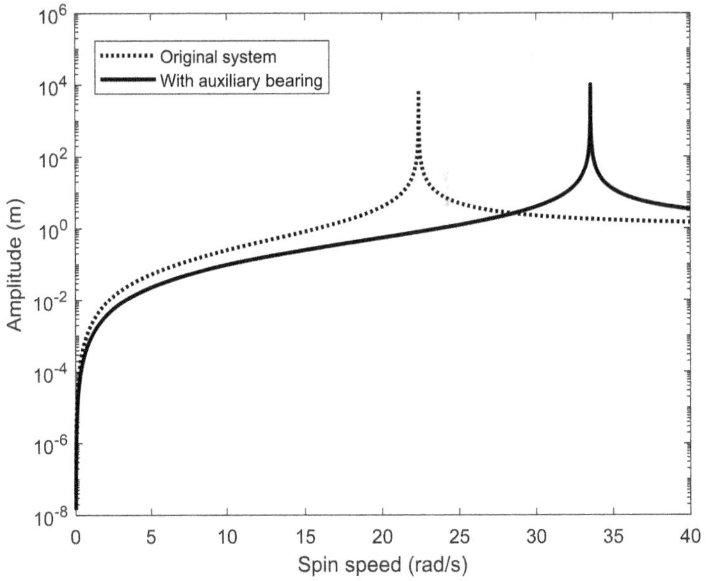

FIGURE 2.23 Plot of rotor response without and with auxiliary bearing (switching frequency will at the intersection of the two curves) (for m =10 kg; k_1 = 5,000 N/m (first rotor system); k_2 = 11,250 N/m (second rotor system with auxiliary bearing)).

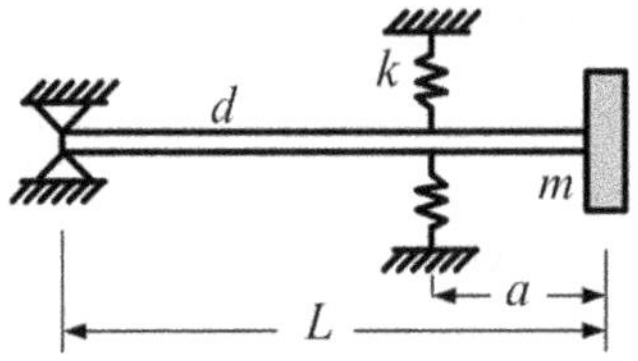

FIGURE 2.24 A rigid shaft supported on a flexible support and a rigid support.

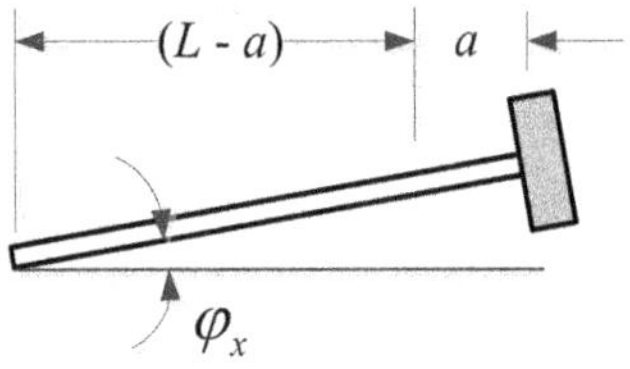

FIGURE 2.25 The shaft in deflected position.

where l is the length of the shaft. The displacement at the spring will be

$$y_s = (l - a)\varphi_x \tag{2.152}$$

The kinetic energy (KE) of the rotor system is given as

$$T = \tfrac{1}{2}m\dot{y}^2 + \tfrac{1}{2}I_d\dot{\varphi}_x^2 = \tfrac{1}{2}ml^2\dot{\varphi}_x^2 + \tfrac{1}{2}I_d\dot{\varphi}_x^2 = \tfrac{1}{2}\left(ml^2 + I_d\right)\dot{\varphi}_x^2 \tag{2.153}$$

The potential energy (PE) of the rotor system is

$$U = \tfrac{1}{2}(2k)y_s^2 = k(l-a)^2\varphi_x^2 \tag{2.154}$$

Similarly, the Lagrange's equation for coordinate φ_x, we have

$$\frac{d}{dt}\left(\frac{\partial T}{\partial \dot{\varphi}_x}\right) - \frac{\partial T}{\partial \varphi_x} + \frac{\partial U}{\partial \varphi_x} = \frac{\partial(\delta W_{nc})}{\partial \varphi_x} \tag{2.155}$$

and $$\frac{d}{dt}\left(\frac{\partial T}{\partial \dot{\varphi}_x}\right) = \left(ml^2 + I_d\right)\ddot{\varphi}_x; \quad \frac{\partial T}{\partial \varphi_x} = 0; \quad \frac{\partial U}{\partial \varphi_x} = 2k(l-a)^2\varphi_x \tag{2.156}$$

Hence, the equation of motion becomes

$$\left(ml^2 + I_d\right)\ddot{\varphi}_x + 2k(l-a)^2\varphi_x = 0 \tag{2.157}$$

Method 2: For a tilt of φ_x, the total spring force can be written as (refer Figure 2.25)

$$F_s = 2k(l-a)\sin\varphi_x \tag{2.158}$$

For a small value of φ_x, we have $\sin\varphi_x \to \varphi_x$. The net spring force reduces to

$$f_s(t) = 2k(l-a)\varphi_x(t) \tag{2.159}$$

Therefore, the total restoring moment acting about the support location will be

$$M(t) = 2k(l-a)^2\varphi_x(t) \tag{2.160}$$

Equations of motion become

$$(I_d + ml^2)\ddot{\varphi}_x + 2k(l-a)^2\varphi_x = 0 \tag{2.161}$$

Hence, the natural frequency is given, as (which is the same as of Eq. (2.157))

$$\omega_{nf} = \sqrt{\frac{2k(l-a)^2}{(I_d + ml^2)}} \tag{2.162}$$

Given data are

$$L = 1 \text{ m},\ a = 0.3 \text{ m},\ k = 1 \text{ kN/m},\ m = 5 \text{ kg and } I_d = 0.02 \text{ kg-m}^2.$$

On substituting numerical values, we get

$$\omega_{nf} = \sqrt{\frac{2\times 1{,}000\times(0.7^2)}{0.02 + 5\times 1^2}} = 13.97 \text{ rad/s}.$$

Exercise 2.22 For a Jeffcott rotor with an offset disc, derive the equations of motion using the energy method (e.g. Lagrange's method - refer to Chapter 7 of Tiwari (2017)) for a single plane. Energy terms are the kinetic energy $T = \tfrac{1}{2}m\dot{y}^2 + \tfrac{1}{2}I_d\dot{\varphi}_x^2$, the potential energy $U = \tfrac{1}{2}k_{11}y^2 + \tfrac{1}{2}k_{22}\varphi_x^2 + k_{12}y\varphi_x$

(with $k_{12} = k_{21}$) and non-conservative virtual work $\delta W_{nc} = \left(me\omega^2 \sin\omega t\right)\delta y$. Here the stiffness terms have been obtained by inverting the influence coefficient matrix. Here δ is the variational operator.

Solution: Given KE and PE energies are

$$T = \tfrac{1}{2}m\dot{y}^2 + \tfrac{1}{2}I_d\dot{\varphi}_x^2 \quad \text{and} \quad U = \tfrac{1}{2}k_{11}y^2 + \tfrac{1}{2}k_{22}\varphi_x^2 + k_{12}y\varphi_x \tag{2.163, 2.164}$$

And the non-conservative virtual work is

$$\delta W_{nc} = (me\omega^2 \sin\omega t)\delta y \tag{2.165}$$

The Lagrange's equation, for coordinate y, is given as

$$\frac{d}{dt}\left(\frac{\partial T}{\partial \dot{y}}\right) + \frac{\partial T}{\partial y} + \frac{\partial U}{\partial y} = \frac{\partial(\delta W_{nc})}{\partial y} \tag{2.166}$$

So that

$$\frac{d}{dt}\left(\frac{\partial T}{\partial \dot{y}}\right) = m\ddot{y}; \quad \frac{\partial T}{\partial y} = 0; \quad \frac{\partial U}{\partial y} = k_{11}y + k_{12}\varphi_x; \quad \frac{\partial(\delta W_{nc})}{\partial y} = me\omega^2 \sin\omega t \tag{2.167}$$

Hence, the equation of motion is given, as

$$m\ddot{y} + k_{11}y + k_{12}\varphi_x = me\omega^2 \sin\omega t \tag{2.168}$$

Similarly, the Lagrange's equation, for coordinate, φ_x, is given as

$$\frac{d}{dt}\left(\frac{\partial T}{\partial \dot{\varphi}_x}\right) + \frac{\partial T}{\partial \varphi_x} + \frac{\partial U}{\partial \varphi_x} = \frac{\partial(\delta W_{nc})}{\partial \varphi_x} \tag{2.169}$$

So that

$$\frac{d}{dt}\left(\frac{\partial T}{\partial \dot{\varphi}_x}\right) = I_d\ddot{\varphi}_x; \quad \frac{\partial T}{\partial \varphi_x} = 0; \quad \frac{\partial U}{\partial \varphi_x} = k_{22}\varphi_x + k_{21}y; \quad \frac{\partial(\delta W_{nc})}{\partial \varphi_x} = 0 \tag{2.170}$$

Hence, the equation of motion becomes

$$I_d\ddot{\varphi}_x + k_{22}\varphi_x + k_{21}y = 0 \tag{2.171}$$

Combining Eqs. (2.168) and (2.171), we get

$$\begin{bmatrix} m & 0 \\ 0 & I_d \end{bmatrix}\begin{Bmatrix} \ddot{y} \\ \ddot{\varphi}_x \end{Bmatrix} + \begin{bmatrix} k_{11} & k_{12} \\ k_{21} & k_{22} \end{bmatrix}\begin{Bmatrix} y \\ \varphi_x \end{Bmatrix} = \begin{Bmatrix} me\omega^2 \sin\omega t \\ 0 \end{Bmatrix} \tag{2.172}$$

Exercise 2.23 In a Jeffcott rotor the disc has an initial tilt by an angle α in transverse plane due to improper assembly, during spinning (ω rad/s) of the rotor, the unbalanced moment due to this would be what? Let I_d be the diametral mass moment of the disc and I_p is the polar mass moment of inertia.

Solution: The unbalance moment M is $(I_p - I_d)\omega^2\alpha$. For thin disc $I_p = 2I_d$, so $M = I_d\omega^2\alpha$. (refer chapter 5 of Tiwari (2017) for more details).

Exercise 2.24 For a single-plane motion (let us consider the vertical plane), obtain the whirl natural frequencies in a closed analytical form of the rotor-bearing system as shown in Figure 2.26. Consider the shaft to be massless and elastic. Both bearings are identical with no mass, and a thin disc is located at a distance of a from left end and b from the right end of the shaft (with $a + b = l$). The expression for the deflection of a simple supported shaft is given in Figure 2.27. Do not consider the gyroscopic effect.

***Solution*:** For the present case, both the shaft and bearings are flexible. We consider only vertical plane motion. The flexible shaft on flexible bearings (Figure 2.28a) have been considered in two cases, **Case 1**: rigid shaft on flexible bearings (Figure 2.28b) and **Case 2**: Flexible shaft on *rigid* bearings (Figure 2.28c).

Case-1: A rigid shaft on flexible bearings: A force P is applied on the shaft (refer to Figure 2.29) and that will be resisted by the spring, the force balance is given as

$$P = k_b y_{b1} + k_b y_{b2} \tag{2.173}$$

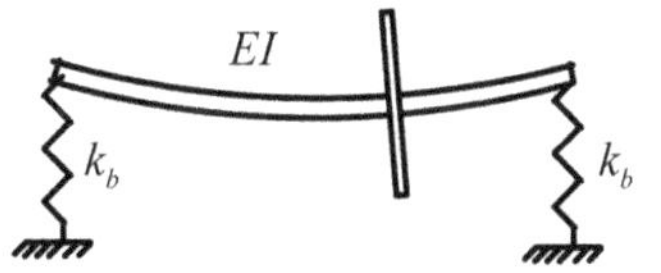

FIGURE 2.26 A flexible shaft on flexible bearings.

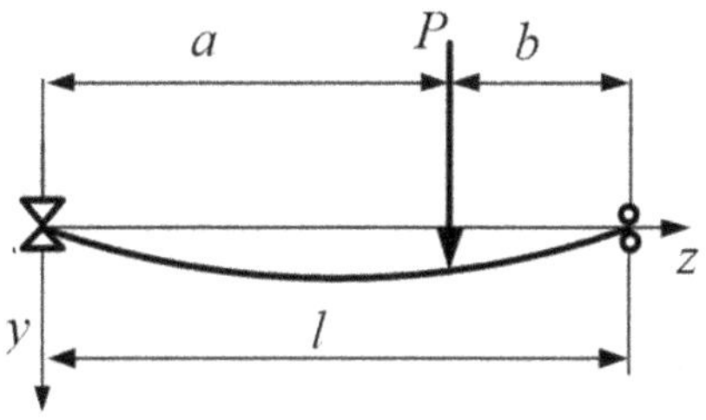

$$y = \frac{Pbz}{6lEI}\left(l^2 - z^2 - b^2\right) \text{ for } \left[0 < z < a\right]$$

$$y = \frac{Pb}{6lEI}\left[\frac{l}{b}(z-a)^3 + \left(l^2 - b^2\right)z - z^3\right] \text{ for } \left[a < z < l\right]$$

FIGURE 2.27 A simply supported shaft deflection due to an offset load.

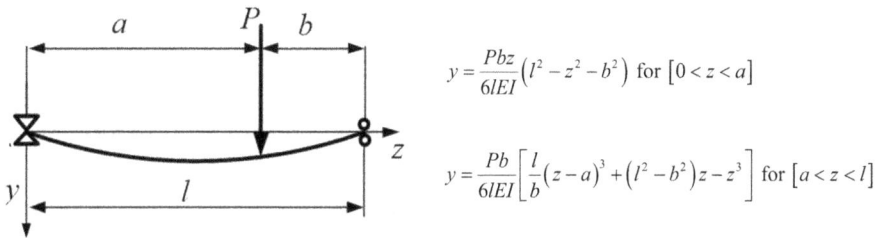

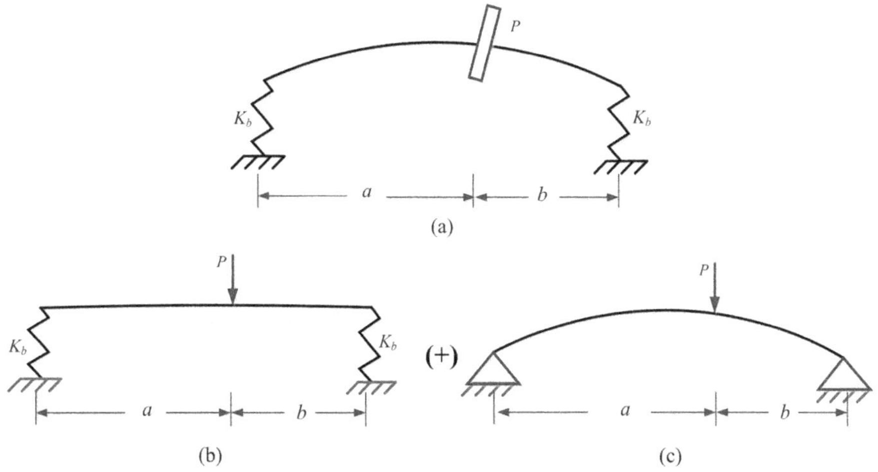

FIGURE 2.28 (a) Flexible shaft on flexible bearings (b) case I: rigid shaft on flexible bearings and (c) case II: flexible shaft on rigid bearings.

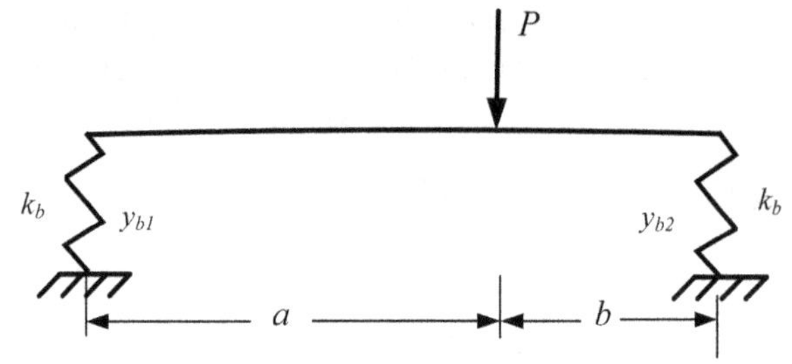

FIGURE 2.29 Rigid shaft on flexible bearings.

where k_b is the bearing stiffness y_{b1} and y_{b2} are shaft end displacements. Also, the moment balance about the location of the applied force P will be given as

$$k_b y_{b1} a = k_b y_{b2} b \tag{2.174}$$

which gives

$$y_{b1} = \frac{b y_{b2}}{a} \tag{2.175}$$

On substituting Eq. (2.175) into Eq. (2.173), we get

$$P = k_b \frac{b y_{b2}}{a} + k_b y_{b2} \quad \Rightarrow P = k_b y_{b2} \frac{l}{a} \quad \Rightarrow y_{b2} = \frac{Pa}{k_b l} \tag{2.176}$$

On substituting Eq. (2.176) into Eq. (2.175), we get

$$y_{b1} = \frac{Pb}{k_b l} \tag{2.177}$$

To derive disc displacement y_{disc1} in terms of y_{b1} and y_{b2}, consider Figure 2.30.

For the case of a deflected rigid shaft, the slope is same along the length, so we have

$$\frac{y_{b2} - y_{disc1}}{b} = \frac{y_{disc1} - y_{b1}}{a} \quad \Rightarrow a y_{b2} - a y_{disc1} = b y_{disc1} - b y_{b1} \quad \Rightarrow y_{disc1} = \frac{a y_{b2} + b y_{b1}}{l} \tag{2.178}$$

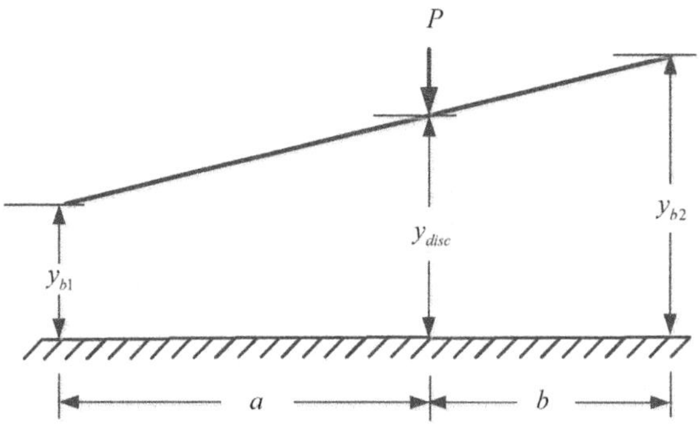

FIGURE 2.30 Deflection of a rigid shaft on flexible supports.

On substituting Eqs. (2.176) and (2.177), we get

$$y_{disc1} = \frac{\dfrac{Pb^2}{k_b l} + \dfrac{Pa^2}{k_b l}}{l} = \frac{P(a^2 + b^2)}{k_b l^2} \tag{2.179}$$

Now we will consider flexible shaft and rigid bearings.

Case-2: Flexible shaft on rigid bearings:

For a flexible shaft on rigid bearings, the deflection at any point along the length, z, is given by

$$y(z) = \frac{Pbz}{6lEI}(l^2 - z^2 - b^2) \quad \text{for} \quad 0 < z < a \tag{2.180}$$

The deflection at offset disc location (Figure 2.31) is given as (for $z = a$, displacement at the location of the force)

$$y_{disc2} = \frac{Pbz}{6lEI}(l^2 - z^2 - b^2) = \frac{Pba}{6lEI}(l^2 - a^2 - b^2) = \frac{Pab}{6lEI}(ab) = \frac{Pa^2b^2}{3lEI} \tag{2.181}$$

Total deflection at the disc, noting Eqs. (2.179) and (2.181), we get

$$y_{disc} = y_{disc1} + y_{disc2} = \frac{P(a^2 + b^2)}{k_b l^2} + \frac{Pa^2b^2}{3lEI} \tag{2.182}$$

Equivalent stiffness (the shaft and bearing stiffness are in series) at the disc location is given as

$$k_{eq} = \frac{P}{y_{disc}} = \frac{1}{\dfrac{1}{k_{beq}} + \dfrac{1}{k_{seq}}} = \frac{1}{\dfrac{(a^2 + b^2)}{k_b l^2} + \dfrac{a^2b^2}{3lEI}} = \frac{3k_b EIl^3}{3EIl(a^2 + b^2) + k_b l^2 a^2 b^2} \tag{2.183}$$

where k_{beq} and k_{seq} are the bearing and shaft equivalent stiffness connected in series. The natural frequency is now given by

$$\omega_{nf} = \sqrt{\frac{k_{eq}}{m}} = \sqrt{\frac{3k_b EIl^3}{3mEIl(a^2 + b^2) + k_b l^2 a^2 b^2}} \tag{2.184}$$

For following rotor parameters, let us obtain the natural frequency of the system

$$m = 14 \text{ kg}; \quad l = 0.4 \text{ m}; \quad d = 0.025 \text{ m}; \quad I = \frac{\pi}{64} d^4 = 1.91 \times 10^{-8} \text{ m}^4;$$

$$E = 2.1 \times 10^{11} \text{ N/m}^2; \quad k_b = 50 \times 10^3 \text{ N/m}; \quad a = 0.2 \text{ m}; \quad b = 0.2 \text{ m};$$

$$k_{beq} = 1.00 \times 10^5 \text{ N/m}; \quad k_{seq} = 3.02 \times 10^6 \text{ N/m}; \quad k_{eq} = 9.68 \times 10^4 \text{ N/m}.$$

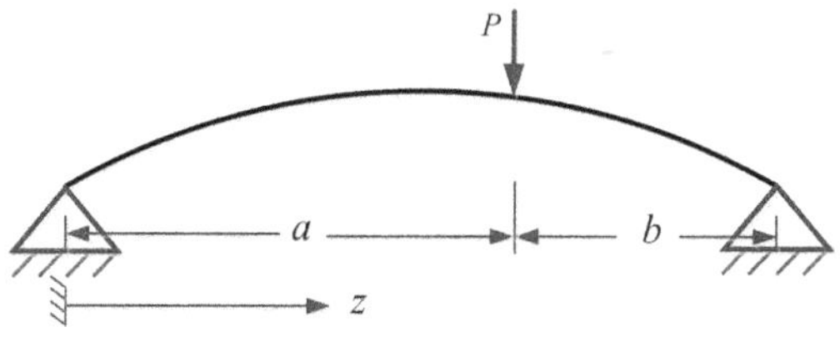

FIGURE 2.31 Defection of a flexible shaft on rigid supports.

Hence, the natural frequency of the system is

$$\omega_{nf} = \sqrt{\frac{k_{eq}}{m}} = \sqrt{\frac{3k_b EIl^3}{3mEIl(a^2+b^2)+k_b l^2 a^2 b^2}} = \sqrt{\frac{9.68\times10^4}{14}} = 83.15 \text{ rad/s}$$

For a more general motion of the disc (i.e. tilting motion) refer to Chapters 4 and 5. In Chapter 12 of Tiwari (2017) using FEM more realistic analysis are made.

Exercise 2.25 Choose a single correct answer from the multiple-choice questions:

i. The critical speed phenomenon of a rotor system is a(n)
   A. free vibration B. forced vibration
   C. transient vibration D. unstable vibration

   Solution: Critical speed phenomenon occurs when unbalanced rotor rotates at its whirl natural frequency. Without unbalance this may not occur since it requires a dynamic force to have resonance condition.

ii. A rigid body is defined as
   A. a body with no deformation B. a body with particles that have fixed distances
   C. both (A) and (B) D. a body with large dimensions

   Solution: A rigid body has no deformation. That means a relative displacement of various particles of the body does not take place. The rigid body assumption is an engineering assumption, based on which we separate the kinematic and dynamic analysis of mechanisms.

iii. A particle has how many degrees of freedom in space?
   A. 1 B. 2 C. 3 D. more than 3

   Solution: A particle is assumed as a point for its dynamic analysis. As a point in space has three coordinates to define its position so a particle will have three degrees-of-freedom (DOFs). A particle has two DOFs in a plane.

iv. A rigid body has how many degrees of freedom in space?
   A. 1 B. 3 C. 6 D. more than 6

   Solution: A rigid body requires three coordinates of a point and three coordinates for its orientation. So, a rigid has six DOFs in space. A rigid body has three DOFs in a plane (two coordinates to define a point on the body and one coordinate for its orientation).

v. A flexible body has how many degrees of freedom in space?
   A. 1 B. 3 C. 6 D. infinite

   Solution: A flexible body has infinite particles and every particle require three DOFs in a space. So, a flexible body has infinite DOFs.

vi. If three particles have fixed relative distances between them, then it represents a system of
   A. a single particle B. a rigid body
   C. a multi-body D. a flexible body

   Solution: Three particles with fixed relative distances between them will represent a rigid body. In a multi-body, each body can have relative motion with other. In a flexible body, every particle in the body will have relative motion.

vii. A system consists of three particles with their relative distances as constant. In space, it has how many degrees of freedom?
A. 1 B. 3 C. 6 D. infinite

Solution: A single particle has three DOFs, one additional particle in the system will add only two more DOF, since it has one constraint to maintain the fixed distance. The third particle will add only one more DOF, since it has two constraints to maintain fixed distances from the first as well as the second particle. Hence, a system consists of three particles with their relative distances as constant have a total of six DOF, i.e., the same as a rigid body.

viii. A perfectly balanced Jeffcott rotor (i.e., a flexible shaft with a disc at midspan) is rotating at a particular speed. If it is perturbed in the transverse plane from its equilibrium, the frequency of whirl would be equal to
A. the shaft spin speed B. the transverse natural frequency
C. more than the transverse natural frequency
D. less than the transverse natural frequency

Solution: A balanced rotor has no external force so the rotor system will not undergo forced vibration. If a perturbation is given to the system then due to this initial disturbance it will have whirling motion with frequency the same as the natural frequency of the system. If gyroscopic effect is present in the system, then the natural frequency of whirl will depend upon the spin speed of the rotor. In such cases, the whirl natural frequency is speed-dependent.

ix. A Jeffcott rotor with an off-set disc (i.e., not at the mid-span) has an initial tilt in the transverse plane. The shaft would experience
A. gyroscopic couple B. an external moment
C. an unbalance force D. A gyroscopic couple and an external moment

Solution: The offset disc would give the gyroscopic couple and an initial tilted disc would give the external moment. The initial tilt of the disc will not give any unbalance force. The unbalanced force comes when the disc has radial eccentricity.

x. The transverse natural frequency of the rotor-bearing system shown in Figure 2.32 would be

A. $\sqrt{\dfrac{(k_b + k_s)}{m}}$ B. $\sqrt{\dfrac{k_b k_s}{m(k_b + k_s)}}$

C. $\sqrt{\dfrac{k_s(k_b + k_s)}{m k_b}}$ D. $\sqrt{\dfrac{k_b(k_b + k_s)}{m k_s}}$

Solution: Because of the symmetry and on considering transverse translational motion only, we can assume that bearings are connected in parallel and its effective stiffness to the shaft will be k_b. Now it can be assumed that the shaft and bearing stiffnesses are in series. So, the effective stiffness of the disc would experience

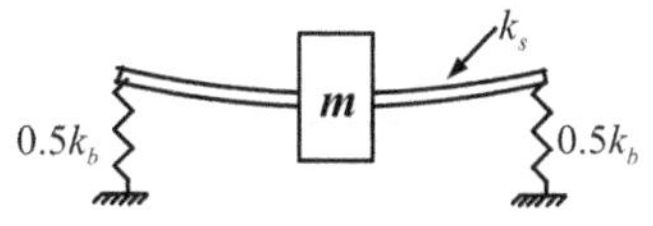

FIGURE 2.32 A symmetric flexible rotor with flexible bearings.

$$k_{eff} = \frac{k_b k_s}{k_b + k_s} \tag{2.185}$$

Hence, the natural frequency of the system is

$$\omega_{nf} = \sqrt{\frac{k_b k_s}{m(k_b + k_s)}} \tag{2.186}$$

In Chapter 4, we will study in more detailed analysis of rotor-bearing interactions including transverse rotational motion.

xi. For a Jeffcott rotor operating at supercritical speed (i.e. well above the critical speed), the rotor deflection would be approaching

A. $\frac{mg}{k}$ B. infinite C. zero D. its eccentricity

Solution: At supercritical region, the rotor deflection will approach to disc eccentricity (refer Figure 2.1). This will take the disc to rotate about its centre of gravity. This phenomenon is called the self-centering of the rotor well above the critical speed. If the disc is perfectly balanced (zero eccentricity) then the rotor will not have any deflection in the post-critical region. Herein, mg/k is the static deflection of the rotor due to its own weight.

xii. For a Jeffcott rotor operating at the critical speed, the rotor response phase with respect to the unbalance force would be approaching

A. 0° B. 90° C. 180°

D. some finite value depending upon the damping value

Solution: At the critical speed the phase will be 90° irrespective of the damping in the system (refer Figure 2.2). Without damping in the system before the critical speed, it will have phase 0°, and after the critical speed, the phase will be 180°.

xiii. The transverse critical speed of a rotor system, as shown in Figure 2.33, is to be fixed at 5.98 rad/s. Take the disc as a point mass with $m = 5$ kg. What is the diameter of the uniform shaft, d? Take shaft length $2a = b = 0.7$ m. Ignore the gyroscopic effect.

A. 4.25 mm B. 4.43 mm C. 5.10 mm D. 5.22 mm

Solution: For the present problem, we have to ignore the gyroscopic effect. Hence, motions in both orthogonal transverse planes are uncoupled. Equations of motion in a single plane can be written, as (refer Tiwari (2017))

$$\begin{bmatrix} m & 0 \\ 0 & I_d \end{bmatrix} \begin{Bmatrix} \ddot{x} \\ \ddot{\varphi}_y \end{Bmatrix} + \begin{bmatrix} \alpha_{xf} & \alpha_{xM} \\ \alpha_{\varphi f} & \alpha_{\varphi M} \end{bmatrix}^{-1} \begin{Bmatrix} x \\ \varphi_y \end{Bmatrix} = \begin{Bmatrix} 0 \\ 0 \end{Bmatrix} \tag{2.187}$$

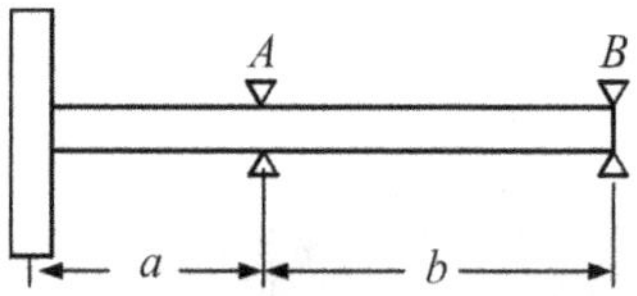

FIGURE 2.33 An overhung rotor.

For free vibration we have,

$$\left(-\omega_{nf}^2\begin{bmatrix} m & 0 \\ 0 & I_d \end{bmatrix}+\begin{bmatrix} \alpha_{xf} & \alpha_{xM} \\ \alpha_{\varphi f} & \alpha_{\varphi M} \end{bmatrix}^{-1}\right)\begin{Bmatrix} x \\ \varphi_y \end{Bmatrix}=\begin{Bmatrix} 0 \\ 0 \end{Bmatrix} \tag{2.188}$$

where ω_{nf} is the natural frequency of the system. For the non-trivial solution, we have

$$\left|-\omega_{nf}^2\begin{bmatrix} m & 0 \\ 0 & I_d \end{bmatrix}+\begin{bmatrix} \alpha_{xf} & \alpha_{xM} \\ \alpha_{\varphi f} & \alpha_{\varphi M} \end{bmatrix}^{-1}\right|=0 \tag{2.189}$$

which gives frequency equation as,

$$mI_d\omega_{nf}^4(\alpha_{xf}\alpha_{\varphi M}-\alpha_{\varphi f}^2)-\omega_{nf}^2(m\alpha_{xf}+I_d\alpha_{\varphi M})+1=0 \tag{2.190}$$

For the point mass ($I_d=0$), the aforementioned equation becomes (refer Appendix 2.1)

$$1-\omega_{nf}^2 m\alpha_{xf}=0 \Rightarrow \quad \omega_{nf}^2=\frac{1}{m\alpha_{xf}} \Rightarrow \quad \omega_{nf}^2=\frac{3EI}{ma^2(a+b)} \Rightarrow \quad \omega_{nf}^2=\frac{8EI}{mb^3} \tag{2.191}$$

We have following rotor parameters

$$m=5\text{ kg};\quad a=0.35\text{ m};\quad b=0.7\text{ m};\quad E=2.1\times10^{11}\text{ N/m}^2;\quad I=\frac{\pi}{64}\times d^4\text{ m}^4.$$

On putting values of parameters in Eq. (2.191), we get

$$5.98^2=\frac{8\times2.1\times10^{11}\times I}{5\times0.7^3} \Rightarrow \quad I=3.65054\times10^{-11}\text{ m}^4;$$

$$\text{or}\quad \frac{\pi}{64}d^4=3.65054\times10^{-11} \Rightarrow \quad d=5.222\times10^{-3}\text{ m} \Rightarrow \quad d=5.222\text{ mm}.$$

xiv. For a disc with mass m, eccentricity e and rotating at ω, the unbalance is defined as
A. $me\omega^2$ B. me C. $me\omega$ D. $e\omega$

Solution: The unbalance is defined as the mass multiplied by eccentricity. For a disc, it will be mass of the disc multiplied by disc eccentricity. For a trial unbalance used in balancing of rotors, it will be a small mass multiplied by its larger radial position.

xv. At resonance, the phase between the force and the displacement of a Jeffcott rotor for different levels of damping would
A. varying between 0° and 90° B. varying between 90° and 180°.
C. 90° D. 180°

Solution: At the critical speed the phase will be 90° irrespective of the damping in the system. With a damping in the system before the critical speed, the phase will vary between 0° and 90° and after the critical speed the phase will vary between 90° and 180°.

xvi. In a Jeffcott rotor, if the shaft is given a constant axial tensile preload at the ends, then the transverse natural frequency of the rotor would

A. increase B. decrease

C. remain constant D. become zero

Solution: Due to a constant axial tensile initial load, the effective transverse stiffness of the shaft will increase. This will increase the critical speed. Initial tension is common in strings and cables, wherein the transverse stiffness develops only due to the initial tension. It should be noted that for a constant axial compressive initial load, the effective transverse stiffness of the shaft will decrease and result in decrease in natural frequency. If shaft is slender, then only by increasing the compressive load gradually the natural frequency will become zero at the critical load of the buckling.

xvii. For a Jeffcott rotor with the vertical shaft, if a constant axial compressive load is applied on the shaft, the transverse natural frequency of the rotor would

A. remain unchanged B. increase

C. decrease D. become zero

Solution: Refer explanation of Exercise 2.25xvi.

xviii. Consider two different rotor systems, the first with a rigid massless shaft and ends supported by two identical flexible bearings, each with a stiffness of, k_b, and the second rotor system with simply-supported bearing conditions and flexible massless shaft (EI is flexural rigidity of the shaft and l is its span). Both rotor systems have a disc of mass, m, which is symmetrically placed with respect to the ends of the shaft. If both rotor systems have the same whirl frequency, then the following relations would prevail

A. $k_b = \dfrac{48EI}{l^3}$ B. $k_b = \dfrac{12EI}{l^3}$

C. $k_b = \dfrac{24EI}{l^3}$ D. $k_b = \dfrac{96EI}{l^3}$

Solution: The first rotor-bearing has the effective stiffness for pure translational motion (two bearings are connected in parallel), as

$$k_{eff} = 2k_b \tag{2.192}$$

The corresponding natural frequency will be

$$\omega_{nf} = \sqrt{\frac{2k_b}{m}} \tag{2.193}$$

The stiffness of a simply-supported rotor with disc at mid-span is given as (refer Appendix 2.1)

$$k_{eff} = \frac{48EI}{l^3} \tag{2.194}$$

The corresponding natural frequency will be

$$\omega_{nf} = \sqrt{\frac{48EI}{ml^3}} \tag{2.195}$$

On equating Eqs. (2.193) and (2.195), we get

$$\omega_{nf} = \sqrt{\frac{2k_b}{m}} = \sqrt{\frac{48EI}{ml^3}} \quad \text{or} \quad k_b = \frac{24EI}{l^3} \tag{2.196}$$

xix. Two unbalances ($m_{b1}e_1$ and $m_{b2}e_2$) are attached to a Jeffcott rotor disc with an angular phase of ϕ_{12} between them with $m_{b2}e_2$ ahead of $m_{b1}e_1$ with respect to the rotation of the rotor. Initially, the phase of $m_{b1}e_1$ is ϕ_0 with x-axis (i.e., horizontal transverse direction) in the direction of rotation of the rotor. The unbalanced force due to spin speed, ω, of the rotor would be

A. $m_{b1}e_1\omega^2\cos(\omega t+\phi_0)+m_{b2}e_2\omega^2\cos(\omega t+\phi_{12})$

B. $m_{b1}e_1\omega^2\cos(\omega t+\phi_0)+m_{b2}e_2\omega^2\cos(\omega t-\phi_0+\phi_{12})$

C. $m_{b1}e_1\omega^2\cos(\omega t+\phi_0+\phi_{12})+m_{b2}e_2\omega^2\cos(\omega t+\phi_0+\phi_{12})$

D. $m_{b1}e_1\omega^2\cos(\omega t+\phi_0)+m_{b2}e_2\omega^2\cos(\omega t+\phi_0+\phi_{12})$

Solution: The initial phase of $m_{b1}e_1$ is ϕ_0, so option (C) is ruled out. The phase of $m_{b2}e_2$ is ahead of phase of $m_{b1}e_1$ by ϕ_{12}, so the phase of $m_{b2}e_2$ from the common reference will be $(\phi_0 + \phi_{12})$. Both option (A) and (B) do not have such term, so option (D) is correct.

xx. For a cantilever shaft with a thin disc at the free end, if the transverse and torsional natural frequencies are same, then the ratio of the length of the shaft to the diameter of the disc would be (take Poisson's ratio as 0.5 for the shaft material)

A. 3/4 B. 4/3 C. 2/3 D. 1/3

Solution: The natural frequency for the transverse vibration of a cantilever rotor is given as

$$\omega_{nf} = \sqrt{\frac{3EI}{ml^3}} \tag{2.197}$$

The torsional natural frequency for a cantilever rotor is given as

$$\omega_{nf} = \sqrt{\frac{GJ}{I_p l}} \tag{2.198}$$

On equating Eqs. (2.197) and (2.198), we get

$$\omega_{nf} = \sqrt{\frac{3EI}{ml^3}} = \sqrt{\frac{GJ}{I_p l}} \Rightarrow \frac{3\times 2G(1+\nu)}{ml^2}\frac{\pi d^4}{64} = \frac{G}{mD^2/8}\frac{\pi d^4}{32} \tag{2.199}$$

$$\text{or} \quad \frac{3\times 1.5}{l^2} = \frac{8}{D^2} \quad \frac{l}{D} = \sqrt{\frac{3\times 3}{8\times 2}} = \frac{3}{4} \tag{2.200}$$

xxi. For a single-degree-of-freedom spring-mass-damper rotor model, the equation of motion is given as $\ddot{x}+2\zeta\omega_{nf}\dot{x}+\omega_{nf}^2 x = e\omega_{nf}^2\cos\omega_{nf}t$. The solution of this equation would have the following form:

A. $x(t)=X(t)\cos\omega_{nf}t$ B. $x(t)=X(t)\sin\omega_{nf}t$

C. $x(t)=X(t)\cos\left(\omega_{nf}t+\phi\right)$ D. $x(t)=X(t)\sin\left(\omega_{nf}t+\phi\right)$

E. $x(t)=X\cos\omega_{nf}t$ F. $x(t)=X\sin\omega_{nf}t$

G. $x(t)=X\cos\left(\omega_{nf}t+\phi\right)$ H. $x(t)=X\sin\left(\omega_{nf}t+\phi\right)$

where z is the damping ratio, ω_{nf} is the undamped natural frequency, ϕ is the phase, X is time-invariant amplitude, and $X(t)$ is time-variant amplitude.

Solution: The given equation is

$$\ddot{x} + 2\zeta\omega_{nf}\dot{x} + \omega_{nf}^2 x = e\omega_{nf}^2 \cos\omega_{nf}t \equiv \mathrm{Re}\left(e\omega_{nf}^2 e^{j\omega_{nf}t}\right) \tag{2.201}$$

Assuming the steady-state solution as

$$x = \mathrm{Re}\left(Xe^{j\omega_{nf}t}\right) \tag{2.202}$$

where X is a complex amplitude. On substituting Eq. (2.202) into Eq. (2.201), we get

$$\begin{gathered}-\omega_{nf}^2 Xe^{j\omega_{nf}t} + 2\zeta\omega_{nf}\left(j\omega_{nf}Xe^{j\omega_{nf}t}\right) + \omega_{nf}^2\left(Xe^{j\omega_{nf}t}\right) = e\omega_{nf}^2 e^{j\omega_{nf}t} \\ \text{or} \quad -\omega_{nf}^2 X + 2j\zeta\omega_{nf}^2 X + \omega_{nf}^2 X = e\omega_{nf}^2 \quad \text{or} \quad X = \frac{e\omega_{nf}^2}{2j\zeta\omega_{nf}^2} = \frac{e}{2j\zeta} = -\frac{ej}{2\zeta}\end{gathered} \tag{2.203}$$

On substituting Eq. (2.203) into Eq. (2.202), we get

$$x = \mathrm{Re}\left(-\frac{ej}{2\zeta}e^{j\omega_{nf}t}\right) = \frac{e}{2\zeta}\sin\omega_{nf}t \tag{2.204}$$

So, the amplitude is time-invariant and has 90° phase with the force, which always remains constant in presence of under-damping at the resonance (since the spin speed of the rotor is equal to the undamped natural frequency). In absence of damping the amplitude will be time-variant (refer eq. (2.19) in chapter 2 of Tiwari (2017)).

xxii. In a Jeffcott rotor, the phase of unbalance force with respect to the undamped response below the critical speed approaches
A. 180° B. 90° C. 45° D. 0°

Solution: For an undamped rotor system, the force and the response are in the same phase (or zero phase) below the critical speed. After critical speed, it is 180°, and at critical speed, it is 90°.

xxiii. In a Jeffcott rotor with a rigid disc having mass and polar mass moment of inertia, when rotating at a nominal speed, the torsional vibration may take place due to
A. only gravity B. gravity with eccentricity
C. only eccentricity D. none of above

Solution: When the disc has an eccentricity, the gravity load gives a moment (refer eq. (2.44) of Tiwari (2017)) and in the presence of an appreciable polar mass moment of inertia of the disc, the torsional vibration takes place. However, for a constant speed, there is no angular acceleration of the disc so the torsional vibration will not be present in such case.

xxiv. In a Jeffcott rotor, when speed is very high ($\omega \gg \omega_{cr}$), the magnitude of rotor displacement is equal to
A. infinity B. zero
C. eccentricity of disc D. static deflection of disc

Solution: For a very high ($\omega \gg \omega_{cr}$), the magnitude of rotor displacement is equal to the eccentricity of the disc (refer Figure 2.1).

xxv. The elastic coupling in the transverse vibration of a circular shaft is
   A. coupling of two orthogonal plane motions but no coupling of the linear and angular displacements
   B. no coupling of two orthogonal plane motions but coupling of the linear and angular displacements
   C. coupling of two orthogonal plane motions and coupling of the linear and angular displacements
   D. no coupling of two orthogonal plane motions and no coupling of the linear and angular displacements

   Solution: The elastic coupling in transverse vibration of a circular (symmetric) shaft has no coupling in two orthogonal plane motions but has coupling of the linear and angular displacements. Because of the shaft symmetry, there is no coupling of two orthogonal plane motions.

xxvi. For the anti-synchronous whirl of a Jeffcott rotor when viewed from the bearing centre
   A. the same face (or mark) of the shaft as that for the synchronous whirl will always be seen
   B. opposite face (or mark) of the shaft as that for the synchronous whirl will always be seen
   C. one of the faces (or marks) will be seen twice in one rotation of the shaft
   D. a different face (or mark) will be visible with no fixed pattern

   Solution: For the anti-synchronous whirl (spin and whirl directions are opposite) of a Jeffcott rotor when viewed from the bearing centre one of the faces (or marks) will be seen twice in one rotation of the shaft (refer Figure 2.34).

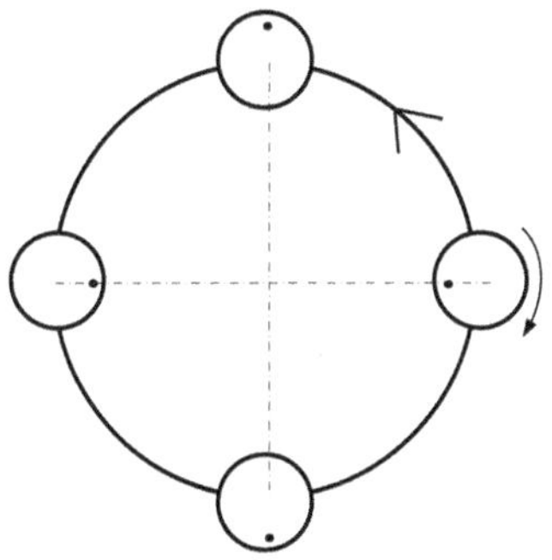

FIGURE 2.34 Anti-synchronous whirl (spin direction CW and whirl direction CCW).

xxvii. The phase between the unbalance force and the response in a Jeffcott rotor at the resonance would be (in radians)
   A. $\pi/2$ B. π C. 0   D. $-\pi$

   Solution: The phase between the unbalance force and the response in a Jeffcott rotor at the resonance would be $\pi/2$ rad.

xxviii. In a cantilever rotor during maintenance, the mild steel shaft is replaced with an aluminum shaft ($G_{Al}/G_S = 1/3$) of the same dimension. The shaft is massless and the disc mass at the free end remains the same. The torsional natural frequency ratio for rotors ($(\omega_{nf\_Al}/(\omega_{nf\_MS})$) would be
   A. 3   B. 1/3   C. $\sqrt{3}$ D. $1/\sqrt{3}$

Solution: The torsional natural frequencies for a cantilever rotor of different materials are given as

$$\omega_{nfMS} = \sqrt{\frac{G_{MS}J}{lI_p}} \quad \text{and} \quad \omega_{nfAl} = \sqrt{\frac{G_{Al}J}{lI_p}} \tag{2.205}$$

On taking the ratio, we get

$$\frac{\omega_{nfAl}}{\omega_{nfMS}} = \sqrt{\frac{G_{Al}}{G_{MS}}} = \sqrt{\frac{1}{3}} \tag{2.206}$$

xxix. For the synchronous whirl of a damped Jeffcott rotor model while rotating the rotor above the critical speed, the correct phase representation between the unbalance force (black circle represents an unbalanced mass) and disc response would be one of the options in Figure 2.35.

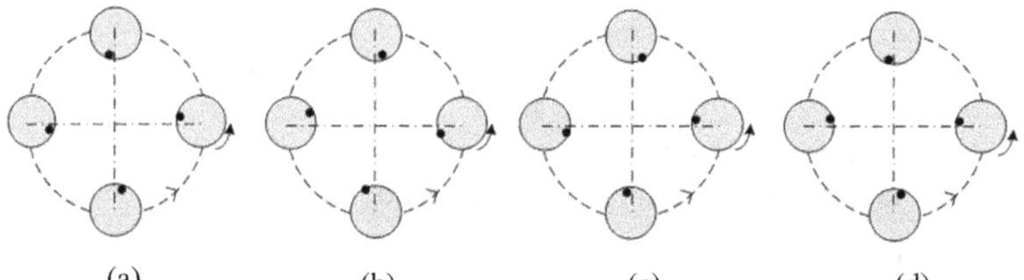

FIGURE 2.35 Shaft whirling and spinning about a bearing centre (a) Option A (b) Option B (c) Option C (d) Option D.

Solution: The phase between the unbalance force and response beyond critical speed approached close to 180°, but with the damping it remains less than 180°. Now the response is radially outward (let us take at 3 o'clock position of option A) and the unbalance force will be radially inward (from shaft centre to the black spot). The angle between the response and the unbalance force can be seen less than 180°. The angle is measured in the direction of the whirl (i.e., CCW). The same is true for Options C and D but not for Option B. Now for 6 o'clock, Option A has less than 180° phase and also for Option D but not for Option C. With the same logic for 9 o'clock, Option A has less than 180° phase and but not for Option D. So only Option A has this condition even at 12 o'clock.

xxx. The correct phase between unbalance force and response after the critical speed with damping in the system will have the following option given in Figure 2.36.
Solution: The phase between the unbalance force and response beyond critical speed approached close to 180°, but with the damping it remains less than 180°. Now the response is radially outward and the unbalance force, *F*, will be radially inward. The angle between the response and the unbalance force should be less than 180°. The angle is measured in the direction of the whirl. So, Options A and D are ruled out. Option B has phase of more than 180° and only Option C has phase of less than 180°. In fact, Option A is at critical speed and Option D is below critical speed.

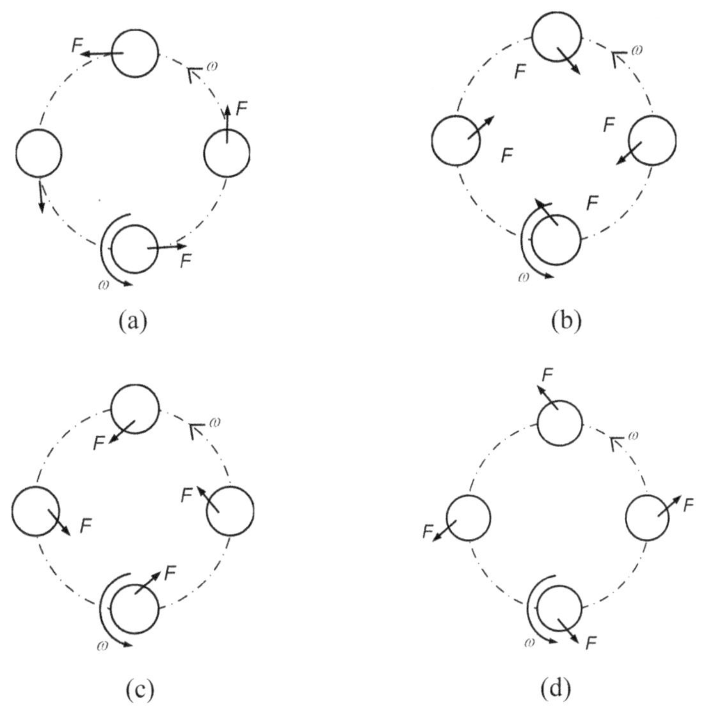

FIGURE 2.36 A shaft orbiting about a bearing centre (a) Option A (b) Option B (c) Option C (d) Option D.

xxxi. The critical speed phenomenon of a rotor is a

A. free vibration B. forced vibration

C. transient vibration D. unstable vibration

Solution: The critical speed phenomenon of a rotor is a forced vibration since it occurs most commonly due to unbalanced force.

xxxii. In a Jeffcott rotor (with a disc at the mid-span), if the disc centre of gravity is at its centre of rotation but the disc is tilted by a small angle with respect to the shaft. The critical speed encountered during rotation of the shaft will be

A. $\sqrt{\dfrac{12EI}{I_d L}}$ B. $\sqrt{\dfrac{48EI}{mL^3}}$ C. $\sqrt{\dfrac{3EI}{I_d L}}$ D. $\sqrt{\dfrac{3EI}{mL^3}}$

Solution: For simply supported case with $a = b = l/2$, we have following influence coefficients (refer Appendix 2.1)

$$\alpha_{11} = \frac{l^3}{48EI}; \quad \alpha_{22} = \frac{-\left(3al - 3a^2 - l^2\right)}{3EIl} = \frac{l}{12EI}; \quad \alpha_{12} = \alpha_{21} = \frac{-\left(3a^2 l - 2a^3 - al^2\right)}{3EIl} = 0 \tag{2.207}$$

The critical speed corresponding to the pure tilting motion will be

$$\omega_{cr} = \sqrt{\frac{1}{I_d \alpha_{22}}} = \sqrt{\frac{12EI}{I_d L}} \tag{2.208}$$

xxxiii. A 2-kg mass of a simply-supported rotor causes a static deflection at the free end of 0.2 mm. What is the natural frequency of the system in rad/s?

A. 221.5 B. 140.1 C. 120.4 D. 220.3

Solution: The rotor data are: Mass of the disc, $m = 2$ kg,
Static deflection of the shaft at the disc centre, $\delta = 0.2 \times 10^{-3}$ m,
The acceleration due to gravity, $g = 9.81$ m/s^2,
The stiffness of the shaft is given as

$$k = \frac{mg}{\delta} = \frac{2 \times 9.81g}{0.2 \times 10^{-3}} = 98{,}100 \text{ N/m} \tag{2.209}$$

Hence, the natural frequency of the rotor system is

$$\omega_{nf} = \sqrt{\frac{k}{m}} = \sqrt{\frac{98{,}100}{2}} = 221.47 \text{ rad/s} \tag{2.210}$$

xxxiv. In a Jeffcott rotor, if the rotor rotates at a very high speed (well above the critical speed), the ratio of shaft deflection and disc eccentricity (absolute value) will be approached towards

A. 0 B. 0.5 C. 1 D. 2

Solution: At the supercritical speed the deflection of the rotor will be approaching asymptotically equal to the disc eccentricity so its ratio will be 1 (refer Figure 2.1).

xxxv. A long rotor supported on flexible bearings and having both radial eccentricity and axial offset of its centre of gravity will produce

A. only force B. only moment
C. both force and moment D. neither force nor moment

Solution: Due to the radial eccentricity and axial offset of its centre of gravity, the rotor will experience the unbalanced force and moment. So, it will experience all critical speeds (pure translational and rotational for a symmetric rotor) by gradually increasing the rotor speed.

xxxvi. The unbalanced force in a rotor system gives

A. a synchronous whirl condition of the response
B. an anti-synchronous whirl condition of the response
C. an asynchronous whirl condition of the response
D. a multi-harmonic synchronous whirl condition of the response

Solution: The unbalanced force in a rotor system gives a synchronous whirl condition of the response. The unbalanced force gives a force of frequency equal to the spin speed of the shaft.

xxxvii. For a Jeffcott rotor with a positive damping, whether the amplitude ratio (rotor displacement/disc eccentricity) can be infinity for a particular non-zero value of positive damping (in terms of the damping ratio) in an unbalanced rotor?

A. 1.414 B. 1 C. −1 D. not feasible

Solution: From Eq. (2.71), we have

$$\bar{R} = \frac{R}{e} = \frac{\bar{\omega}^2}{\sqrt{\left(1-\bar{\omega}^2\right)^2 + \left(2\zeta\bar{\omega}\right)^2}} \tag{2.211}$$

Whether denominator can be zero? So that the amplitude ratio becomes infinity. Let us see the condition for it

$$\left(1-\bar{\omega}^2\right)^2+\left(2\zeta\bar{\omega}\right)^2=0 \quad \Rightarrow \left(1-\bar{\omega}^2\right)^2+4\zeta^2\bar{\omega}^2=0 \quad \Rightarrow \zeta^2=-\frac{\left(1-\bar{\omega}^2\right)^2}{4\bar{\omega}^2} \tag{2.212}$$

It can be seen that the right-side term is always negative. So the damping ratio will be complex, for this condition to satisfy. So it is not feasible to have infinite amplitude ratio with damping in the system. But for negative damping, the system may go to unstable condition but it is not a resonance condition in fact it is a free vibration instability phenomenon.

xxxviii. In a Jeffcott rotor with an offset disc, the coupling is present in
A. two orthogonal transverse translational displacements
B. two orthogonal transverse rotational displacements
C. the transverse translational and rotational displacements in the same plane
D. the transverse translational and rotational displacements in the orthogonal plane

Solution: In a Jeffcott rotor with an offset disc, the coupling is present in the transverse translational and rotational displacements in the same plane. Due to symmetry of the shaft, there will not be any coupling in two orthogonal planes.

xxxix. The influence coefficient α_{ij} is defined as
A. the displacement at the *i*th station due to a unit force at the *j*th station keeping all other forces at zero.
B. the force at the *i*th station due to a unit displacement at the *j*th station keeping all other displacements at zero.
C. the force at the *j*th station due to a unit displacement at the *i*th station keeping all other displacements at zero.
D. the displacement at the *j*th station due to a unit force at the *i*th station keeping all other forces at zero.

Solution: The influence coefficient α_{ij} is defined as the displacement at the *i*th station due to a unit force at the *j*th station keeping all other forces at zero.

xl. Obtain the bending natural frequency (in rad/s) for the synchronous motion of a cantilever rotor. Take the mass of the disc m = 1 kg. The shaft is assumed to be massless, and its length and diameter are 0.3 m and 0.01 m, respectively. Take Young's modulus of the shaft as $E = 2.1 \times 10^{11}$ N/m^2.
A. 104.4 B. 196.6 C. 121.4 D. 88.3

Solution: The rotor data are

Young's modulus, $E = 2 \times 10^{11}$ N/m^2; Mass of the disc, m = 1 kg,
The diameter of the shaft, d = 0.01 m; The length of the shaft, l = 0.3 m

The second moment of area is $I=\dfrac{\pi d^4}{64}=\dfrac{\pi\times 0.01^4}{64}=4.9087\times 10^{-10}$ m^4

The stiffness is given as $k=\dfrac{3EI}{l^3}=\dfrac{3\times 2.1\times 10^{11}}{0.3^3}=1.0908\times 10^4$ N/m

Hence, the natural frequency of the system is $\omega_{nf}=\sqrt{\dfrac{k}{m}}=\sqrt{\dfrac{1.0908\times 10^4}{1}}=104.44$ rad/s.

xli. For a simply-supported rotor with a point mass, m, at the mid-span. The natural frequency of the system is ω_{nf}. Let l and d be the length and the diameter, respectively, of the shaft, and E be Young's modulus. The diameter, d, of the shaft expression is given as

A. $\left(\frac{4ml^3\omega_{nf}^2}{3E\pi}\right)^{(1/4)}$ B. $\left(\frac{64ml^3\omega_{nf}^2}{3E\pi}\right)^{(1/4)}$

D. $\left(\frac{4ml^3\omega_{nf}^2}{3E\pi}\right)^{(1/2)}$ D. $\left(\frac{64ml^3\omega_{nf}^2}{3E\pi}\right)^{(1/2)}$

Solution: The stiffness of a simply-supported rotor with disc at mid-span is given as

$$k = \frac{48EI}{l^3} = \frac{48E\pi d^4}{64l^3} = \frac{3E\pi d^4}{4l^3} \tag{2.213}$$

So, the natural frequency of the system is

$$\omega_{nf}^2 = \frac{k}{m} = \frac{3E\pi d^4}{4ml^3} \tag{2.214}$$

which gives

$$d = \left(\frac{4ml^3\omega_{nf}^2}{3E\pi}\right)^{(1/4)} \tag{2.215}$$

REFERENCE

Tiwari, R., 2017, *Rotor Systems: Analysis and Identification.* Boca Raton, FL: CRC Press.

FINAL REMARKS

In this chapter, a very simple exercise problems on rotor vibration analysis have been presented. Mostly, the rotor system with the single or two degrees-of-freedom (DOFs) are considered for transverse vibration and very few cases for torsional vibration also. For two DOFs with elastic shaft, mostly the influence coefficients method is popular and is used extensively in this chapter. Various types of boundary conditions are considered, including cantilever, simply-supported and over-hung rotor with intermediate support. Few cases with support flexibility, stepped-shaft, two-rotor train and coupling effect are considered with very basic steps to give overall understanding of such simple rotor system analysis. Mainly, free and forced vibration analyses are covered. A large number of MCQs helps in understanding very fundamental concepts and small analysis steps for deeper understanding of rotor terminology and phenomena.

APPENDIX 2.1: USEFUL FORMULAS

S.N.	Formulas	Descriptions
1	For thin disc (about centre of gravity) $I_p = 2I_d = mk^2 = \frac{1}{2}mr^2$ with $k = r/\sqrt{2}$ $I_p = 2I_d = \frac{1}{8}mD^2$	Herein, k is the radius of gyration, r is the radius of thin disc, D is the disc diameter, m is the mass of the disc, I_d is the diametral mass moment of inertia and I_p is the polar mass moment of inertia
2	For long cylinder (about centre of gravity) $I_p = mk^2 = \frac{1}{2}mr^2 = \frac{1}{8}mD^2$ $I_d = \frac{1}{16}mD^2 + \frac{1}{12}ml^2$	D is the diameter of cylinder, m is the mass of the cylinder, l is the length of cylinder and r is the radius of the cylinder.
3	$\begin{bmatrix} a & b \\ c & d \end{bmatrix}^{-1} = \frac{1}{ad-bc}\begin{bmatrix} d & -b \\ -c & a \end{bmatrix}$	Inverse of 2 × 2 matrix
4	$[M] = \begin{bmatrix} a_r + ja_i & b_r + jb_i \\ c_r + jc_i & d_r + jd_i \end{bmatrix}$ $[N] = \begin{bmatrix} \mathrm{Re}[M] & \mathrm{Im}[M] \\ -\mathrm{Im}[M] & \mathrm{Re}[M] \end{bmatrix} = \begin{bmatrix} a_r & b_r & a_i & b_i \\ c_r & d_r & c_i & d_i \\ -a_i & -b_i & a_r & b_r \\ -c_i & -d_i & c_r & d_r \end{bmatrix},$ $[P] = [N]^{-1} = \begin{bmatrix} \mathrm{Re}[inv(M)] & \mathrm{Im}[inv(M)] \\ -\mathrm{Im}[inv(M)] & \mathrm{Re}[inv(M)] \end{bmatrix} = \begin{bmatrix} P_{11} & P_{12} \\ P_{21} & P_{22} \end{bmatrix}$ $[M]^{-1} = [P_{11} + jP_{12}] = \left[\mathrm{Re}[inv(M)] + j\,\mathrm{Im}[inv(M)]\right]$	Inverse of a complex matrix (2 × 2) or (N × N). *Example*: Obtain inverse of $[M]_{1\times 1} = (a + jb)$ such that $[M][M]^{-1} = (a + jb)(c + jd) = 1$. $[N] = \begin{bmatrix} a & b \\ -b & a \end{bmatrix},$ $[P] = [N]^{-1} = \frac{1}{a^2+b^2}\begin{bmatrix} a & -b \\ b & a \end{bmatrix},$ $[M]^{-1} = [P_{11} + jP_{12}] = \left[\frac{a}{a^2+b^2} + j\frac{-b}{a^2+b^2}\right]$
5	$A^{-1} = \frac{1}{\lvert A\rvert}\begin{bmatrix} \begin{vmatrix} a_{22} & a_{23} \\ a_{32} & a_{33} \end{vmatrix} & \begin{vmatrix} a_{13} & a_{12} \\ a_{33} & a_{32} \end{vmatrix} & \begin{vmatrix} a_{12} & a_{13} \\ a_{22} & a_{23} \end{vmatrix} \\ \begin{vmatrix} a_{23} & a_{21} \\ a_{33} & a_{31} \end{vmatrix} & \begin{vmatrix} a_{11} & a_{13} \\ a_{31} & a_{33} \end{vmatrix} & \begin{vmatrix} a_{13} & a_{11} \\ a_{23} & a_{21} \end{vmatrix} \\ \begin{vmatrix} a_{21} & a_{22} \\ a_{31} & a_{32} \end{vmatrix} & \begin{vmatrix} a_{12} & a_{11} \\ a_{32} & a_{31} \end{vmatrix} & \begin{vmatrix} a_{11} & a_{12} \\ a_{21} & a_{22} \end{vmatrix} \end{bmatrix}.$	It is inverse of a 3 × 3 matrix $A = \begin{bmatrix} a_{11} & a_{12} & a_{13} \\ a_{21} & a_{22} & a_{23} \\ a_{31} & a_{32} & a_{33} \end{bmatrix}$
6	Influence coefficients for simply supported shaft (refer Figure 2.37): $\begin{bmatrix} \alpha_{yf} & \alpha_{yM} \\ \alpha_{\varphi f} & \alpha_{\varphi M} \end{bmatrix} = \begin{bmatrix} \frac{a^2b^2}{3EIl} & \frac{-(3a^2l - 2a^3 - al^2)}{3EIl} \\ \frac{ab(b-a)}{3EIl} & \frac{-(3al - 3a^2 - l^2)}{3EIl} \end{bmatrix};$ $l = a + b.$	Herein, α_{ij} represent the displacement (subscript: y - translational and ϕ - rotational) at ith station due to a unit force (subscript: f - force and M - moment) at jth station keeping all other forces/moments to zero. 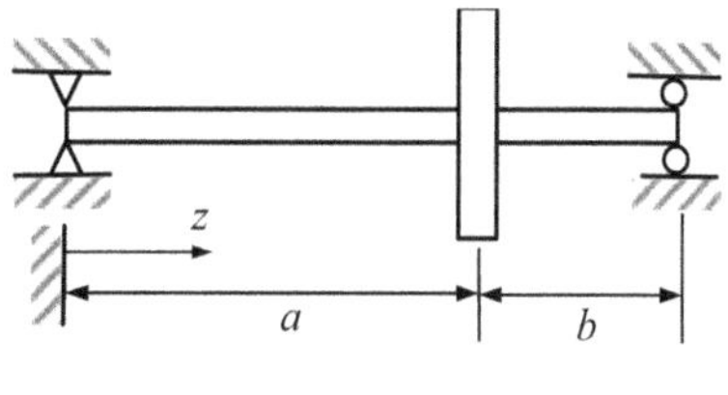 **FIGURE 2.37** A simply supported shaft.

(*Continued*)

S.N.	Formulas	Descriptions

7 Influence coefficients for simply supported shaft (Jeffcott rotor) with $a = b = 0.5l$ (refer Figure 2.38):

$$\begin{bmatrix} \alpha_{yf} & \alpha_{yM} \\ \alpha_{\varphi f} & \alpha_{\varphi M} \end{bmatrix} = \begin{bmatrix} \dfrac{l^3}{48EI} & 0 \\ 0 & \dfrac{l}{12EI} \end{bmatrix}$$

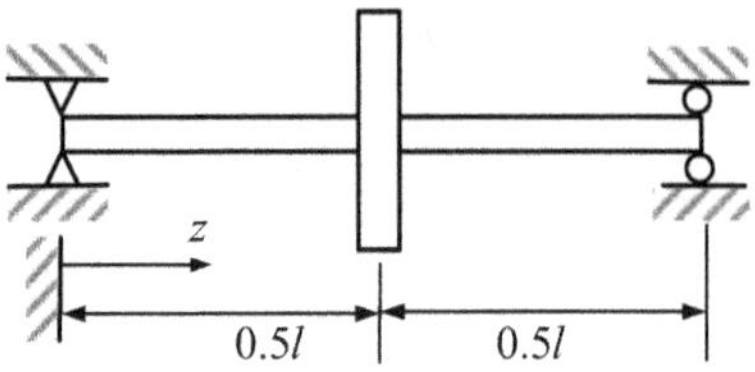

FIGURE 2.38 A simply-supported (Jeffcott) rotor system.

8 Influence coefficients for cantilever shaft (refer Figure 2.39):

$$\begin{bmatrix} \alpha_{yf} & \alpha_{yM} \\ \alpha_{\varphi f} & \alpha_{\varphi M} \end{bmatrix} = \begin{bmatrix} \dfrac{l^3}{3EI} & \dfrac{l^2}{2EI} \\ \dfrac{l^2}{2EI} & \dfrac{l}{EI} \end{bmatrix}$$

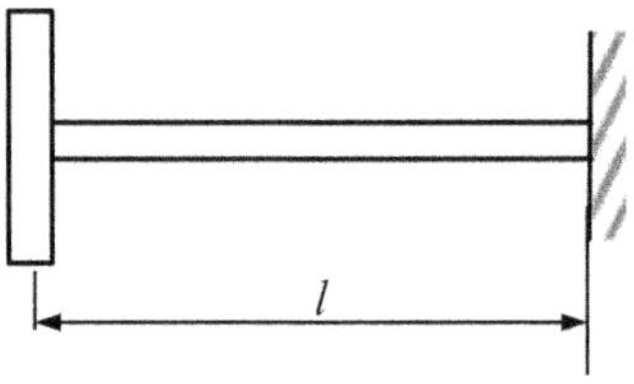

FIGURE 2.39 A cantilever rotor system.

9 Influence coefficients for an overhung shaft with roller supports (refer Figure 2.40) (tilting considered):

$$\begin{bmatrix} \alpha_{yf} & \alpha_{yM} \\ \alpha_{\varphi f} & \alpha_{\varphi M} \end{bmatrix} = \begin{bmatrix} \dfrac{a^2(a+b)}{3EI} & \dfrac{a(3a+2b)}{6EI} \\ \dfrac{a(3a+2b)}{6EI} & \dfrac{(3a+b)}{3EI} \end{bmatrix}$$

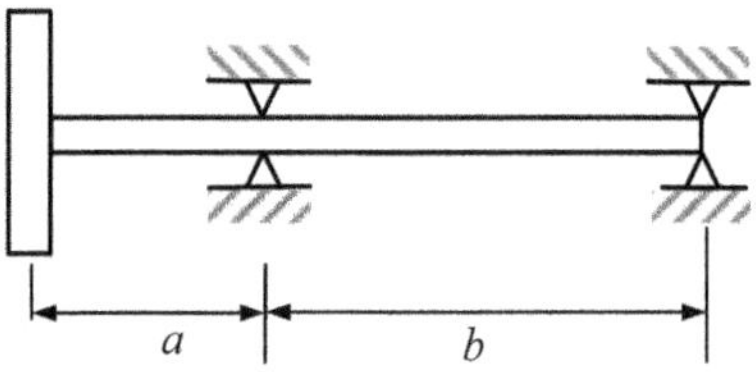

FIGURE 2.40 An overhung shaft with roller supports.

10. Influence coefficients for an overhung shaft with a fixed support (refer Figure 2.41) (tilting considered):

$$\begin{bmatrix} \alpha_{yf} & \alpha_{yM} \\ \alpha_{\varphi f} & \alpha_{\varphi M} \end{bmatrix} = \begin{bmatrix} \dfrac{a^2(4a+3b)}{12EI} & \dfrac{a(2a+b)}{4EI} \\ \dfrac{a(2a+b)}{4EI} & \dfrac{(4a+b)}{4EI} \end{bmatrix}$$

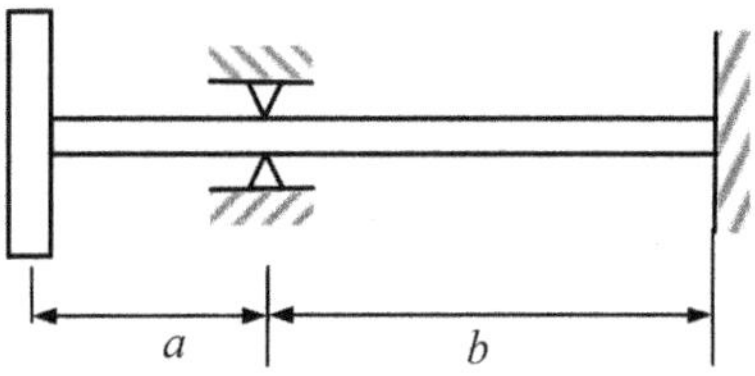

FIGURE 2.41 An overhung rotor system with a fixed support.

11. For a flexible shaft on simply-supported bearings (refer Figure 2.42), the deflection at any point along the length, z, is given by

$$y(z) = \frac{Pbz}{6lEI}(l^2 - z^2 - b^2) \quad \text{for} \quad 0 < z < a$$

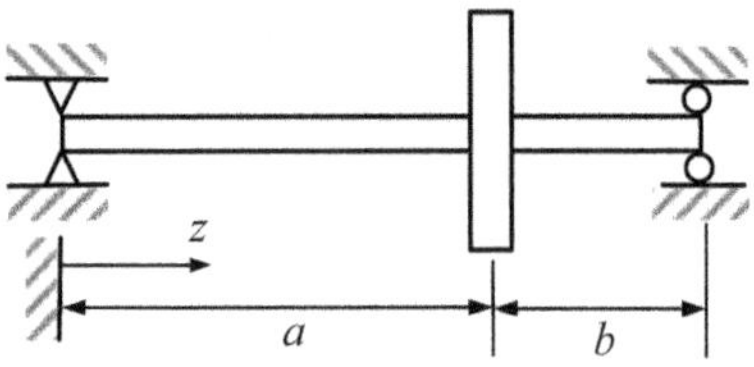

FIGURE 2.42 A simply supported shaft.

12. For fixed-fixed support (refer Figure 2.43)

$$\begin{bmatrix} \alpha_{yf} & \alpha_{yM} \\ \alpha_{\varphi f} & \alpha_{\varphi M} \end{bmatrix} = \begin{bmatrix} \dfrac{a^3b^3}{3EIl^3} & \dfrac{a^4b^2(b-a)}{2EIl^3} \\ \dfrac{a^4b^2(b-a)}{2EIl^3} & \dfrac{a^2b^2}{3EIl^3} \end{bmatrix}$$

$$\alpha_{yf} = \frac{a^3b^3}{3EIl^3} \text{ with } l = a + b$$

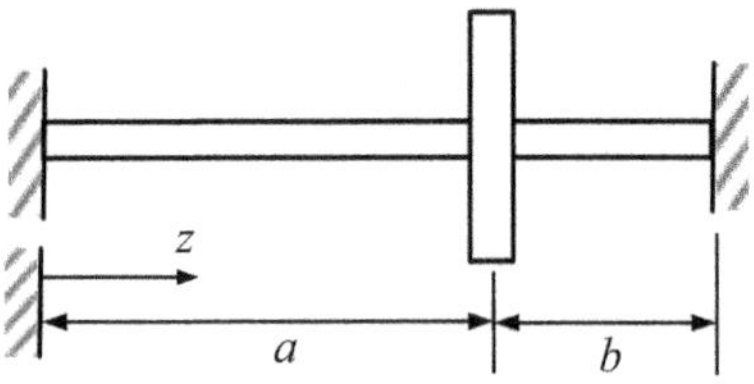

FIGURE 2.43 A fixed-fixed support rotor system.

(Continued)

S.N.	Formulas	Descriptions
13.	For fixed-fixed support with load at mid-span (Figure 2.44) $$\begin{bmatrix} \alpha_{yf} & \alpha_{yM} \\ \alpha_{\varphi f} & \alpha_{\varphi M} \end{bmatrix} = \begin{bmatrix} \dfrac{l^3}{192EI} & 0 \\ 0 & \dfrac{l}{24EI} \end{bmatrix}$$	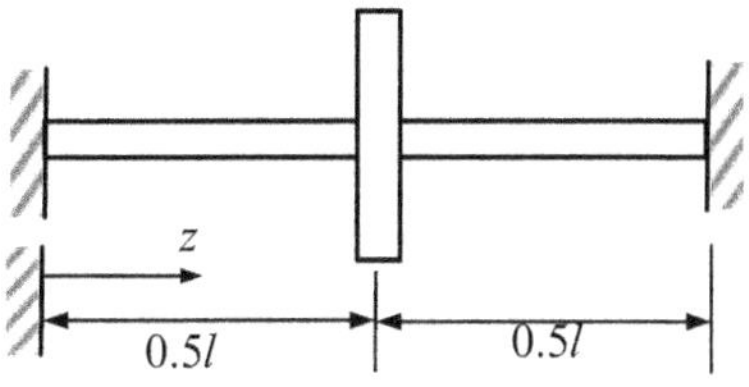 FIGURE 2.44 A fixed-fixed support with load at mid-span.
14.	For fixed-fixed support, (refer case 12 also) the deflection with load not at middle (refer Figure 2.45) $$y = \frac{Pb^2z^2}{6EIl^3}\{3al - z(3a+b)\} \text{ for } 0<z<a$$ $$y = \frac{Mbz^2}{2EIl^2}(2a-b) - \frac{Mabz^3}{EIl^3} \text{ for } 0<z<a$$ with $\varphi_x = \dfrac{dy}{dz}$ $$\begin{bmatrix} \alpha_{yf} & \alpha_{yM} \\ \alpha_{\varphi f} & \alpha_{\varphi M} \end{bmatrix} = \begin{bmatrix} \dfrac{a^3b^2}{6EIl^3}\{3l-(3a+b)\} & \dfrac{a^2b^2}{2EIl^3}\{2a^2l - a^2(3a+b)\} \\ \dfrac{a^2b^2}{2EIl^3}\{2a^2l - a^2(3a+b)\} & -\dfrac{ab(a^2+b^2-ab)}{EIl^3} \end{bmatrix}$$	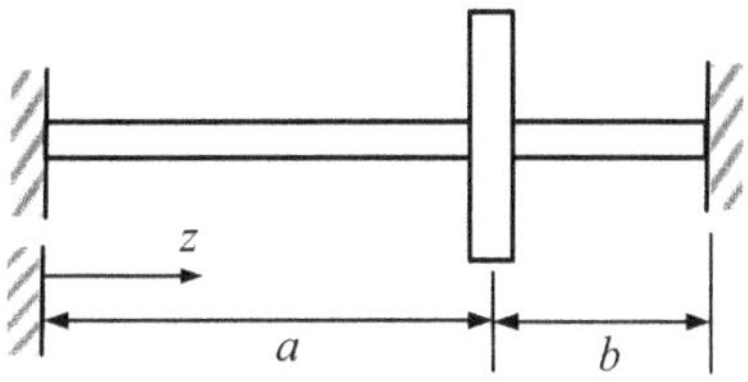 FIGURE 2.45 A fixed-fixed support rotor system.
15.	For disc locations 1 and 2 (refer Figure 2.46) (tilting ignored): $$\begin{bmatrix} \alpha_{y_1f_1} & \alpha_{y_1f_2} \\ \alpha_{y_2f_1} & \alpha_{y_2f_2} \end{bmatrix} = \begin{bmatrix} \dfrac{l^3}{8EI} & \dfrac{l^3}{32EI} \\ \dfrac{l^3}{32EI} & \dfrac{l^3}{48EI} \end{bmatrix}$$ where l is the length as shown in adjacent figure and 1 and 2 refer to disc location.	Herein, α_{ij} represent the displacement (subscript: y - translational) at ith station due to a unit force (subscript: f - force) at jth station keeping all other forces to zero. 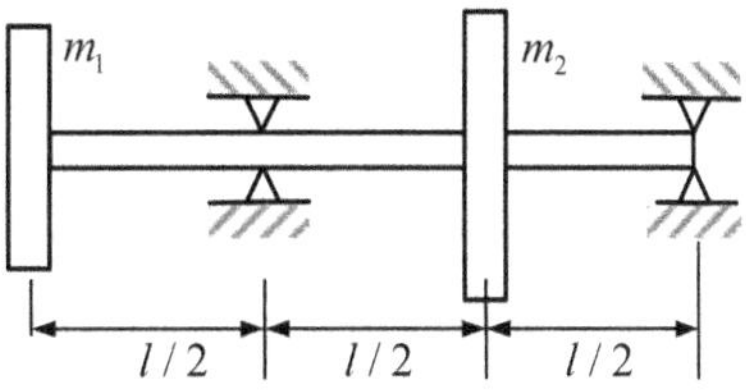 FIGURE 2.46 An overhung rotor system.
16.	Tapered shaft (refer Figure 2.47) Diameter: $d(z) = d_1 + \dfrac{d_2 - d_1}{L}z$ Second polar moment of area: $$J(z) = \frac{\pi d^4(z)}{32} = \frac{\pi}{32}\left(d_1 + \frac{d_2 - d_1}{L}z\right)^4$$ Stiffness: $k_t = \dfrac{3\pi G(d_2d_1)^3}{32L(d_2^2 + d_1d_2 + d_1^2)}$ Maximum shear stress at smaller end ($d = d_1$): $\tau_{max} = \dfrac{16T}{\pi d_1^3}$	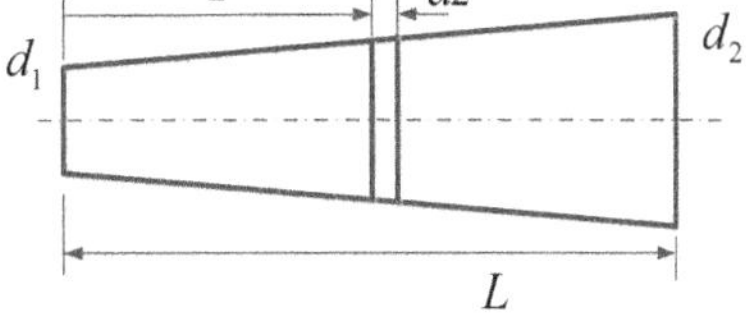 FIGURE 2.47 A tapered shaft.

ANSWERS TO MCQs

2.25

i. B	ii. C	iii. C	iv. C	v. D	vi. B
vii. C	viii. B	ix. D	x. B	xi. D	xii. B
xiii. D	xiv. B	xv. C	xvi. A	xvii. C	xviii. C
xix. D	xx. A	xxi. F	xxii. B	xxiii. B	xxiv. C
xxv. B	xxvi. C	xxvii. A	xxviii. D	xxix. A	xxx. C
xxxi. B	xxxii. A	xxxiii. A	xxxiv. C	xxxv. C	xxxvi. A
xxxvii. D	xxxviii. C	xxxix. A	xxxl. A	xxxli. A	

3 Rotordynamic Parameters of Bearings

Exercise 3.1 List the parameters on which the stiffness of rolling element bearings depends.

***Solution*:** There are several parameters that affects the stiffness of the rolling element bearings. It can be divided into macro- and micro-geometries. The macro-geometries include the standard boundary (outer) dimensions and non-standard internal dimensions. The boundary dimension includes the outer diameter, bore diameter and width. Internal dimensions include the dimension of rolling elements (the diameter for the ball, the diameter and the length for the cylindrical roller, the diameters at ends and the length for the tapered roller and the curvature radius and the length for the spherical roller), raceways dimensions (the inner raceway (ring) diameter, the outer raceway (ring) diameter), the curvatures of raceways (the inner and outer raceway curvature for ball bearings), contact angles (at inner and outer raceways with rolling element) and number of rolling elements. Micro-geometry includes the tolerances (ball/roller, and inner and outer raceways), fits (between shaft and bearing bore and between bearing housing and bearing outer diameters) and roughness of various rolling element parts (ball/roller, and inner and outer raceways). Macro-geometry is fixed for a particular bearing once it is manufactured, but micro-geometries may change during operation. The main parameter that changes with the bearing micro-geometries and affects the stiffness of the rolling bearing is the preload (or the clearance). In fact, the preload can change due to various other parameters such as temperature, speed, centrifugal forces in rings and rolling elements, gyroscopic effect in rolling elements (it changes contact angle), lubrication, etc. In fact, when lubrication comes into play, one more very important micro-geometric parameter that affects it is the elasto-hyrodynamic lubrication (EHL) minimum film thickness. So, it can be seen that stiffness of the rolling element bearing is very complex to calculate considering all such practical aspects of its operating conditions, and because of this, the stiffness of the rolling bearing is often estimated experimentally in actual working conditions (refer chapter 14 for more details in Tiwari (2017)).

Exercise 3.2 List the parameters on which the stiffness and damping coefficients of fluid-film hydrodynamic bearings and seals depend.

***Solution*:** The stiffness and damping parameters in hydrodynamic bearings and seals depend on their macro- and micro-geometry, lubrication properties, operating speed, operating load, operating clearance, etc. Since varied geometry of types of bearings and seals are available, describing all of them one by one will be difficult. In hydrodynamic bearing, the lubrication inlet/outlet hole positions and the shape of the bearing (cylindrical, number of lobes, number of tilting pads, etc.) play an important role. For seals, apart from complex internal geometries, even the texture of inner surface plays an important role in the stiffness and damping coefficients. Again, due to complexity, practice engineers often rely on experimental estimation of the stiffness and damping coefficients (refer Chapter 14 of Tiwari (2017)).

Exercise 3.3 The specifications of the ball and roller bearings are given in Table 3.1. Plot the variation of the bearing radial stiffness versus the elastic radial deformation. Show the variation for different radial clearances. Take increments in the load of 100 N and the radial clearance as 0.005 mm for plotting the variation of the radial stiffness.

DOI: 10.1201/9781032638218-3

TABLE 3.1
Geometrical and Load Parameters of the Bearing

		Parameters of the Bearing	
S.N.	**Type of the Geometry**	**Ball**	**Roller**
1	Basic bore diameter (mm)	40	20
2	Basic outside diameter (mm)	68	47
3	Inner groove radii (mm)	4.95	–
4	Outer groove radii (mm)	4.96	–
5	Ball/roller diameter (mm)	9.53	6.50
6	Roller length (mm)	–	6.50
7	Radial clearance range (mm)	0.005–0.020	0.005–0.020
8	Radial loads (N)	300–1,000	300–1,000
9	Width (mm)	15	14

***Solution*:** Case I: Ball Bearing

In general, the internal geometric dimensions of the bearings are not provided in the *Bearing Manufacture's Catalogues* (e.g., SKF General Catalogue). These dimensions can either be measured after disassembling the bearing (these may be different for same bearing number by different manufacturers) or they can be approximately calculated by the expressions in terms of boundary dimensions of the bearing provided in the *Bearing Catalogues*, for obtaining the stiffness of the bearing. Alternatively, more advanced optimum design can be attempted (Gupta et al., 2007; Rao and Tiwari, 2007). Given parameters of the ball bearing is:

The diameter of the ball $D_b = 9.53$ mm.
The inner ring groove radius of curvatures, $r_i = 4.95$ mm.
The outer ring groove radius of curvatures, $r_o = 4.96$ mm.

General proportions for ball bearings are obtained as follows (PSG, 1982):

$$\text{The pitch diameter or the mean diameter: } D_m \approx \frac{1}{2}(D+d) = 54 \text{ mm}. \tag{3.1}$$

$$\text{The radial thickness of rings, } s = \frac{1}{2}\left\{\frac{1}{2}(D-d) - D_b\right\} = 2.2350 \text{ mm}. \tag{3.2}$$

$$\text{The number of ball, } Z \approx \frac{\pi D_m}{D_b} = 17.80. \text{ Hence } Z = 18 \text{ can be taken.} \tag{3.3}$$

For a zero radial clearance, the inner and outer ring diameters at raceway grooves are:

$$d_0 = D_m + D_b = 54 + 9.53 = 63.53 \text{ mm}; \quad d_i = D_m - D_b = 54 - 9.53 = 44.47 \text{ mm} \tag{3.4}$$

For a ball bearing of the bore diameter 40 mm; from the data given, we have radial clearance of 0.005–0.020 mm. For the present example, let us take the maximum clearance, hence $g = -0.020$ mm (where g is the preload, so for clearance a negative sign is used).

Finally, the bearing internal dimensions are given/obtained as:

$$r_i = 4.95 \text{ mm}; \quad r_o = 4.96 \text{ mm}; \quad D_b = 9.53 \text{ mm}; \quad Z = 18$$
$$d_i = 44.47 \text{ mm}; \quad d_0 = 63.53 \text{ mm} \quad D_m = 54 \text{ mm}; \quad g = -0.020 \text{ mm}.$$

Based on these dimensions, the following parameters, which will be used for finding the bearing stiffness, can be estimated

$$\gamma = \frac{D_b}{D_m} = \frac{9.53}{54} = 0.1765; \quad f_i = \frac{r_i}{D_b} = \frac{4.95}{9.53} = 0.5194; \quad f_o = \frac{r_o}{D_b} = \frac{4.96}{9.53} = 0.5205 \tag{3.5}$$

$$\Sigma\rho_i = \frac{1}{D_b}\left(4 - \frac{1}{f_i} + \frac{2\gamma}{1-\gamma}\right) = \frac{1}{6}\left(4 - \frac{1}{0.5194} + \frac{2\times 0.1765}{1-0.1765}\right) = 0.2627 \text{ mm}^{-1} \tag{3.6}$$

$$\Sigma\rho_o = \frac{1}{D_b}\left(4 - \frac{1}{f_o} - \frac{2\gamma}{1+\gamma}\right) = \frac{1}{6}\left(4 - \frac{1}{0.5205} - \frac{2\times 0.1765}{1+0.1765}\right) = 0.1866 \text{ mm}^{-1} \tag{3.7}$$

$$F(\rho)_i = \frac{\dfrac{1}{f_i} + \dfrac{2\gamma}{1-\gamma}}{4 - \dfrac{1}{f_i} + \dfrac{2\gamma}{1-\gamma}} = \frac{\dfrac{1}{0.5194} + \dfrac{2\times 0.1765}{1-0.1765}}{4 - \dfrac{1}{0.5194} + \dfrac{2\times 0.1765}{1-0.1765}} = 0.9403 \tag{3.8}$$

Corresponding to the curvature difference of $F(\rho_i) = 0.9403$, the non-dimensional relative displacement is given as $\delta_i^* = 0.6013$ (refer Table 3.2).

TABLE 3.2
Dimensionless Contact Parameters

$F(\rho)$	a^*	b^*	δ^*
0.0000	1.0000	1.0000	1.0000
0.1075	1.0760	0.9318	0.9974
0.3204	1.2623	0.8114	0.9761
0.4795	1.4556	0.7278	0.9429
0.5916	1.6440	0.6687	0.9077
0.6716	1.8258	0.6245	0.8733
0.7332	2.0110	0.5881	0.8394
0.7948	2.2650	0.5480	0.7961
0.8350	2.4940	0.5186	0.7602
0.8737	2.8000	0.4863	0.7169
0.9100	3.2330	0.4499	0.6636
0.9366	3.7380	0.4166	0.6112
0.9574	4.3950	0.3830	0.5551
0.9729	5.2670	0.3490	0.4960
0.9838	6.4480	0.3150	0.4352
0.9909	8.0620	0.2814	0.3745
0.9951	10.2220	0.2497	0.3176
0.9973	12.7890	0.2232	0.2705
0.9982	14.8390	0.2072	0.2427
0.9989	17.9740	0.1882	0.2106
0.9995	23.5500	0.1644	0.1717
0.9995	37.3800	0.1305	0.1199
1.0000	∞	0.0000	0.0000

$$F(\rho_o)=\frac{\dfrac{1}{f_o}+\dfrac{2\gamma}{1+\gamma}}{4-\dfrac{1}{f_o}-\dfrac{2\gamma}{1+\gamma}}=\frac{\dfrac{1}{0.5205}-\dfrac{2\times 0.1765}{1+0.1765}}{4-\dfrac{1}{0.5205}-\dfrac{2\times 0.1765}{1+0.1765}}=0.9116 \tag{3.9}$$

Corresponding to the curvature difference of $F(\rho_i)=0.9116$, the non-dimensional relative displacement is given as $\delta_o^*=0.6605$ (refer Table 3.2). The load-deflection constant at the inner and outer raceway contacts are given as (refer Eq. (3.32) of Tiwari (2017))

$$K_{pi}=2.15\times 10^5\left(\Sigma\rho_i\right)^{-1/2}\left(\delta_i^*\right)^{-3/2}=2.15\times 10^5\times 0.2627^{-1/2}\times 0.6013^{-3/2}=8.9975\times 10^5\ \text{N/mm}^{1.5} \tag{3.10}$$

and $K_{po}=2.15\times 10^5\left(\Sigma\rho_o\right)^{-1/2}\left(\delta_o^*\right)^{-3/2}=2.15\times 10^5\times 0.1866^{-1/2}\times 0.6605^{-3/2}=9.272\times 10^5\ \text{N/mm}^{1.5}$ (3.11)

The load deflection factor for a single ball is given as

$$K_{Pio}=\left\{\frac{1}{(1/K_{Pi})^{2/3}+(1/K_{Po})^{2/3}}\right\}=\left\{\frac{1}{\left(1/8.9975\times 10^5\right)^{0.667}+\left(1/9.272\times 10^5\right)^{0.667}}\right\}$$
$$=3.2289\times 10^5\ \text{N/mm}^{1.5} \tag{3.12}$$

The bearing stiffness can be expressed as

$$k(x)=a+bx^2 \quad \text{with} \quad a=k(0) \quad \text{and} \quad b=\frac{k(g)-k(0)}{g^2} \tag{3.13}$$

The bearing stiffness is given as (Tiwari, 2017)

$$k(x)=1.5K_{pio}\sum_{i=1}^{Z}\left(g+x\cos\psi_i\right)^{0.5}\left(\cos\psi_i-\frac{C}{B}\sin\psi_i\right)\cos\psi_i \tag{3.14}$$

with $n=3/2$, $Z=18$, $g=-0.02$ mm.

When the load direction is between two rolling elements, then

$$\psi_i=\frac{\pi}{18}(2i-1),\quad i=1,2,\ldots,18 \quad \text{so that} \quad \psi_i=\frac{\pi}{18},\frac{3\pi}{18},\ldots,\frac{35\pi}{18} \tag{3.15}$$

$$B=\sum_{i=1}^{Z}\left(g+x\cos\psi_i\right)^{0.5}\sin^2\psi_i \quad \text{and} \quad C=\sum_{i=1}^{Z}\left(g+x\cos\psi_i\right)^{0.5}\sin\psi_i\cos\psi_i \tag{3.16}$$

At $x=0$, we have (with clearance, there is no contact of rolling elements with rings so both B and C are zero)

$$B=\sum_{i=1}^{18}(-0.02)^{0.5}\sin^2\psi_i=(-0.02)^{0.5}\left(\sin^2\frac{\pi}{18}+\sin^2\frac{3\pi}{18}+\cdots+\sin^2\frac{35\pi}{18}\right)=0 \qquad (3.17)$$

$$C=\sum_{i=1}^{18}(-0.02)^{0.5}\sin\psi_i\cos\psi_i=0 \qquad (3.18)$$

$$\text{and} \quad k(0)=1.5K_{pio}\sum_{i=1}^{18}(-0.02)^{0.5}(\cos\psi_i-0)\cos\psi_i=0 \qquad (3.19)$$

When there is no contact of rolling element with rings, the stiffness is also zero. In fact, $k(0)$ will not be required in the estimation of stiffness as we will see in subsequent steps.

At $x=2|g|=2|-0.02|=0.04$ mm (so that once the clearance is filled the rolling elements have compression equal to the clearance level), from Eqs. (3.16) and (3.14), we have

$$B=\sum_{i=1}^{Z}(g+x\cos\psi_i)^{0.5}\sin^2\psi_i=\sum_{i=1}^{18}(-0.02+0.04\cos\psi_i)^{0.5}\sin^2\psi_i=0.1576 \qquad (3.20)$$

$$C=\sum_{i=1}^{Z}(g+x\cos\psi_i)^{0.5}\sin\psi_i\cos\psi_i=\sum_{i=1}^{18}(-0.02+0.04\cos\psi_i)^{0.5}\sin\psi_i\cos\psi_i=0 \qquad (3.21)$$

and

$$\begin{aligned} k(2|g|)&=1.5K_{pio}\sum_{i=1}^{Z}(g+x\cos\psi_i)^{0.5}\left(\cos\psi_i-\frac{C}{B}\sin\psi_i\right)\cos\psi_i \\ &=1.5K_{pio}\sum_{i=1}^{18}(-0.02+(0.04)\cos\psi_i)^{0.5}\cos^2\psi_i=2.4898\times10^5\ \text{N/m} \end{aligned} \qquad (3.22)$$

Hence, the stiffness will be zero during displacement $x=0$ to $|g|$, i.e.

$$k(x=0\ \text{to}\ |g|)=0 \qquad (3.23)$$

Once the rolling elements are in contact with the rings, then its stiffness will be given by

$$k(x)=k(2|g|)(2|g|-|g|)^{-2/3}x^{2/3}=2.4898\times10^5\times(0.02)^{-2/3}x^{2/3}=3.3792\times10^6x^{2/3}\ \text{N/mm} \qquad (3.24)$$

Figure 3.1 shows the variation of stiffness of the ball bearing with the displacement for different clearances (or negative preloads).

Case II: Roller Bearing

The analysis of cylindrical roller bearings is similar to deep-groove roller bearing, except for a few differences in the load-deflection exponent, the curvature sum and curvature difference expressions.

A more accurate determination of the stiffness of rolling contact bearings can be performed as follows (Harris, 2001). For the steel roller and the steel raceway contact, we have

$$K_L=7.86\times10^4 l^{8/9}\ \text{N/mm}^{10/9} \qquad (3.25)$$

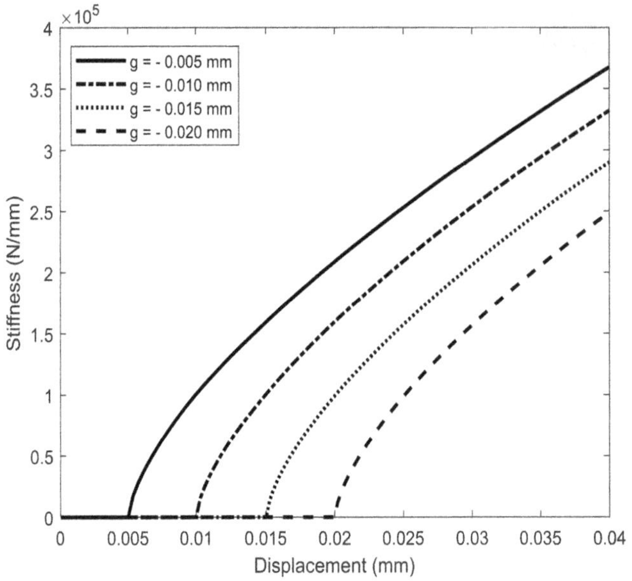

FIGURE 3.1 Variation of stiffness of ball bearing with displacement for different clearances (or negative preloads).

where l is the effective length of the roller in mm. Herein, it can be observed that, unlike ball bearing, the load-deflection proportionately constant depends on the length of the roller, apart from material properties. It can be written for the inner and outer ring contacts, as

$$K_{Li} = K_{Lo} = 7.86\times 10^4 l^{8/9}\ \text{N/mm}^{10/9} \tag{3.26}$$

so that $K_{Lio} = \left\{ \dfrac{1}{(1/K_{Li})^{9/10} + (1/K_{LO})^{9/10}} \right\}^{10/9}$ N/mm$^{10/9}$

For a given external radial load, F_r, on a roller bearing (Figure 3.2), the total elastic force on the line of contact of the ith roller with the inner and outer races is expressed as (Ragulskis et al. 1974; Tiwari 2017)

$$f_i = K_{Lio}\left(g + x\cos\psi_i + y\sin\psi_i\right)^{10/9} \tag{3.27}$$

with $\psi_i = \phi + (\text{i} - 1)v$ and $i = 1,2,3,\ldots Z$

where g is the radial preload or negative of radial clearance ($g = -c_r$) between the roller and the races, x and y are the displacements of the moving ring in the direction of the radial load and perpendicular to the direction of the radial load, respectively; ψ_i is the angle between the lines of action of the radial load (direction of displacement of the moving ring) and the radius passing through the centre of the ith roller, K_{Lio} is a coefficient of proportionality depending on the geometric and material properties of the bearing for line contact (the value of K_{Lio}, for the bearing with its specifications, can be estimated by the method suggested by Harris (2001)), v and ϕ are angles (see Figure 3.2) with $v = 2\pi/Z$, and Z is number of rolling elements. The projection of f_i along the line of action of the applied force (x-axis direction) is

$$f_i = K_{Lio}\left(g + x\cos\psi_i + y\sin\psi_i\right)^{10/9}\cos\psi_i \tag{3.28}$$

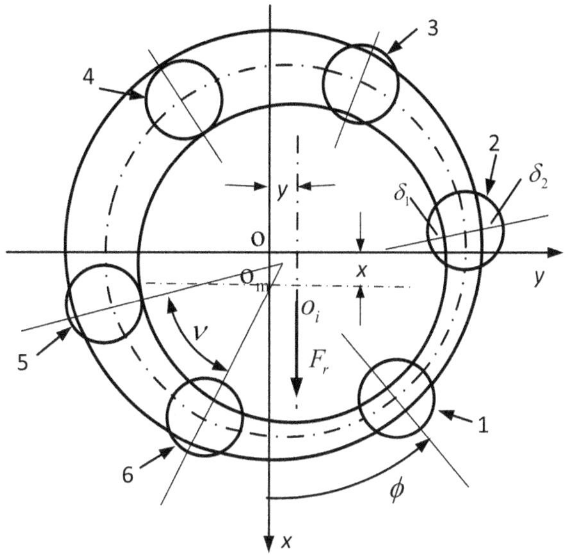

FIGURE 3.2 A line diagram of a radially loaded bearing.

The total elastic force in the direction of the applied force (x-axis) is

$$F_x = \sum_{i=1}^{Z} f_i = \sum_{i=1}^{Z} K_{Lio}\left(g + x\cos\psi_i + y\sin\psi_i\right)^{10/9}\cos\psi_i \tag{3.29}$$

where Z is the total number of rollers in the bearing. It should be noted that in the aforementioned equation within bracket only positive value need to be considered. If a negative term comes, then that terms should be ignored. Negative term indicates that the rolling element is not loaded (i.e., it is in an unloaded zone), and it will not participate in the load sharing.

Using the condition of zero elastic force in the direction perpendicular to the external load, the deformation, y, perpendicular to the radial force line is obtained as

$$F_y = \sum_{i=1}^{Z} f_i = \sum_{i=1}^{Z} K_{Lio}\left(g + x\cos\psi_i + y\sin\psi_i\right)^{10/9}\sin\psi_i = 0 \tag{3.30}$$

Expanding the aforementioned equation in a McLaurin series (Taylor series expansion) in terms of power y and limiting up to two terms, we get

$$F_y = 0 = F_y\Big|_{y=0} + \frac{\partial F}{\partial y}\Big|_{y=0} y + \frac{1}{2!}\frac{\partial^2 F}{\partial y^2}\Big|_{y=0} y^2 + \cdots \tag{3.31}$$

or $$0 = \sum_{i=1}^{Z} K_{Lio}\left\{\left(g + x\cos\psi_i + y\sin\psi_i\right)^{10/9}\Big|_{y=0} + \frac{\partial}{\partial y}\left(g + x\cos\psi_i + y\sin\psi_i\right)^{10/9}\Big|_{y=0} y + \cdots\right\}\sin\psi_i = 0$$

or $$0 = \sum_{i=1}^{Z} K_{Lio}\left\{\left(g + x\cos\psi_i\right)^{10/9} + \frac{10}{9}\left(g + x\cos\psi_i + y\sin\psi_i\right)^{1/9}\sin\psi_i\Big|_{y=0} y + \cdots\right\}\sin\psi_i = 0$$

$$\text{or}\quad 0=\sum_{i=1}^{Z}K_{Lio}\left\{\left(g+x\cos\psi_i\right)^{10/9}+\frac{10}{9}y\left(g+x\cos\psi_i\right)^{`1/9}\sin\psi_i+\cdots\right\}\sin\psi_i=0$$

$$\text{or}\quad K_{Lio}\sum_{i=1}^{Z}\left(g+x\cos\psi_i\right)^{10/9}\sin\psi_i+\frac{10}{9}yK_{Lio}\sum_{i=1}^{Z}\left\{\left(g+x\cos\psi_i\right)^{`1/9}\sin^2\psi_i\right\}=0$$

$$\text{or}\quad \frac{10}{9}y\sum_{i=1'}^{Z}\left(g+x\cos\psi_i\right)^{1/9}\sin^2\psi_i=-\sum_{i=1}^{Z}\left(g+x\cos\psi_i\right)^{10/9}\sin\psi_i$$

$$\text{or}\quad y=-\frac{\sum_{i=1}^{Z}\left(g+x\cos\psi_i\right)^{10/9}\sin\psi_i}{\frac{10}{9}\sum_{i=1}^{Z}\left(g+x\cos\psi_i\right)^{1/9}\sin^2\psi_i}\tag{3.32}$$

Equations (3.28) and (3.32) are used in Eq. (3.29) and the bearing stiffness is determined as a function of the deformation x as

$$k(x)=\frac{\partial F_x}{\partial x}\tag{3.33}$$

On substituting Eq. (3.29) in Eq. (3.33), considering Eq. (3.32), the bearing stiffness is expressed as a function of deformation as

$$k(x)=\frac{\partial F_x}{\partial x}=\frac{\partial}{\partial x}\left\{\sum_{i=1}^{Z}K_{Lio}\left(g+x\cos\psi_i+y\sin\psi_i\right)^{`10/9}\cos\psi_i\right\}$$

$$\text{or}\quad k(x)=\frac{\partial F_x}{\partial x}=\frac{\partial}{\partial x}\left[\sum_{i=1}^{Z}K_{Lio}\left\{g+x\cos\psi_i-\frac{\sum_{i=1}^{Z}\left(g+x\cos\psi_i\right)^{10/9}\sin\psi_i}{\frac{10}{9}\sum_{i=1}^{Z}\left(g+x\cos\psi_i\right)^{1/9}\sin^2\psi_i}\sin\psi_i\right\}^{`10/9}\cos\psi_i\right]$$

$$\text{or}\quad k(x)=\frac{\partial F_x}{\partial x}=K_{Lio}\frac{10}{9}\left[\begin{array}{l}\sum_{i=1}^{Z}\left\{g+x\cos\psi_i-\frac{\sum_{i=1}^{Z}\left(g+x\cos\psi_i\right)^{10/9}\sin\psi_i}{\frac{10}{9}\sum_{i=1}^{Z}\left(g+x\cos\psi_i\right)^{1/9}\sin^2\psi_i}\sin\psi_i\right\}^{`1/9}\\ \frac{\partial}{\partial x}\left\{g+x\cos\psi_i-\frac{\sum_{i=1}^{Z}\left(g+x\cos\psi_i\right)^{10/9}\sin\psi_i}{\frac{10}{9}\sum_{i=1}^{Z}\left(g+x\cos\psi_i\right)^{1/9}\sin^2\psi_i}\sin\psi_i\right\}\cos\psi_i\end{array}\right]$$

$$\text{or}\quad k(x)=\frac{\partial F_x}{\partial x}=K_{Lio}\frac{10}{9}\left[\begin{array}{l}\left[\sum_{i=1}^{Z}\left\{g+x\cos\psi_i-\dfrac{\sum_{i=1}^{Z}\left(g+x\cos\psi_i\right)^{10/9}\sin\psi_i}{\frac{10}{9}\sum_{i=1}^{Z}\left(g+x\cos\psi_i\right)^{1/9}\sin^2\psi_i}\sin\psi_i\right\}\right]^{`1/9}\\ \left\{\cos\psi_i-\dfrac{9}{10}\dfrac{\partial}{\partial x}\dfrac{\sum_{i=1}^{Z}\left(g+x\cos\psi_i\right)^{10/9}\sin\psi_i}{\sum_{i=1}^{Z}\left(g+x\cos\psi_i\right)^{1/9}\sin^2\psi_i}\sin\psi_i\right\}\cos\psi_i\end{array}\right]$$

$$\text{or}\quad k(x)=\frac{\partial F_x}{\partial x}=K_{Lio}\frac{10}{9}\left[\begin{array}{l}\left[\sum_{i=1}^{Z}\left\{g+x\cos\psi_i-\dfrac{\sum_{i=1}^{Z}\left(g+x\cos\psi_i\right)^{10/9}\sin\psi_i}{\frac{10}{9}\sum_{i=1}^{Z}\left(g+x\cos\psi_i\right)^{1/9}\sin^2\psi_i}\sin\psi_i\right\}\right]^{`1/9}\\ \left\{\begin{array}{l}\cos\psi_i-\dfrac{9}{10}\dfrac{\dfrac{\partial}{\partial x}\left\{\sum_{i=1}^{Z}\left(g+x\cos\psi_i\right)^{10/9}\sin\psi_i\right\}\left\{\sum_{i=1}^{Z}\left(g+x\cos\psi_i\right)^{1/9}\sin^2\psi_i\right\}}{\left\{\sum_{i=1}^{Z}\left(g+x\cos\psi_i\right)^{1/9}\sin^2\psi_i\right\}^2}\sin\psi_i\\ +\dfrac{9}{10}\dfrac{\left\{\sum_{i=1}^{Z}\left(g+x\cos\psi_i\right)^{10/9}\sin\psi_i\right\}\dfrac{\partial}{\partial x}\left\{\sum_{i=1}^{Z}\left(g+x\cos\psi_i\right)^{1/9}\sin^2\psi_i\right\}}{\left\{\sum_{i=1}^{Z}\left(g+x\cos\psi_i\right)^{1/9}\sin^2\psi_i\right\}^2}\sin\psi_i\end{array}\right\}\end{array}\right]$$

$$\text{or}\quad k(x)=\frac{\partial F_x}{\partial x}=K_{Lio}\frac{10}{9}\left[\sum_{i=1}^{Z}\left\{g+x\cos\psi_i-\frac{\sum_{i=1}^{Z}(g+x\cos\psi_i)^{10/9}\sin\psi_i}{\frac{10}{9}\sum_{i=1}^{Z}(g+x\cos\psi_i)^{1/9}\sin^2\psi_i}\sin\psi_i\right\}^{1/9}\left\{\begin{array}{l}\cos\psi_i-\dfrac{9}{10}\dfrac{\left\{\frac{10}{9}\sum_{i=1}^{Z}(g+x\cos\psi_i)^{1/9}\sin\psi_i\cos\psi_i\right\}\left\{\sum_{i=1}^{Z}(g+x\cos\psi_i)^{1/9}\sin^2\psi_i\right\}}{\left\{\sum_{i=1}^{Z}(g+x\cos\psi_i)^{1/9}\sin^2\psi_i\right\}^2}\sin\psi_i\\[2ex]+\dfrac{9}{10}\dfrac{\left\{\sum_{i=1}^{Z}(g+x\cos\psi_i)^{10/9}\sin\psi_i\right\}\left\{\frac{1}{9}\sum_{i=1}^{Z}(g+x\cos\psi_i)^{-8/9}\sin^2\psi_i\right\}}{\left\{\sum_{i=1}^{Z}(g+x\cos\psi_i)^{1/9}\sin^2\psi_i\right\}^2}\sin\psi_i\end{array}\right\}\cos\psi_i\right]$$

$$
\text{or} \quad k(x) = \frac{\partial F_x}{\partial x} = 1.11 K_{Lio} \left[\sum_{i=1}^{Z} \left\{ g + x\cos\psi_i - \frac{\sum_{i=1}^{Z} (g + x\cos\psi_i)^{10/9} \sin\psi_i}{1.11 \sum_{i=1}^{Z} (g + x\cos\psi_i)^{1/9} \sin^2\psi_i} \sin\psi_i \right\}^{1/9} \left\{ \begin{array}{l} \cos\psi_i - \dfrac{\left\{ 1.11 \sum_{i=1}^{Z} (g + x\cos\psi_i)^{1/9} \sin\psi_i \cos\psi_i \right\} \left\{ \sum_{i=1}^{Z} (g + x\cos\psi_i)^{1/9} \sin^2\psi_i \right\}}{1.11 \left\{ \sum_{i=1}^{Z} (g + x\cos\psi_i)^{1/9} \sin^2\psi_i \right\}^2} \sin\psi_i \\ + \dfrac{\left\{ \sum_{i=1}^{Z} (g + x\cos\psi_i)^{10/9} \sin\psi_i \right\} \left\{ 0.11 \sum_{i=1}^{Z} (g + x\cos\psi_i)^{-8/9} \sin^2\psi_i \right\}}{1.11 \left\{ \sum_{i=1}^{Z} (g + x\cos\psi_i)^{1/9} \sin^2\psi_i \right\}^2} \sin\psi_i \end{array} \right\} \cos\psi_i \right]
$$

or $$k(x)=1.11K_{Lio}\sum_{i=1}^{Z}\left\{g+x\cos\psi_i-\left(\frac{A}{1.11B}\right)\sin\psi_i\right\}^{1/9}\left\{\cos\psi_i-\frac{1.11CB-0.11AD}{1.11B^2}\sin\psi_i\right\}\cos\psi_i \tag{3.34}$$

with
$$A=\sum_{i=1}^{Z}(g+x\cos\psi_i)^{10/9}\sin\psi_i;\qquad B=\sum_{i=1}^{Z}(g+x\cos\psi_i)^{1/9}\sin^2\psi_i;$$
$$C=\sum_{i=1}^{Z}(g+x\cos\psi_i)^{1/9}\sin\psi_i\cos\psi_i;\qquad D=\sum_{i=1}^{Z}(g+x\cos\psi_i)^{-8/9}\sin^2\psi_i\cos\psi_i \tag{3.35}$$

It should be noted that the aforementioned equation is valid for the case when the elastic force in y-direction (perpendicular to load direction) is zero during slow rotation of the bearing (i.e., neglecting the centrifugal forces and gyroscopic effects). However, it is still valid for any orientation of rolling elements by suitably changing the angle in the expression. In the aforementioned equations, when the load-deflection exponent is expressed as n (where $n = 3/2$ for ball bearing and $n = 10/9$ for roller bearing), it then becomes

$$k(x)=nK_{io}\sum_{i=1}^{Z}\left\{g+x\cos\psi_i-\left(\frac{A}{nB}\right)\sin\psi_i\right\}^{n-1}\left\{\cos\psi_i-\frac{nCB-(n-1)AD}{nB^2}\sin\psi_i\right\}\cos\psi_i \tag{3.36}$$

with
$$A=\sum_{i=1}^{Z}(g+x\cos\psi_i)^{n}\sin\psi_i;\qquad B=\sum_{i=1}^{Z}(g+x\cos\psi_i)^{n-1}\sin^2\psi_i;$$
$$C=\sum_{i=1}^{Z}(g+x\cos\psi_i)^{n-1}\sin\psi_i\cos\psi_i;\qquad D=\sum_{i=1}^{Z}(g+x\cos\psi_i)^{n-2}\sin^2\psi_i\cos\psi_i \tag{3.37}$$

For a special case, $\phi = v/2$, (i.e., when the radial load is passing through the bisector of the two neighbouring rolling elements) and $y = 0$ (i.e., the coupling effect has been neglected, which is often the case for rolling bearings) formulae for $k(x)$ is appreciably simplified. For $y = 0$, from Eq. (3.32), we get

$$\sum_{i=1}^{Z}(g+x\cos\psi_i)^{10/9}\sin\psi_i=0\equiv A \tag{3.38}$$

Hence, for $y = 0$; $A = 0$. The stiffness expression (3.34) can be simplified to

$$k(x)=1.11K_{Lio}\sum_{i=1}^{Z}\left\{g+x\cos\psi_i-\left(\frac{A}{1.11B}\right)\sin\psi_i\right\}^{1/9}\left\{\cos\psi_i-\frac{1.11CB-0.11AD}{1.11B^2}\sin\psi_i\right\}\cos\psi_i$$

or $$k(x)=1.11K_{Lio}\sum_{i=1}^{Z}\{g+x\cos\psi_i\}^{1/9}\left\{\cos\psi_i-\frac{1.11CB}{1.11B^2}\sin\psi_i\right\}\cos\psi_i$$

$$\text{or} \quad k(x) = 1.11K_{Lio}\sum_{i=1}^{Z}\left(g - x\cos\psi_i\right)^{1/9}\left\{\cos\eta_i - \frac{C}{B}\sin\psi_i\right\}\cos\psi_i \tag{3.39}$$

$$\text{with} \quad \psi_i = \frac{\pi}{Z}(2i-1); \quad i = 1,2,3,\ldots,Z \tag{3.40}$$

$$B = \sum_{i=1}^{Z}(g + x\cos\psi_i)^{1/9}\sin^2\psi_i; \quad C = \sum_{i=1}^{Z}(g + x\cos\psi_i)^{1/9}\sin\psi_i\cos\psi_i \tag{3.41}$$

The preloading is given as

$$g = -\frac{1}{2}P_d = D - \frac{1}{2}(d_o\text{-}d_i) \tag{3.42}$$

where P_d is the diametral clearance ($P_d = 2c_r$). When $x = 0$, let $k(x) = k(0) = k_o$. From Eq. (3.39), we have

$$k_o = K_{Lio}\sum_{i=1}^{Z}g^{1/9}\left(\cos\psi_i - \frac{C}{1.11B}\sin\psi_i\right)\cos\psi_i \tag{3.43}$$

$$B = \sum_{i=1}^{Z}g^{1/9}\sin^2\psi_i; \quad C = \sum_{i=1}^{Z}g^{1/9}\sin\psi_i\cos\psi_i \tag{3.44}$$

We can have different levels of preloading conditions as follows:

Case I: $g > 0$ (Positive preloading). A simplified mathematical expression of the bearing stiffness is taken as

$$k(x) = a - bx^2 \quad \text{when} \quad |x| \le g \tag{3.45}$$

$$\text{and} \quad k(x) = b_1 x^{2/3} \quad \text{when} \quad |x| \ge g \tag{3.46}$$

Let $k(x = g) = k(g) = k_g$. At $x = 0$, from Eq. (3.45), we get $a = k_o$; and at $x = g$, we get $b = \dfrac{k_o - k_g}{g^2}$. So that the stiffness can be expressed as

$$k(x) = k_o - \frac{k_g - k_o}{g^2}x^2 \quad \text{when} \quad |x| \le g \tag{3.47}$$

Similarly, from Eq. (3.46) from information at $x = g$, we get

$$k(x) = k_g g^{-2/3}x^{2/3} \quad \text{when} \quad |x| \ge g \tag{3.48}$$

Case 2: $g = 0$ (No clearance). A simplified mathematical expression of the bearing stiffness is taken as

$$k(x) = b_2 x^{2/3} \tag{3.49}$$

At $x = c$ (c is a constant), we get $b_2 = k_c c^{-2/3}$ with $k_c = k(c)$. So that the stiffness can be expressed as

$$k(x) = k_c c^{-2/3} x^{2/3} \tag{3.50}$$

Case 3: $g < 0$ (pre-clearance or negative preload). A simplified mathematical expression of the bearing stiffness is taken as

$$\begin{aligned} k(x) &= 0 && \text{when} \quad |x| \le g \\ &= b_3 x^{2/3} && \text{when} \quad |x| > g \end{aligned} \tag{3.51}$$

For $x = d$ ($d > |g|$, where d is a constant), we get $b_3 = k_d (d)^{-3/2}$ with $k_d = k(d)$. So that the stiffness can be expressed as

$$k(x) = k_d d^{-3/2} x^{3/2} \quad \text{when} \quad x > |g| \tag{3.52}$$

Equations (3.47), (3.48), (3.50) and (3.52) are basically simplified expressions for the bearing stiffness expression given by Eq. (3.39). These forms of bearing stiffness can be used to analyse the nonlinear behaviour of rotor bearing systems (Tiwari and Vyas, 1995). It should be noted that the analysis presented here can be used to ball bearings by taking load-deflection exponent equals to $n = 3/2$ (or 1.5). Various geometrical relations for roller bearings are summarised again as

$$\sum \rho_i = \frac{1}{D_r}\left(\frac{2}{1-\gamma}\right); \quad \sum \rho_o = \frac{1}{D_r}\left(\frac{2}{1+\gamma}\right); \quad \gamma = \frac{D_r \cos 0^\circ}{D_m} \tag{3.53}$$

$$\text{and} \quad F(\rho)_i = F(\rho)_o = 1 \tag{3.54}$$

where D_r is the nominal roller diameter.

Numerical solution: Given parameters of the roller bearing is

The diameter of the roller $D_b = 6.5$ mm; The length of the roller $L = 6.5$ mm.
The inner ring groove radius of curvatures, $r_i = \infty$ mm.
The outer ring groove radius of curvatures, $r_o = \infty$ mm.

General proportions for roller bearings, as shown in Figure 3.3, are obtained as follows (PSG, 1982):

$$\text{The pitch diameter or the mean diameter: } D_m \approx \frac{1}{2}(D+d) = 33.5000 \text{ mm}. \tag{3.55}$$

$$\text{The radial thickness of rings, } s = \frac{1}{2}\left\{\frac{1}{2}(D-d) - D_b\right\} = 3.5000 \text{ mm}. \tag{3.56}$$

$$\text{The number of ball, } Z \approx \frac{\pi D_m}{D_b} = 16.19. \text{ Hence } Z = 16 \text{ can be taken.} \tag{3.57}$$

For a zero radial clearance, the inner and outer ring diameters at raceway grooves are:

$$d_0 = D_m + D_b = 33.5 + 6.5 = 40 \text{ mm}; \quad d_i = D_m - D_b = 33.5 - 6.5 = 27 \text{ mm} \tag{3.58}$$

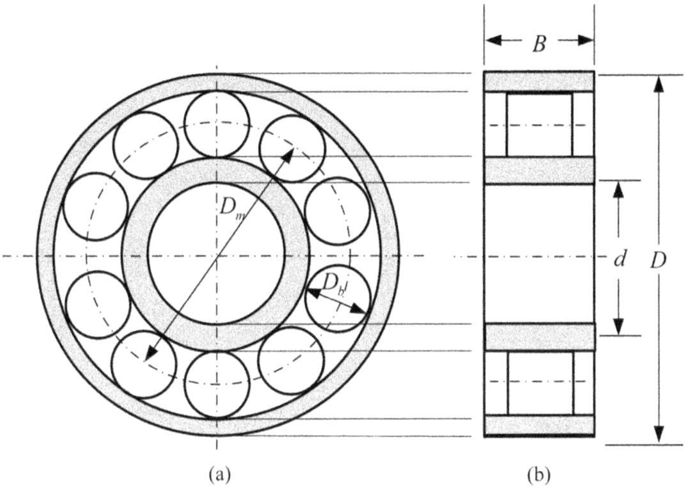

FIGURE 3.3 Internal geometries of a cylindrical roller bearing (a) end view (b) side view.

For a roller bearing of the bore diameter 20 mm; from the data given we have radial clearance of 0.005–0.020 mm. Let us for the present example take the maximum clearance, hence $g = -0.020$ mm (where g is the preload, so for clearance a negative sign is used). Finally, the bearing internal dimensions are given/obtained as:

$$r_i = \infty \text{ mm}; \quad r_o = \infty \text{ mm}; \quad D_b = 6.5 \text{ mm}; \quad Z = 16$$
$$d_i = 27.00 \text{ mm}; \quad d_0 = 40.00 \text{ mm}; \quad D_m = 33.50 \text{ mm}; \quad g = -0.020 \text{ mm}.$$
$$L = 6.5 \text{ mm}.$$

Based on these dimensions now the following parameters, which will be used for finding the bearing stiffness, can be estimated

$$\gamma = \frac{D_b}{D_m} = \frac{6.5}{33.5} = 0.1940; \quad f_i = \frac{r_i}{D_b} = \frac{\infty}{6.5} = \infty; \quad f_o = \frac{r_o}{D_b} = \frac{\infty}{6.5} = \infty \tag{3.59}$$

$$\Sigma\rho_i = \frac{1}{D_b}\left(\frac{2}{1-\gamma}\right) = \frac{1}{6.5}\left(\frac{2}{1-0.1940}\right) = 0.3818 \text{ mm}^{-1} \tag{3.60}$$

$$\Sigma\rho_o = \frac{1}{D_b}\left(\frac{2}{1+\gamma}\right) = \frac{1}{6.5}\left(\frac{2}{1+0.1940}\right) = 0.2577 \text{ mm}^{-1} \tag{3.61}$$

$$F(\rho)_i = 1.00 \tag{3.62}$$

$$F(\rho_o) = 1.00 \tag{3.63}$$

The load-deflection constant at the inner and outer raceway contacts are given as (refer Eq. (3.26))

$$K_{Li} = K_{Lo} = 7.86 \times 10^4 l^{8/9} = 7.86 \times 10^4 (6.5)^{8/9} = 3.6348 \times 10^5 \text{ N/mm}^{10/9} \tag{3.64}$$

So that, the load deflection factor for a single roller contact is given as

$$K_{Lio} = \left\{\frac{1}{(1/K_{Li})^{9/10} + (1/K_{LO})^{9/10}}\right\}^{10/9} = 1.6827\times10^{5}\ \text{N/mm}^{10/9} \tag{3.65}$$

The bearing stiffness is given as

$$k(x) = 1.11K_{Pio}\sum_{i=1}^{Z}(g + x\cos\psi_i)^{1/9}\left(\cos\psi_i - \frac{C}{B}\sin\psi_i\right)\cos\psi_i \tag{3.66}$$

with $n = 10/9$, $Z = 16$, $g = -0.02$ mm.

When the load direction is between two rolling elements, then

$$\psi_i = \frac{\pi}{16}(2i-1),\quad i = 1,2,\ldots,16 \quad \text{so that} \quad \psi_i = \frac{\pi}{16},\frac{3\pi}{16},\ldots,\frac{31\pi}{16} \tag{3.67}$$

$$B = \sum_{i=1}^{Z}(g + x\cos\psi_i)^{1/9}\sin^2\psi_i \quad \text{and} \quad C = \sum_{i=1}^{Z}(g + x\cos\psi_i)^{1/9}\sin\psi_i\cos\psi_i \tag{3.68}$$

At $x = 0$, we have (with clearance no contact of rolling elements with rings so both B and C are zero)

$$B = \sum_{i=1}^{16}(-0.02)^{1/9}\sin^2\psi_i = (-0.02)^{1/9}\left(\sin^2\frac{\pi}{16} + \sin^2\frac{3\pi}{16} + \cdots + \sin^2\frac{31\pi}{16}\right) \equiv 0 \tag{3.69}$$

$$C = \sum_{i=1}^{16}(-0.02)^{1/9}\sin\psi_i\cos\psi_i \equiv 0 \tag{3.70}$$

$$\text{and} \quad k(0) = K_{L_{io}}\sum_{i=1}^{16}(-0.02)^{1/9}(\cos\psi_i - 0)\cos\psi_i \equiv 0 \tag{3.71}$$

The stiffness also will be zero when there is no contact of rolling element with rings. In fact, $k(0)$ will not be required in the estimation of stiffness (as in the case of preload) as we will see in subsequent steps.

At $x = 2|g| = 2|-0.02| = 0.04$ mm (so that once clearance is filled the rolling elements have compression equal to the clearance level), we have

$$B = \sum_{i=1}^{Z}(g + x\cos\psi_i)^{1/9}\sin^2\psi_i = \sum_{i=1}^{16}(-0.02 + 0.04\cos\psi_i)^{1/9}\sin^2\psi_i = 1.1323 \tag{3.72}$$

$$C = \sum_{i=1}^{Z}(g + x\cos\psi_i)^{1/9}\sin\psi_i\cos\psi_i = \sum_{i=1}^{16}(-0.02 + 0.04\cos\psi_i)^{1/9}\sin\psi_i\cos\psi_i = 0 \tag{3.73}$$

and

$$k(2|g|) = 1.11K_{Pio}\sum_{i=1}^{Z}(g + x\cos\psi_i)^{1/9}\left(\cos\psi_i - \frac{C}{B}\sin\psi_i\right)\cos\psi_i$$
$$= 1.11K_{Pio}\sum_{i=1}^{16}(-0.02 + (0.04)\cos\psi_i)^{1/9}\cos^2\psi_i = 4.5034\times10^5 \text{ N/m} \tag{3.74}$$

Hence, the stiffness will be zero during displacement $x = 0$ to $|g|$, i.e.

$$k(x = 0 \text{ to } |g|) = 0 \tag{3.75}$$

Once the rolling elements are in contact with the rings, then its stiffness will be given by

$$k(x) = k(2|g|)(2|g| - |g|)^{-2/3}x^{2/3} = 4.5034\times10^5\times(0.02)^{-2/3}x^{2/3} = 6.1120\times10^6 x^{2/3} \text{ N/mm} \tag{3.76}$$

Figures 3.4 and 3.5 show the variation of stiffness of roller bearing with displacement for different clearances (or negative preloads) and preloads, respectively.

Exercise 3.4 For Exercise 3.3, obtain the variation of stiffness of the ball and roller bearings for different orientations of the rolling elements (i.e., for different values of ϕ). Discuss the results obtained.

***Solution*:** Herein, different angular position of rolling element with line of action of the force is plotted by varying the relative phase.

For ball bearing: For a given external radial load, F_r, on a ball bearing (Figure 3.6), the total elastic force on the line of contact of the *i*th roller with the inner and outer races is expressed as (Ragulskis et al., 1974; Tiwari 2017)

$$f_i = K_{pio}(g + x\cos\psi_i + y\sin\psi_i)^{3/2} \tag{3.77}$$

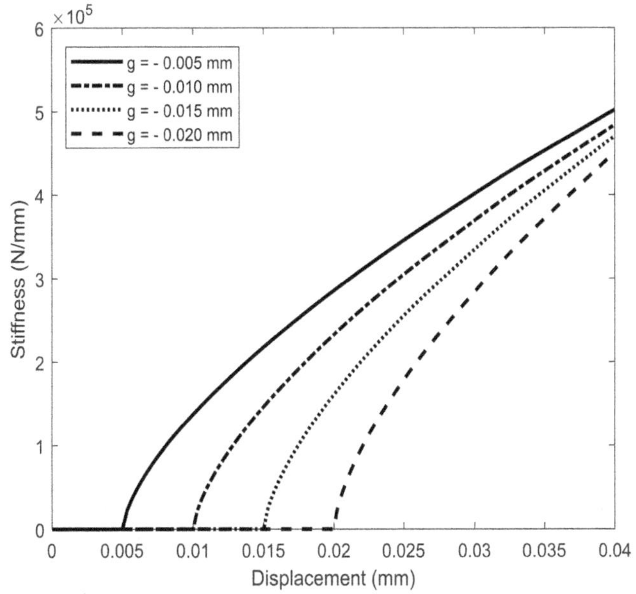

FIGURE 3.4 Variation of stiffness of roller bearing with displacement for different clearances ($x > |g|$).

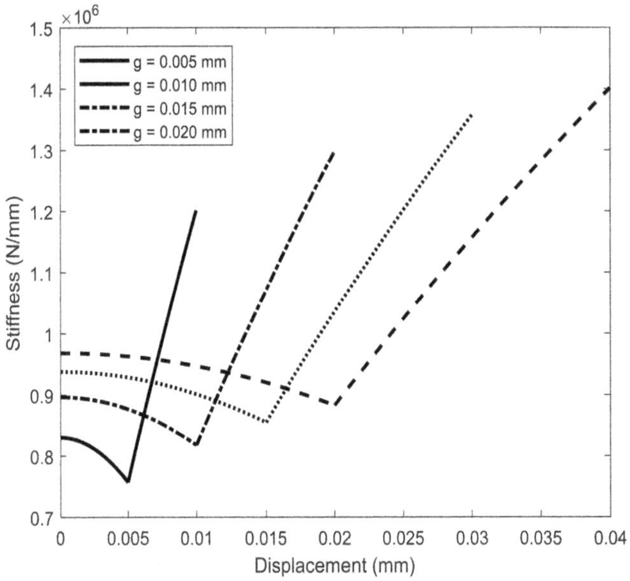

FIGURE 3.5 Variation of stiffness of roller bearing with displacement for different preloads ($x > g$).

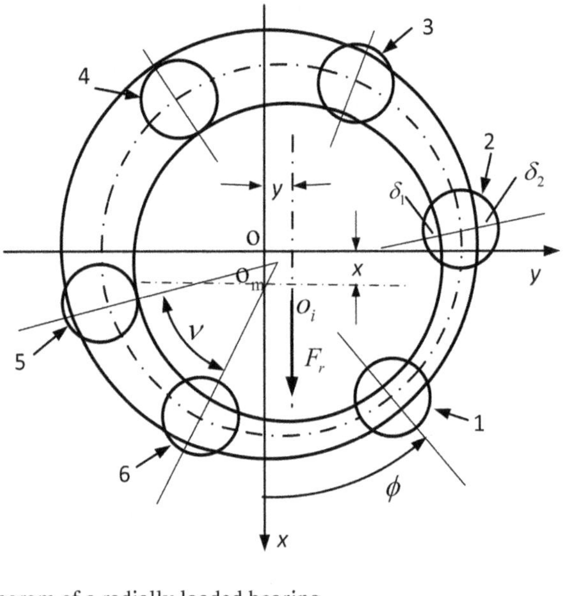

FIGURE 3.6 A line diagram of a radially loaded bearing.

with $\psi_i = \phi + (i-1)\nu$, and i = 1, 2, 3, …, Z

$$\text{and } \psi_i = \phi + \frac{\pi}{18}(2i-1), \quad i = 1,2,\ldots,18 \text{ so that } \psi_i = \frac{\pi}{18}, \frac{3\pi}{18}, \ldots, \frac{17\pi}{6}$$

where ϕ is the phase and it is varied from 0 (the force line action along midway of two rolling elements) to $4\pi/Z$ (where $2\pi/Z$ is the included angle between two neighbouring rolling elements). When a rolling element is along the line of action, least stiffness is seen as compared to when it is along the midway of two rolling elements. Plot in Figure 3.7 shows the variation of stiffness with phase. Figure 3.8 shows the variation of stiffness with displacement for different phases.

For roller bearing: For a given external radial load, F_r, on a roller bearing (Figure 3.6), the total elastic force on the line of contact of the ith roller with the inner and outer races is expressed as (Ragulskis et al. 1974; Tiwari 2017)

$$f_i = K_{Lio}\left(g + x\cos\psi_i + y\sin\psi_i\right)^{10/3} \tag{3.78}$$

with $\psi_i = \phi + (i-1)v$ and $i = 1, 2, 3, \ldots, Z$

where ϕ is the phase and it is varied from 0 (roller along force line action) to $4\pi/Z$ (where $2\pi/Z$ is the included angle between two neighbouring rolling elements). Plot in Figure 3.9 shows the variation of stiffness with phase, which has quite different characteristics as compared to the ball bearing (refer Figure 3.7). Figure 3.10 shows the variation of stiffness with displacement for different phases, which has a similar trend as that of ball bearing (refer Figure 3.8).

Exercise 3.5 Obtain the plots of the dimensionless stiffness and damping coefficients versus the Sommerfeld number (between 0.01 and 2) for the short bearing assumption of a hydrodynamic bearing.

***Solution*:** The eight linearised stiffness and damping coefficients depend on the steady-state operating conditions of the journal, and in particular upon the angular speed. For the short bearing, the dimensionless bearing stiffness and damping coefficients, $\bar{k}_{ij} = k_{ij}c_r/W$, $\bar{c}_{ij} = c_{ij}c_r/W$, and $i, j = x, y$, as a function of the steady eccentricity ratio, ε, of the bearing are given as (Smith, 1969)

$$\bar{k}_{xy} = \frac{\pi\left\{\pi^2 - 2\pi^2\varepsilon^2 - \left(16-\pi^2\right)\varepsilon^4\right\}Q(\varepsilon)}{\varepsilon\sqrt{1-\varepsilon^2}}; \quad \bar{k}_{yy} = \frac{4\left\{\pi^2 + \left(32+\pi^2\right)\varepsilon^2 + 2\left(16-\pi^2\right)\varepsilon^4\right\}Q(\varepsilon)}{\left(1-\varepsilon^2\right)};$$

$$\bar{k}_{yx} = \frac{-\pi\left\{\pi^2 + \left(32+\pi^2\right)\varepsilon^2 + 2\left(16-\pi^2\right)\varepsilon^4\right\}Q(\varepsilon)}{\varepsilon\sqrt{1-\varepsilon^2}}; \quad \bar{k}_{xx} = 4\left\{2\pi^2 + (16-\pi^2)\varepsilon^4\right\}Q(\varepsilon)$$

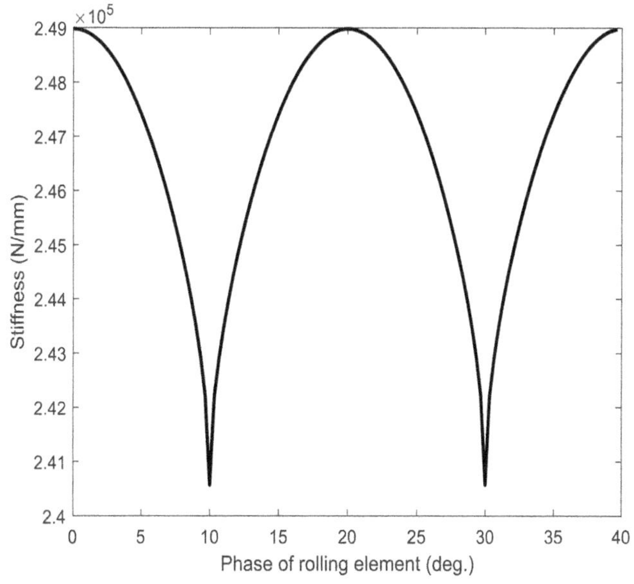

FIGURE 3.7 Variation of stiffness at $x = g$ with different phases of rolling element ($g = -0.02$ mm).

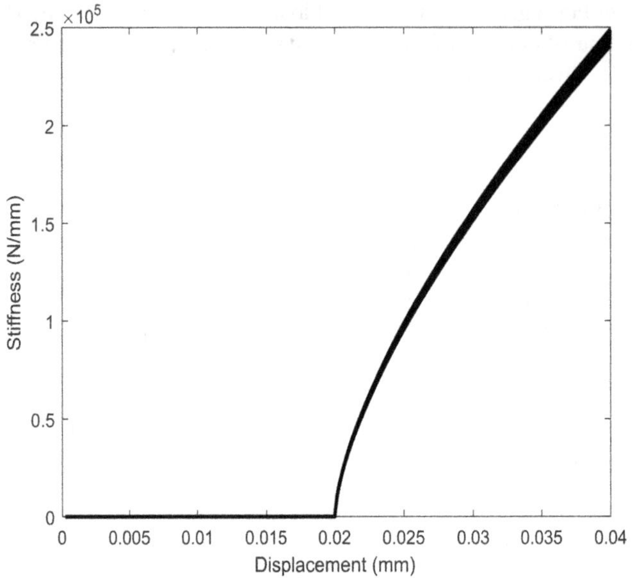

FIGURE 3.8 Variation of stiffness with displacement of bearing for different phases ($g = -0.02$ mm, phase 0°–40°: two rotation of balls).

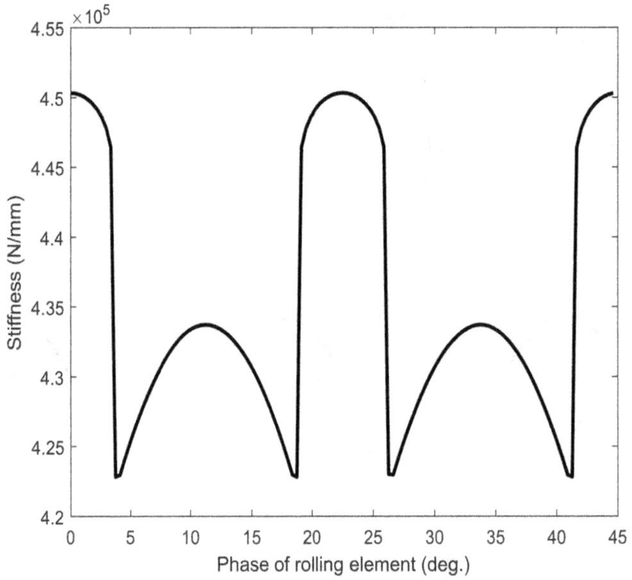

FIGURE 3.9 Variation of stiffness at $x = g$ with different phases of rolling element ($g = -0.02$ mm).

$$\bar{c}_{xx} = \frac{2\pi\sqrt{1-\varepsilon^2}\left\{\pi^2 + 2\left(\pi^2 - 8\right)\varepsilon^2\right\}Q(\varepsilon)}{\varepsilon}; \quad \bar{c}_{xx} = \bar{c}_{yx} = -8\left\{\pi^2 + 2(\pi^2 - 8)\varepsilon^2\right\}Q(\varepsilon)$$

$$\bar{c}_{yy} = \frac{2\pi\left\{\pi^2 + 2\left(24 - \pi^2\right)\varepsilon^2 + \pi^2\varepsilon^4\right\}Q(\varepsilon)}{\varepsilon\sqrt{1-\varepsilon^2}} \tag{3.79}$$

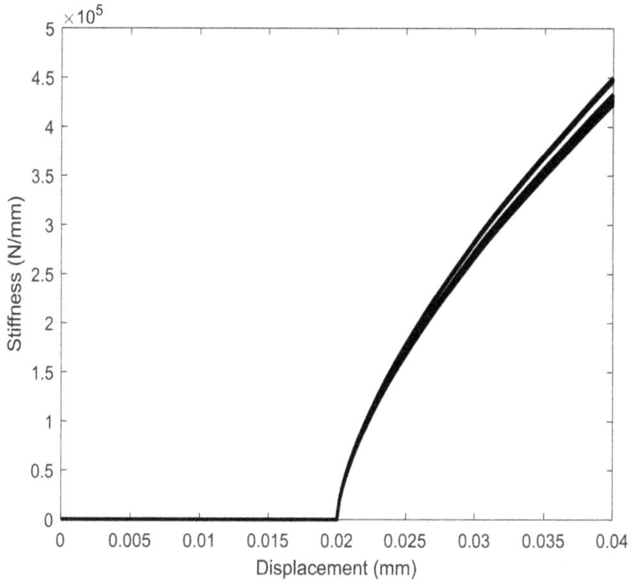

FIGURE 3.10 Variation of stiffness with displacement of bearing for different phases ($g = -0.02$ mm, phase 0°–40°: two rotation of rollers).

with

$$Q(\varepsilon) = \frac{1}{\left\{\pi^2 + (16 - \pi^2)\varepsilon^2\right\}^{3/2}} \quad \text{and} \quad \varepsilon = \frac{e_r}{c_r} \tag{3.80}$$

where W is the journal weight, e_r is the journal eccentricity and c_r is the radial clearance. It should be noted that $k_{xy} \neq k_{yx}$ whereas $c_{xy} = c_{yx}$. To determine the stiffness and damping coefficients of a short bearing, the Sommerfeld number or bearing characteristic number

$$S = \frac{\mu DLN}{W}\left(\frac{r}{c_r}\right)^2 \tag{3.81}$$

is first determined, where W is the load on the bearing, r is the bearing radius, D is the journal diameter, L is the length of bearing, μ is the viscosity of lubricant at operating temperature, $\Omega = 2\pi N$) the angular speed of journal in rad/s, and N is the number of revolutions per second. We can then determine the eccentricity ratio under steady-state operating conditions by

$$S\left(\frac{L}{D}\right)^2 = \frac{\left(1-\varepsilon^2\right)^2}{\pi\varepsilon\sqrt{\pi^2\left(1-\varepsilon^2\right)+16\varepsilon^2}} \tag{3.82}$$

Variations of the dimensionless stiffness and damping coefficients with Sommerfeld number for the case of short bearing approximation are shown in Figures 3.11 and 3.12, respectively. Variation of the eccentricity ratio with the Sommerfeld number is given in Figure 3.13. It should be noted that the eccentricity ratio and the Sommerfeld number have a unique relationship. Based on the Sommerfeld number of a bearing from Figures 3.11–3.13 the dimensionless stiffness coefficients, dimensionless damping coefficients and eccentricity ratio can be obtained.

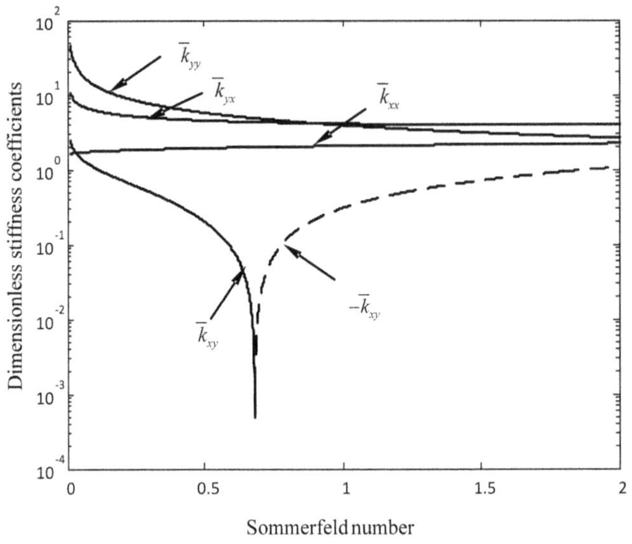

FIGURE 3.11 Variations of dimensionless stiffness coefficients for short bearings.

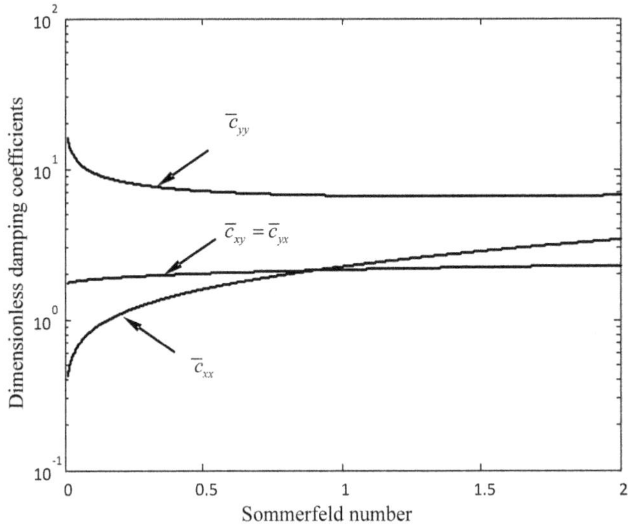

FIGURE 3.12 Variations of dimensionless damping coefficients for short bearings.

Exercise 3.6 Choose a single answer from the multiple-choice questions

i. A high-speed hydrodynamic bearing is modelled by linearised dynamic coefficients, these coefficients are

A. stiffness and damping B. virtual or added mass
C. both (A) and (B) D. None of the above

Solution: In the high-speed application of hydrodynamic bearings, the fluid inertia is significant as compared to the stiffness and damping forces in bearings. Similar to the stiffness and damping linearised coefficients, the fluid inertia gives linearised virtual or added-mass coefficients. Apart from hydrodynamic bearings, seals also in high-speed application impart virtual or added mass effects.

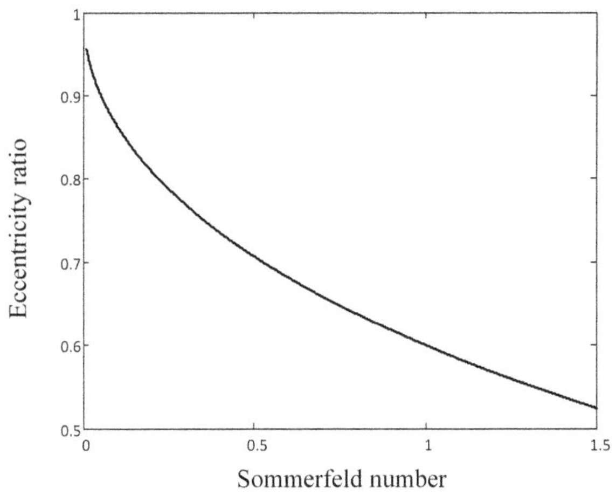

FIGURE 3.13 Variation of the eccentricity ratio with the Sommerfeld number.

ii. The hydrodynamic fluid-film bearing is governed by the
   A. Bernoulli equation
   B. Reynolds equation
   C. Navier-Strokes equation
   D. Maxwell equation

   Solution: The hydrodynamic fluid-film bearing is governed by the Reynolds equation. Section 3.3.2 (Tiwari, 2017) gives details of assumptions involved in the derivation of the hydrodynamic theory based on the Reynolds equation.

iii. For variable speed and load conditions, the most appropriate bearing would be
   A. Squeeze-film bearing
   B. Fluid-film damper
   C. Gas bearing
   D. Rolling element bearing

   Solution: For variable speed and load conditions, the most appropriate bearing would be the rolling element bearing. The fluid-film and gar bearings are designed to operate the rotor at some fixed eccentricity for a given load and speed so their performance may deteriorate on changing the speed and load conditions.

iv. For the attenuation of rotor vibrations at the resonance, the machine element used is
   A. Seal
   B. Fluid-film bearing
   C. Rolling element bearing
   D. Squeeze-film damper

   Solution: At resonance damping plays a vital role in containing the sharp rise in vibration amplitude. For the attenuation of rotor vibrations at the resonance, the machine element used is the squeeze-film damper, which provides damping in the rotor system. Often it is combined with rolling element bearings by providing a gap filled with fluid at the outer ring.

v. For the short-bearing approximation in the fluid-film bearing analysis
   A. the radial and circumferential pressure gradients are negligible
   B. the radial and axial pressure gradients are negligible
   C. the radial pressure gradient is negligible
   D. the circumferential and axial pressure gradients are negligible

   Solution: For the short-bearing approximation in the fluid-film bearing analysis the radial and circumferential pressure gradients are negligible. Due to its short axial length, the variation of the pressure in the axial direction is appreciable as compared to the other two directions.

vi. In a rotor system, the cross-coupling of two orthogonal transverse motions due to the bearing spring and damping forces are predominant in the case of the
A. bush or sleeve bearing
B. journal bearing
C. rolling bearing
D. magnetic bearings

Solution: In a rotor system the cross-coupling of two orthogonal transverse motions due to the bearing spring and damping forces are predominant in the case of the journal bearing. Sleeve, magnetic and rolling bearings impart nearly isotropic stiffness properties with negligible cross-coupled terms.

vii. Due to flexible bearings the rotor critical speeds as compared to the rigid supports are expected to
A. increase
B. decrease
C. have no effect
D. increase or decrease

Solution: Due to flexible bearings the rotor critical speeds as compared to the rigid supports are expected to decrease due to the flexibility of the shaft and bearings are in series. This will decrease the natural frequency of the rotor system.

viii. In a fluid-film bearing for the long bearing approximate solution of the Reynolds equation
A. the pressure variation along the axial direction and the radial direction is neglected
B. the pressure variation along the radial direction is neglected
C. the pressure variation along the axial direction is neglected
D. the pressure variation in any direction is not neglected

Solution: In fluid-film bearing for the long bearing approximate solution of the Reynolds equation the pressure variation along the axial direction and the radial direction is neglected. In long bearing axial length is so long that the pressure distribution can be assumed uniform in most of its length.

(ix) For a deep groove ball bearing the radial clearance is 20 μm and a radial load is applied such that the inner ring centre gets displaced with respect to the outer ring centre by 40 μm in the direction of radial load, *the load zone* would be
A. 30°
B. 60°
C. 90°
D. 120°

Solution: We have the radial clearance $c_r = 20$ μm. The inner ring centre gets displaced with respect to the outer ring centre, $x_m = 40$ μm. Thus, the load zone angle is

$$\varphi = \cos^{-1}\left(\frac{c_r}{x_m}\right) = \cos^{-1}\left(\frac{20}{40}\right) = 60° \tag{3.83}$$

x. The contact angle in rolling bearings is related to the
A. static capacity of bearing
B. dynamic capacity of bearing
C. radial load capacity of bearing
D. axial load capacity of bearing

Solution: The contact angle in rolling bearings is related to the axial load capacity of bearing. Deep groove ball bearings have a lesser contact angle as compared to the angular contact ball bearings. In inner ring split ball bearing due to three-point contact, it can take bi-axial loads and its application is very common in aerospace bearings.

xi. For a long-bearing approximation in the fluid-film hydrodynamic bearing
A. the radial pressure gradient is appreciable
B. the axial pressure gradient is appreciable
C. the circumferential pressure gradient is appreciable
D. the axial and radial pressure gradients are appreciable

Solution: For a long-bearing approximation in the fluid-film hydrodynamic bearing the circumferential pressure gradient is appreciable as compared to the axial and radial pressure gradients.

xii. The most stable rotor operation is expected in a(n)
A. cylindrical journal bearing
B. four lobe journal bearing
C. off-set halves journal bearing
D. titling-pad journal bearing

Solution: The most stable rotor operation is expected in a titling-pad journal bearing. However, its load-carrying capacity is the lowest. On the other hand, the cylindrical journal bearing is most common to have instability but has the highest load-carrying capacity.

xiii. In hydrodynamic bearings, the pressure in fluid is generated mainly due to
A. external pressure
B. motion of journal
C. squeezing action of fluid
D. heating

Solution: In hydrodynamic bearings, the pressure in fluid is generated mainly due to the motion of the journal. During motion of the journal, the fluid is forced into a narrow wedge area and due to this a very high hydrodynamic pressure is generated.

xiv. In rolling bearing due to dry contact, the load deformation relation is governed by
A. Hooke's law
B. Newton's law
C. Hertz's law
D. the Reynolds equation

Solution: In rolling bearing due to dry contact, the load deformation relation is governed by Hertz's law. In such cases, direct contact between two elastic bodies are involved and stresses develop in a very localised area. In ball bearing the contact is ideally for no load a point contact and for roller bearing it is a line contact. For a loading condition, the point contact becomes a small ellipse and line contact becomes a rectangular area. While lubricant is at contact point then elasto-hydrodynamic theory prevail and it predicts very high localised pressure at minimum film thickness of lubricant.

xv. For very high-speed applications, seals provide the following to the rotor system (herein the added-mass is the fluid mass)
A. damping and added-mass
B. stiffness and damping
C. stiffness and added mass
D. stiffness, damping and added-mass

Solution: For very high-speed applications, seals provide the stiffness, damping and added-mass coefficients to the rotor system. Often seals lead to instability in the rotor if the seal design is not proper.

xvi. The journal centre position, in a fluid-film bearing for a particular speed and without unbalance force occupies
A. a fixed position
B. a variable position
C. the centre of the bearing bore
D. a point vertically downward from the centre of the bearing bore

Solution: The journal centre position, in a fluid-film bearing for a particular speed and without unbalance force occupies a fixed position. However, with some disturbance, it will oscillate about the fixed point (i.e. the static equilibrium position). On changing the speed of the rotor, the equilibrium position changes and the trace is called the equilibrium locus of journal bearing.

xvii. Bearing dynamic coefficients are obtained based on linearisation at the

A. bearing centre B. shaft centre
C. static equilibrium position D. dynamic equilibrium position

Solution: Bearing dynamic coefficients are obtained based on linearisation of bearing fluid-film forces at the static equilibrium position. It is assumed that the rotor has small oscillations about its static equilibrium position and based on these fluid-film forces are linearised.

REFERENCES

Gupta, S., Tiwari, R., and Nair, S.B., 2007, Multi-objective design optimization of rolling bearings using genetic algorithms, *Mechanism and Machine Theory*, **42**, 1418–1443.

Harris, T.A., 2001, *Rolling Bearing Analysis*. New York, NY: Wiley.

PSG, 1982, Faculty of Mechanical Engineering, PSG College of Technology, *Design Data*. Coimbatore, India: DPV Printers.

Ragulskis, K.M., Jurkauskas, A.Y., Atstupenas, V.V., Vitkute, A.Y., and Kulvec, A.P., 1974, *Vibration of Bearings*. Vilnyus: Mintis Publishers.

Rao, B.R. and Tiwari, R., 2007, Optimum design of rolling element bearings using genetic algorithms, *Mechanism and Machine Theory*, **42**(2), 233–250.

Smith, D.M., 1969, *Journal Bearings in Turbomachinery*. London: Chapman and Hall.

Tiwari, R., 2017, *Rotor Systems: Analysis and Identification*. Boca Raton, FL: CRC Press.

Tiwari, R., and Vyas, N. S., 1995, Estimation of non-linear stiffness parameters of rolling element bearings from random response of rotor-bearing systems, *Journal of Sound and Vibration*, **187**(2), 229–239.

GENERAL REMARKS

This chapter briefly covers mainly rolling element bearing, fluid-film bearings, seals and dampers. Basic functions and principles of these important elements, which allow relative motion between the rotor and the stator are introduced. These topics themselves deserve a separate book. But just to give an idea about how these elements affect the dynamics of rotors, which will be covered in Chapter 4, this introductory chapter has been written through numerical, descriptive and MCQs.

ANSWERS TO MCQs

3.6

i. C	ii. B	iii. D	iv. D	v. A	vi. B
vii. B	viii. A	ix. B	x. D	xi. C	xii. D
xiii. B	xiv. C	xv. D	xvi. A	xvii. C	

4 Transverse Vibrations of Simple Rotor-Bearing-Foundation Systems

Exercise 4.1 Obtain the bending critical speeds of a rotor as shown in Figure 4.1. It consists of a massless rigid shaft (with 1 m span and 0.7 m from the disc to the left bearing), a rigid disc (5 kg mass and 0.1 kg-m^2 diametral mass moment of inertia), and it is supported on two identical flexible bearings (1 kN/m of stiffness for each bearing). Consider motion in the vertical plane only. Is there any difference in critical speeds when the disc is placed at the centre of the rotor? If not, then justify the same, and if yes, then obtain the same. Ignore the gyroscopic couple.

***Solution*:** In the present case, the shaft is assumed to be rigid and has a single plane motion. This leads to having the shaft and disc together as a rigid member, and two generalised coordinates will be enough to define its motion while ignoring axial vibration. Let the left and right ends of the shaft be represented as end A and end B, respectively. Consider that the motion of the disc (along with shaft) is displaced (refer Figure 4.2) vertically upwards y and tilted by an angle φ_x. We get the equation of motion as follows:

$$m\ddot{y} = -k_A(y + l_A\varphi_x) - k_B(y - l_B\varphi_x) \quad \text{or} \quad m\ddot{y} + (k_A + k_B)y + (k_A l_A - k_B l_B)\varphi_x = 0 \tag{4.1}$$

and $$I_d\ddot{\varphi}_x = -k_A(y + l_A\varphi_x)l_A + k_B(y - l_B\varphi_x)l_B \quad \text{or} \quad I_d\ddot{\varphi}_x + (k_A l_A^2 + k_B l_B^2)\varphi_x + (k_A l_A - k_B l_B)y = 0 \tag{4.2}$$

where distances are up to the disc from respective shaft ends (i.e., A or B), as shown in Figure 4.2.

Equations of motion (Eqs. (4.1) and (4.2)) can be written in a matrix form, as

$$\begin{bmatrix} m & 0 \\ 0 & I_d \end{bmatrix}\begin{Bmatrix} \ddot{y} \\ \ddot{\varphi}_x \end{Bmatrix} + \begin{bmatrix} k_A + k_B & k_A l_A - k_B l_B \\ k_A l_A - k_B l_B & k_A l_A^2 + k_B l_B^2 \end{bmatrix}\begin{Bmatrix} y \\ \varphi_x \end{Bmatrix} = \begin{Bmatrix} 0 \\ 0 \end{Bmatrix} \tag{4.3}$$

Similarly, from Figure 4.3, we get

$$m\ddot{x} = -k_A(x - l_A\varphi_y) - k_B(x + l_B\varphi_y) \quad \text{or} \quad m\ddot{x} + (k_B + k_A)x + (k_B l_B - k_A l_A)\varphi_x = 0 \tag{4.4}$$

and $$I_d\ddot{\varphi}_y = -k_B(x + l_B\varphi_y)l_B + k_A(x - l_A\varphi_y)l_A \quad \text{or} \quad I_d\ddot{\varphi}_y + (k_B l_B^2 + k_A l_A^2)\varphi_y + (k_B l_B - k_A l_A)x = 0 \tag{4.5}$$

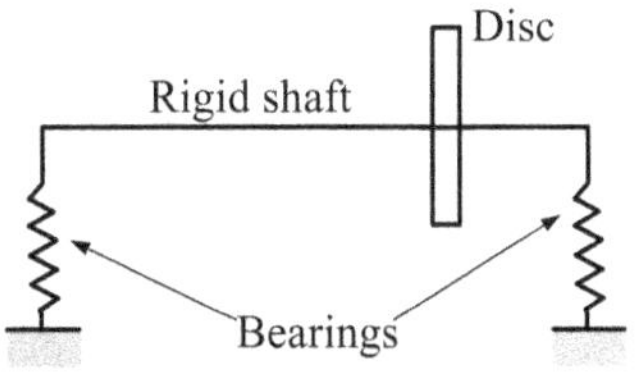

FIGURE 4.1 A rigid rotor supported on flexible bearings.

DOI: 10.1201/9781032638218-4

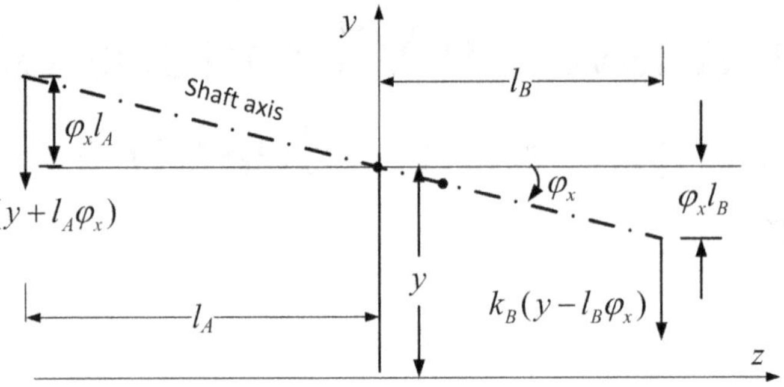

FIGURE 4.2 Free-body diagram of the shaft for simultaneous translational and rotational motion in y–z plane.

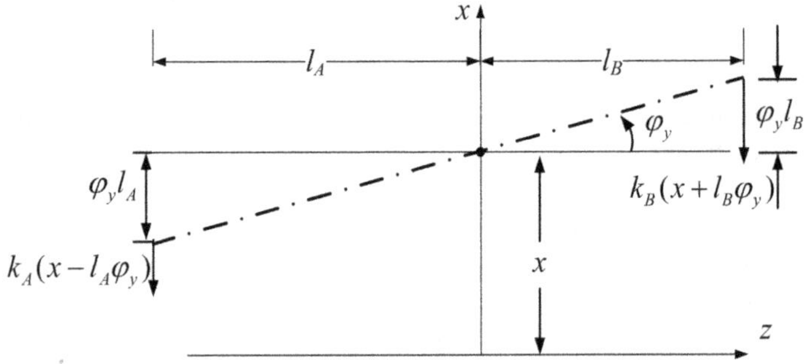

FIGURE 4.3 Free-body diagram of the shaft for simultaneous translational and rotational motion z–x plane.

which can be combined in a matrix, as

$$\begin{bmatrix} m & 0 \\ 0 & I_d \end{bmatrix}\begin{Bmatrix} \ddot{x} \\ \ddot{\varphi}_y \end{Bmatrix}+\begin{bmatrix} (k_B+k_A) & (k_Bl_B-k_Al_A) \\ (k_Bl_B-k_Al_A) & \left(k_Bl_B^2+k_Al_A^2\right) \end{bmatrix}\begin{Bmatrix} x \\ \varphi_y \end{Bmatrix}=\begin{Bmatrix} 0 \\ 0 \end{Bmatrix} \tag{4.6}$$

It should be noted that the mass matrix is uncoupled (off-diagonal terms are zero) and the stiffness matrix is coupled and symmetric. This means the translational and angular displacements are coupled. However, it can be seen that when the disc is in the middle and if both bearings have identical stiffness properties, then their diagonal terms will be zero. This means the translational and angular displacements will be uncoupled or the translational and angular motions will be independent.

Case I: Disc at offset position

The following properties of the rotor-bearing system are given:

$$l_A=0.7\text{ m},\quad l_B=0.3\text{ m},\quad m=5\text{ kg},\quad I_d=0.1\text{ kg-m}^2,\quad k_A=k_B=1{,}000\text{ N/m}$$

For the free vibration and by substituting values of the given parameters in Eq. (4.3), we get

$$-\omega_{nf}^2\begin{bmatrix} 5 & 0 \\ 0 & 0.1 \end{bmatrix}\begin{Bmatrix} y \\ \varphi_x \end{Bmatrix}+\begin{bmatrix} 2{,}000 & 400 \\ 400 & 580 \end{bmatrix}\begin{Bmatrix} y \\ \varphi_x \end{Bmatrix}=\begin{Bmatrix} 0 \\ 0 \end{Bmatrix} \tag{4.7}$$

Defining

$$\mathbf{D} = \mathbf{M}^{-1}\mathbf{K} = \begin{bmatrix} 0.2 & 0 \\ 0 & 10 \end{bmatrix} \begin{bmatrix} 2{,}000 & 400 \\ 400 & 580 \end{bmatrix} = \begin{bmatrix} 400 & 80 \\ 4{,}000 & 5{,}800 \end{bmatrix} \tag{4.8}$$

Using Characteristic Polynomial Method:

Now for the non-trial solution, we have

$$\left| \mathbf{D} - \omega_{nf}^2 \mathbf{I} \right| = 0 \tag{4.9}$$

From the aforementioned equation and substituting the values of matrices, we get

$$\begin{vmatrix} 400 - \omega_{nf}^2 & 80 \\ 4{,}000 & 5{,}800 - \omega_{nf}^2 \end{vmatrix} = 0 \tag{4.10}$$

$$\text{or} \quad \left(400 - \omega_{nf}^2\right)\left(5{,}800 - \omega_{nf}^2\right) - 80 \times 4{,}000 = 0; \quad \text{or} \quad \omega_{nf}^4 - 6{,}200\omega_{nf}^2 + 200{,}000 = 0 \tag{4.11}$$

Solving the aforementioned equation, we get

$$\omega_{nf1}^2 = 341.3772 \Rightarrow \omega_{nf1} = 18.48 \text{ rad/s} \quad \text{and} \quad \omega_{nf2}^2 = 5{,}858.6 \Rightarrow \omega_{nf2} = 76.54 \text{ rad/s}$$

Using Eigenvalue Problem Approach:

Formulating the eigenvalue problem from Eq. (4.7), as

$$(\mathbf{K} - \lambda \mathbf{M})\eta = \mathbf{0} \quad \text{with} \quad \lambda = \omega_{nf}^2 \tag{4.12}$$

Which can be written as

$$(\mathbf{D} - \lambda \mathbf{I})\eta = \mathbf{0} \quad \text{with} \quad \bar{\mathbf{D}} = \bar{\mathbf{M}}^{-1}\mathbf{K} \quad \text{and} \quad \eta = \begin{Bmatrix} y \\ \varphi_x \end{Bmatrix} \tag{4.13}$$

On obtaining eigenvalues, **V**, and eigenvectors, **U**, of matrix **D** given in Eq. (4.8), we get

$$\mathbf{V} = \begin{bmatrix} 341.4 & 0 \\ 0 & 5858.6 \end{bmatrix}; \quad \text{and} \quad \mathbf{U} = \begin{bmatrix} -0.8066 & -0.0147 \\ 0.5911 & -0.9999 \end{bmatrix} \tag{4.14}$$

which in normalised form of eigenvector matrix is

$$\tilde{\mathbf{U}} = \begin{bmatrix} 1 & 1 \\ -0.7328 & 68.2328 \end{bmatrix} \tag{4.15}$$

Natural frequencies, from eigenvalue matrix **V**, can be obtained

$$\omega_{nf1}^2 = v_{11} = 341.3772 \Rightarrow \omega_{nf1} = 18.48 \text{ rad/s} \tag{4.16}$$

$$\text{and} \quad \omega_{nf2}^2 = v_{22} = 5{,}858.6 \Rightarrow \omega_{nf2} = 76.54 \text{ rad/s} \tag{4.17}$$

The eigenvector matrix **U** indicates relative displacements of generalised coordinates chosen (i.e., y and φ_x). For better visualisation of mode shapes, it will be better if we transform the eigenvectors such that we get shaft end translatory displacements (y_A, y_B) (refer Figure 4.2), as

$$y_A = y + l_A\varphi_x; \quad y_B = y - l_B\varphi_x \tag{4.18}$$

which gives

$$\begin{Bmatrix} y_A \\ y_B \end{Bmatrix} = \begin{bmatrix} 1 & l_A \\ 1 & -l_B \end{bmatrix} \begin{Bmatrix} y \\ \varphi_x \end{Bmatrix} \quad \text{with} \quad \mathbf{T} = \begin{bmatrix} 1 & l_A \\ 1 & -l_B \end{bmatrix} \tag{4.19}$$

where **T** is the transformation matrix. On transforming eigenvector matrix, we get

$$\mathbf{U}_n = \mathbf{TU} = \begin{bmatrix} 1 & 0.7 \\ 1 & -0.3 \end{bmatrix} \begin{bmatrix} -0.8066 & 0.0147 \\ -0.5911 & -0.9999 \end{bmatrix} = \begin{bmatrix} -0.3929 & -0.7146 \\ -0.9839 & 0.2853 \end{bmatrix} \tag{4.20}$$

On normalising the aforementioned eigenvectors, we get

$$\tilde{\mathbf{U}}_n = \begin{bmatrix} 1 & 1 \\ 2.5045 & -0.3993 \end{bmatrix} \tag{4.21}$$

In the first column above, the eigenvector (or relative displacements of shaft ends) corresponds to the first mode of vibration, and the second one corresponds to the second mode. The first mode is in-phase motion (both ends have same up (or down) motion simultaneously) and the second is anti-phase motion (one end has opposite up (or down) motion with respect to the other at a particular instance).

Transformation of equations of motion in y_A and y_B, we need the following transformation

$$\begin{Bmatrix} y_A \\ y_B \end{Bmatrix} = \begin{bmatrix} 1 & l_A \\ 1 & -l_B \end{bmatrix} \begin{Bmatrix} y \\ \varphi_x \end{Bmatrix} \Rightarrow \begin{Bmatrix} y \\ \varphi_x \end{Bmatrix} = \begin{bmatrix} 1 & l_A \\ 1 & -l_B \end{bmatrix}^{-1} \begin{Bmatrix} y_A \\ y_B \end{Bmatrix} = \frac{1}{l} \begin{bmatrix} l_B & l_A \\ 1 & -1 \end{bmatrix} \begin{Bmatrix} y_A \\ y_B \end{Bmatrix}$$

$$\text{or} \quad \begin{Bmatrix} y \\ \varphi_x \end{Bmatrix} = \bar{\mathbf{T}} \begin{Bmatrix} y_A \\ y_B \end{Bmatrix} \quad \text{with} \quad \bar{\mathbf{T}} = \frac{1}{l} \begin{bmatrix} l_B & l_A \\ 1 & -1 \end{bmatrix} = \begin{bmatrix} 0.3 & 0.7 \\ 1 & -1 \end{bmatrix} \tag{4.22}$$

Another interesting part of the transformation matrix $\bar{\mathbf{T}}$ is that if we want the EOMs (i.e. the mass and stiffness matrices) in the new coordinate system then it can be used, as

$$\bar{\mathbf{M}} = \bar{\mathbf{T}}^T \mathbf{M} \bar{\mathbf{T}} = \begin{bmatrix} 0.3 & 1 \\ 0.7 & -1 \end{bmatrix} \begin{bmatrix} 5 & 0 \\ 0 & 0.1 \end{bmatrix} \begin{bmatrix} 0.3 & 0.7 \\ 1 & -1 \end{bmatrix} = \begin{bmatrix} 0.5500 & 0.9500 \\ 0.9500 & 2.5500 \end{bmatrix} \tag{4.23}$$

$$\text{and} \quad \bar{\mathbf{K}} = \bar{\mathbf{T}}^T \mathbf{K} \bar{\mathbf{T}} = \begin{bmatrix} 0.3 & 1 \\ 0.7 & -1 \end{bmatrix} \begin{bmatrix} 2{,}000 & -400 \\ -400 & 580 \end{bmatrix} \begin{bmatrix} 0.3 & 0.7 \\ 1 & -1 \end{bmatrix} = \begin{bmatrix} 1{,}000 & 0 \\ 0 & 1{,}000 \end{bmatrix} \tag{4.24}$$

Formulating the eigenvalue problem, as

$$\left(\bar{\mathbf{K}} - \lambda \bar{\mathbf{M}}\right)\eta = \mathbf{0} \quad \text{with} \quad \lambda = \omega_{nf}^2 \tag{4.25}$$

Which can be written as

$$\left(\bar{\mathbf{D}} - \lambda \mathbf{I}\right)\bar{\eta} = \mathbf{0} \quad \text{with} \quad \bar{\mathbf{D}} = \bar{\mathbf{M}}^{-1}\bar{\mathbf{K}} \quad \text{and} \quad \bar{\eta} = \begin{Bmatrix} y_A \\ y_B \end{Bmatrix} \tag{4.26}$$

On obtaining eigenvalues, $\mathbf{V}_n$, and eigenvectors, $\mathbf{U}_n$, of matrix $\bar{\mathbf{D}}$, we get

$$\mathbf{V}_n = \begin{bmatrix} 341.4 & 0 \\ 0 & 5{,}858.6 \end{bmatrix} \text{ and } \mathbf{U}_n = \begin{bmatrix} 0.3708 & 0.9287 \\ 0.9287 & -0.3708 \end{bmatrix} \tag{4.27}$$

Natural frequencies are same as the previous case. The normalising eigenvectors, we get

$$\tilde{\mathbf{U}}_n = \begin{bmatrix} 1 & 1 \\ 2.5045 & -0.3993 \end{bmatrix} \tag{4.28}$$

which is same as obtained by the transformation of eigenvectors in Eq. (4.21). The equations of motion in y_A and y_B as generalised coordinates have the following form

$$\frac{1}{l^2}\begin{bmatrix} ml_B^2 + I_d & ml_A l_B - I_d \\ ml_A l_B - I_d & ml_A^2 + I_d \end{bmatrix}\begin{Bmatrix} \ddot{y}_A \\ \ddot{y}_B \end{Bmatrix} + \begin{bmatrix} k_A & 0 \\ 0 & k_B \end{bmatrix}\begin{Bmatrix} y_A \\ y_B \end{Bmatrix} = \begin{Bmatrix} 0 \\ 0 \end{Bmatrix} \tag{4.29}$$

which can be obtained by following potential and kinetic energies by Lagrange's equation.

$$U = \tfrac{1}{2}k_A y_A^2 + \tfrac{1}{2}k_B y_B^2 \tag{4.30}$$

$$\text{and} \quad T = \tfrac{1}{2}m\dot{y}^2 + \tfrac{1}{2}I_d\dot{\varphi}_x^2 \tag{4.31}$$

$$\text{with} \quad y = y_A - l_A\varphi_x; \quad \varphi_x = \frac{y_A - y_B}{l} \quad \text{or} \quad y = y_A - l_A\frac{y_A - y_B}{l}; \quad \varphi_x = \frac{y_A - y_B}{l} \tag{4.32}$$

Therefore, the KE can be written as

$$T = \tfrac{1}{2}m\left(\dot{y}_A - l_A\frac{\dot{y}_A - \dot{y}_B}{l}\right)^2 + \tfrac{1}{2}I_d\left(\frac{\dot{y}_A - \dot{y}_B}{l}\right)^2 \tag{4.33}$$

It is to be noted that Eq. (4.29) has the mass matrix coupled and stiffness matrix uncoupled. The stiffness matrix is uncoupled since we took generalised coordinates at bearing locations where stiffness is present. In Eq. (4.3), the mass matrix is uncoupled since, in that case, the generalised coordinates are taken at centre of gravity of the rotor system. It is left to the reader to obtain EOMs in which the mass and stiffness matrices will be same as obtained in Eqs. (4.23) and (4.24), and rest of free vibration analysis will give the same natural frequencies and mode shapes as obtained earlier.

Case II: Disc at mid-span

If the disc is at the centre, the stiffness matrix, $\mathbf{K}$, will be uncoupled, as

$$\mathbf{K} = \begin{bmatrix} 2{,}000 & 0 \\ 0 & 500 \end{bmatrix}$$

$$\text{So that} \quad \mathbf{A} = \mathbf{M}^{-1}\mathbf{K} = \begin{bmatrix} 0.2 & 0 \\ 0 & 10 \end{bmatrix}\begin{bmatrix} 2{,}000 & 0 \\ 0 & 500 \end{bmatrix} = \begin{bmatrix} 400 & 0 \\ 0 & 5{,}000 \end{bmatrix}$$

Using Characteristic Polynomial Method:

So, for the free vibration, we have

$$\left|\mathbf{A} - \omega^2 \mathbf{I}\right| = 0$$

From the aforementioned equation and substituting the values of matrices we get

$$\begin{vmatrix} 400 - \omega_{nf}^2 & 0 \\ 0 & 5{,}000 - \omega_{nf}^2 \end{vmatrix} = 0$$

$$\text{or} \quad \left(400 - \omega_{nf}^2\right)\left(5{,}000 - \omega_{nf}^2\right) = 0$$

Solving the aforementioned equation, we get

$$\omega_{nf1}^2 = 400 \Rightarrow \omega_{nf1} = 20.00 \text{ rad/s} \quad \text{and} \quad \omega_{nf2}^2 = 5{,}000 \Rightarrow \omega_{nf2} = 70.71 \text{ rad/s}$$

In this case, the normalised eigenvector matrix becomes

$$\tilde{\mathbf{U}}_n = \begin{bmatrix} 1 & 1 \\ 1 & -1 \end{bmatrix}$$

So, in this case of the mode of vibration, either will have the same displacement; however, in the first mode in the same direction and in the second case in the opposite direction.

So there are differences in both natural frequencies, the first one increased ($\omega_{nf1} = 18.48$–20.00 rad/s) and the second one decreased $\omega_{nf2} = 76.54$–70.71 rad/s. This is due to the fact that the first mode is easy to excite when disc is at the middle and the second mode is easy to excite when the disc is at offset position. Any guess as to what will happen if the disc is at one of bearing locations? Readers may try and check the same.

Exercise 4.2 Consider a long, rigid rotor, R, supported on two identical bearings, B_1 and B_2, as shown in Figure 4.4. The direct stiffness coefficients of both bearings in the horizontal and vertical directions are equal (i.e. k). Ignore the direct damping and the cross-coupled stiffness and damping coefficients of both bearings. The mass of the rotor is m, the span of the rotor is l and the diametral mass moment of inertia is I_d. Derive equations of motion for the system and obtain natural frequencies of whirl. Ignore the gyroscopic effect.

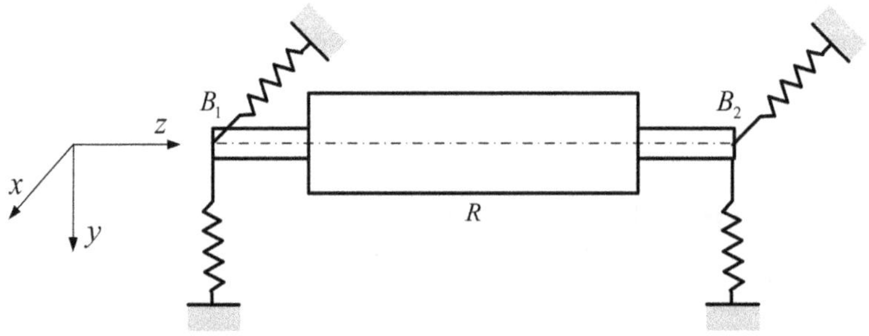

FIGURE 4.4 A long rigid rotor supported on flexible bearings.

***Solution*:** Considering no axial and radial unbalances in the rotor, i.e. C and G are coincident. For the pure transverse translational displacement, on balancing the forces in the vertical upward y direction from free the body diagram of the shaft (Figure 4.5a), we get

$$ky + ky = -m\ddot{y} \quad \text{or} \quad m\ddot{y} + 2ky = 0 \tag{4.34}$$

Similarly, on balancing the forces in the horizontal x-direction (Figure 4.5b), we get

$$kx + kx = -m\ddot{x} \quad \text{or} \quad m\ddot{x} + 2kx = 0 \tag{4.35}$$

Now, considering a pure tilting motion about the horizontal and vertical axes (Figure 4.6), the moment equilibriums, we get

$$-0.25k\varphi_x l^2 - 0.25k\varphi_x l^2 = I_d\ddot{\varphi}_x \quad \text{or} \quad 0.5kl^2\varphi_x + I_d\ddot{\varphi}_x = 0 \tag{4.36}$$

$$\text{and} \quad -0.25k\phi_y l^2 - 0.25k\phi_y l^2 = I_d\ddot{\phi}_y \quad \text{or} \quad 0.5kl^2\phi_y + I_d\ddot{\phi}_y = 0 \tag{4.37}$$

Now, combining the Eqs. (4.34) through (4.37) in a matrix form, we get

$$\begin{bmatrix} m & 0 & 0 & 0 \\ 0 & m & 0 & 0 \\ 0 & 0 & I_d & 0 \\ 0 & 0 & 0 & I_d \end{bmatrix} \begin{Bmatrix} \ddot{x} \\ \ddot{y} \\ \ddot{\varphi}_y \\ \ddot{\varphi}_x \end{Bmatrix} + \begin{bmatrix} 2k & 0 & 0 & 0 \\ 0 & 2k & 0 & 0 \\ 0 & 0 & 0.5kl^2 & 0 \\ 0 & 0 & 0 & 0.5kl^2 \end{bmatrix} \begin{Bmatrix} x \\ y \\ \varphi_y \\ \varphi_x \end{Bmatrix} = \begin{Bmatrix} 0 \\ 0 \\ 0 \\ 0 \end{Bmatrix} \tag{4.38}$$

Both the mass and stiffness matrices are uncoupled, so in fact, all four equations can be solved independently of each other. For a simple harmonic vibration, we have

$$\ddot{x} = \omega_{nf}^2 x; \quad \ddot{y} = \omega_{nf}^2 y; \quad \ddot{\varphi}_x = -\omega_{nf}^2\varphi_x \quad \text{and} \quad \ddot{\varphi}_y = -\omega_{nf}^2\varphi_y \tag{4.39}$$

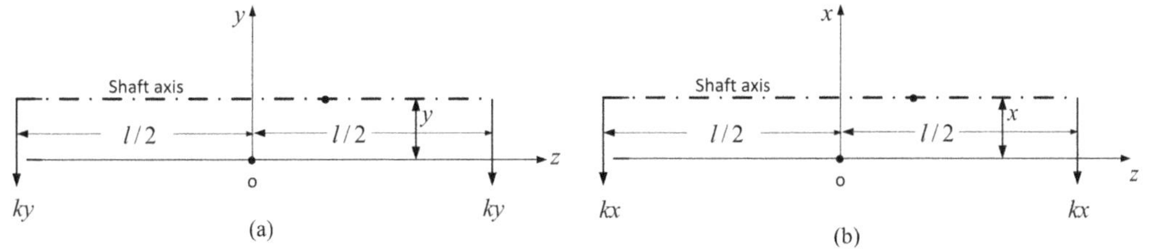

FIGURE 4.5 A long rigid rotor supported on flexible bearings in pure translational motion (a) y–z plane (b) z–x plane.

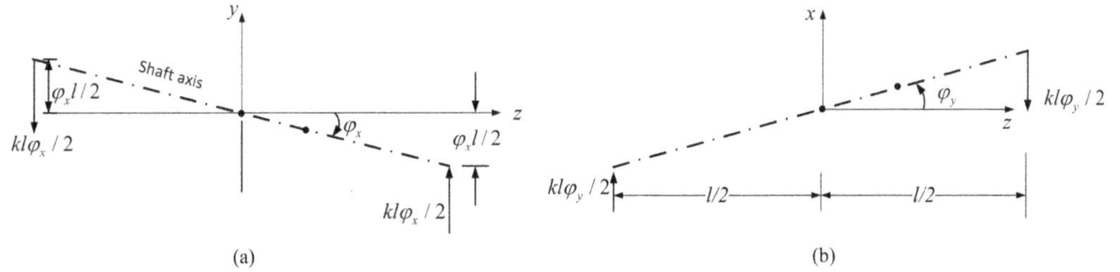

FIGURE 4.6 A long rigid rotor supported on flexible bearings in pure rotational motion (a) y–z plane (b) z–x plane.

Substituting (4.39) in the matrix Eq. (4.38) (Eqs. (4.34) through (4.37)) and on solving, we get

$$\omega_{nf1} = \sqrt{\frac{2k}{m}}; \quad \omega_{nf2} = \sqrt{\frac{2k}{m}}; \quad \omega_{nf3} = \sqrt{\frac{0.5kl^2}{I_d}}; \quad \omega_{nf4} = \sqrt{\frac{0.5kl^2}{I_d}} \tag{4.40}$$

Since the rotor is symmetric and the bearings are identical and isotropic so effectively, we have only two natural frequencies. Herein, two modes shapes will be similar to Exercise 4.1 for the case when the disc was at mid-span, i.e., both ends will have same displacements, and only their direction with respect to each other will change depending upon the first mode (pure translator mode) or second mode (pure tilting mode).

Exercise 4.3 Find critical speeds of the rotor-bearing system shown in Figure 4.7. The shaft is rigid and massless. The mass of the disc is m_d=1 kg with negligible diametral mass moment of inertia. Bearings B_1 and B_2 are identical bearings and have the following properties: k_{xx}=1.1 kN/m, k_{yy}=1.8 kN/m, k_{xy}=0.2 kN/m and k_{yx}=0.1 kN/m. Let B_1D=75 mm and DB_2=50 mm.

***Solution*:** Since in the present case, the diametral mass moment of inertia is negligible, so corresponding tilting vibration will not take place. So only transverse translatory motion will take place. However, due to the presence of cross-coupled stiffness terms, the two orthogonal motions will be coupled and need to be analysed simultaneously. Equations of motion can be obtained by considering the free body diagram in y–z and x–z planes (refer Figures 4.8 and 4.9), respectively, as

$$-2(k_{yy}y + k_{yx}x) = m\ddot{y} \Rightarrow m\ddot{y} + 2(k_{yy}y + k_{yx}x) = 0 \tag{4.41}$$

and
$$-2(k_{xx}x + k_{xy}y) = m\ddot{x} \Rightarrow m\ddot{x} + 2(k_{xx}x + k_{xy}y) = 0 \tag{4.42}$$

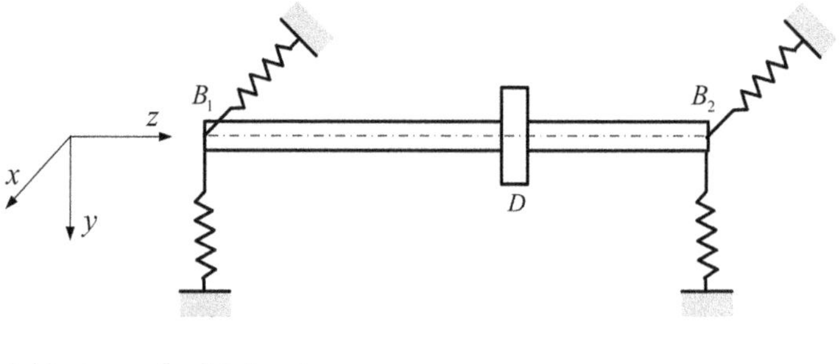

FIGURE 4.7 A rigid rotor on flexible bearings.

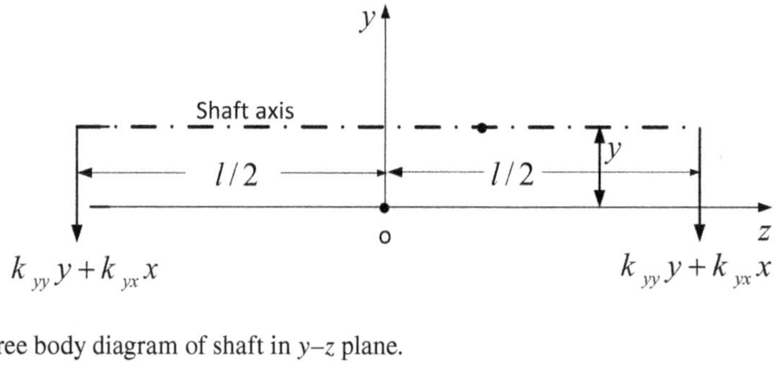

FIGURE 4.8 Free body diagram of shaft in y–z plane.

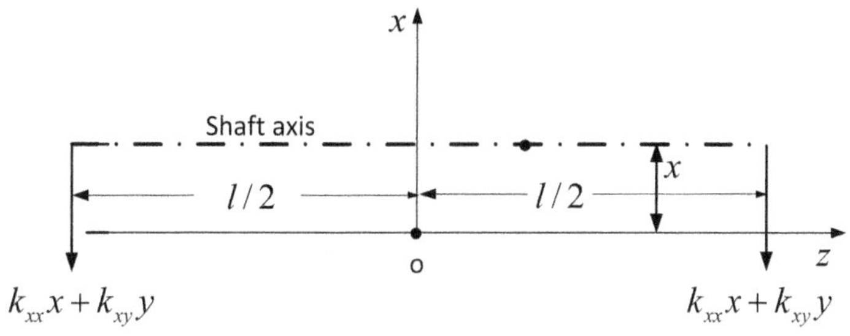

FIGURE 4.9 Free body diagram of shaft in z–x plane.

We are considering only the pure translational motion as it is given that the rotor have a negligible diametral mass moment of inertia. Hence, the pure tilting motion will not contribute towards the critical speed. Arranging Eq. (4.41) and Eq. (4.42) in the matrix form, as

$$\begin{bmatrix} m & 0 \\ 0 & m \end{bmatrix}\begin{Bmatrix} \ddot{x} \\ \ddot{y} \end{Bmatrix}+\begin{bmatrix} 2k_{xx} & 2k_{xy} \\ 2k_{yx} & 2k_{yy} \end{bmatrix}\begin{Bmatrix} x \\ y \end{Bmatrix}=\begin{Bmatrix} 0 \\ 0 \end{Bmatrix} \tag{4.43}$$

Herein, the mass matrix is uncoupled and the stiffness matrix is coupled. The cross-coupled stiffness gives coupling of motions in two orthogonal directions. For the free vibration, it takes the following form,

$$\left(-\omega_{nf}^2\begin{bmatrix} m & 0 \\ 0 & m \end{bmatrix}+\begin{bmatrix} 2k_{xx} & 2k_{xy} \\ 2k_{yx} & 2k_{yy} \end{bmatrix}\right)\begin{Bmatrix} x \\ y \end{Bmatrix}=\begin{Bmatrix} 0 \\ 0 \end{Bmatrix} \tag{4.44}$$

where ω_{nf} is the natural frequency. Taking a determinant equal to zero, this gives the frequency equation, as

$$m^2\omega_{nf}^4 - 2m\omega_{nf}^2(k_{xx}+k_{yy})+4(k_{xx}k_{yy}-k_{xy}k_{yx})=0 \tag{4.45}$$

The following rotor-bearing data are given

Mass of the disc = 1 kg; $k_{xx} = 1.1$ kN/m; $k_{yy} = 1.8$ kN/m; $k_{xy} = 0.20$ kN/m;

$k_{yx} = 0.10$ kN/m.

Substituting values, we get

$$\omega_{nf}^4 - 2\omega_{nf}^2(1.1+1.8)\times 1{,}000 + 4(1.1\times 1.8 - 0.2\times 0.1)\times 10^6 = 0 \tag{4.46}$$

which gives,

$$\omega_{nf1} = 46.31 \text{ rad/s} \quad \text{and} \quad \omega_{nf2} = 60.45 \text{ rad/s} \tag{4.47}$$

Herein, mode shapes have not been obtained; however, they will represent the relative displacement of the whole shaft in the vertical and horizontal directions. When both have the same sign then it will be forward whirl and when opposite sign then it will have backward whirl. The eigenvector of the present case is (detail of obtaining this already explained in great detail in Exercise 4.1)

$$\mathbf{U}=\begin{bmatrix} 0.2651 & 0.9907 \\ 0.9642 & -0.1362 \end{bmatrix} \tag{4.48}$$

which means that during the transition to the first critical speed, the rotor will have the backward whirl and on crossing the second critical speed it will have the forward whirl. Before the critical speed, the rotor has always forward whirl. So, during the speed range of the first critical speed and second critical speed, it will have backward whirl and the remaining range it will forward whirl.

Exercise 4.4 For Exercise 4.3, take 25 g-mm of the unbalance in the disc at a phase of 38° from a shaft reference point. Plot the disc response amplitude and phase versus the spin speed of the rotor to show all critical speeds. Plot the variation of bearing forces with the spin speed of rotor.

***Solution*:** The equations of motion for the given rotor system (refer Figures 4.10 and 4.11), considering both translational and rotational displacements, are

$$m\ddot{x} + 2\left(k_{xx}x + k_{xy}y\right) = f_x(t) \tag{4.49}$$

$$m\ddot{y} + 2\left(k_{yx}x + k_{yy}y\right) = f_y(t) \tag{4.50}$$

$$I_d\ddot{\varphi}_y + k_{xx}\left(0.5l^2\varphi_y\right) + k_{xy}\left(0.5l^2\varphi_x\right) = M_{xz}(t) \tag{4.51}$$

$$I_d\ddot{\varphi}_x + k_{yx}\left(0.5l^2\varphi_y\right) + k_{yy}\left(0.5l^2\varphi_x\right) = M_{yz}(t) \tag{4.52}$$

in the x, y, φ_y, and φ_x directions, respectively. Here f and M represent the external force and moment (e.g., due to an radial and axial eccentricity). Herein, the coupling of motions has not been considered among the translational and rotational motions.

The external force and moment can be written as

$$f_x = me\omega^2\cos(\omega t + \phi) \equiv me\omega^2 e^{j(\omega t+\phi)} \text{ with } F_x = me\omega^2 e^{j\phi} \tag{4.53}$$

$$f_y = me\omega^2\sin(\omega t + \phi) \equiv -jme\omega^2 e^{j(\omega t+\phi)} \text{ with } F_y = -jme\omega^2 e^{j\phi} \tag{4.54}$$

$$M_{xz} = me\omega^2 e_z\cos(\omega t + \phi) \equiv mee_z\omega^2 e^{j(\omega t+\phi)} \text{ with } \bar{M}_{xz} = me\omega^2 e_z e^{j\phi} \tag{4.55}$$

$$M_{yz} = me\omega^2 e_z\sin(\omega t + \phi) \equiv -jmee_z\omega^2 e^{j(\omega t+\phi)} \text{ with } \bar{M}_{yz} = -jme\omega^2 e_z e^{j\phi} \tag{4.56}$$

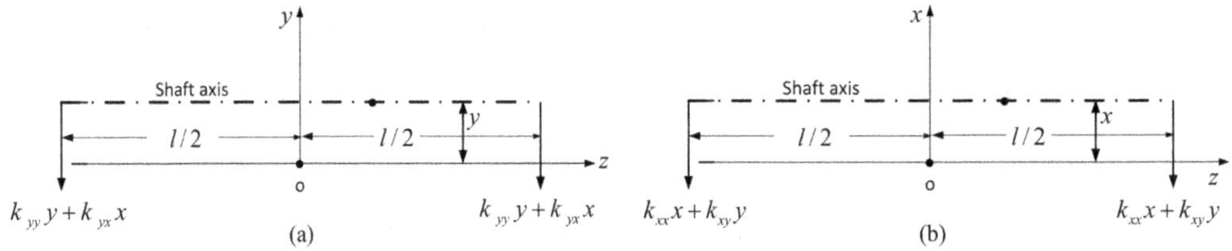

FIGURE 4.10 Pure translational motion of the rigid shaft in (a) y–z plane (b) z–x plane.

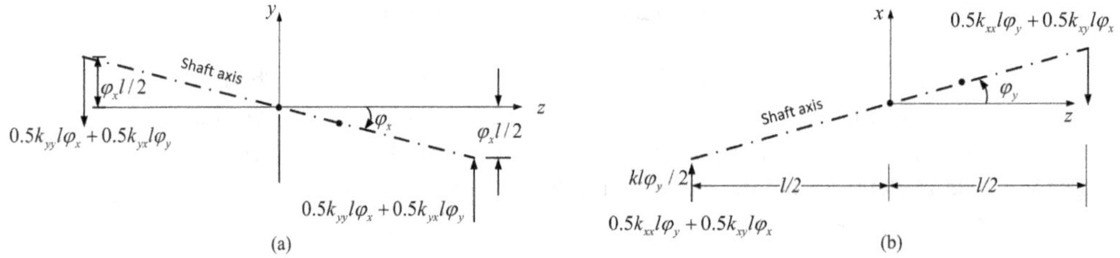

FIGURE 4.11 Pure rotational motion of the rigid shaft in (a) y–z plane (b) z–x plane.

where F_x and F_y are complex unbalance forces (which contain the amplitude and phase information) along the x and y axes, respectively; and $\bar{M}_{xz}$ and $\bar{M}_{yz}$ are complex unbalance moments (which contain the amplitude and phase information) about the y and x axes, respectively. Herein, ϕ is the phase of radial eccentricity, e, with respect to x-axis at time $t = 0$, and e_z is the axial eccentricity, which has additional magnitude only, the phase is governed by the radial eccentricity only. The unbalance responses are assumed as

$$x = Xe^{j\omega t},\; y = Ye^{j\omega t},\; \varphi_y = \Phi_y e^{j\omega t},\; \varphi_x = \Phi_x e^{j\omega t} \tag{4.57}$$

where X, Y, Φ_y and Φ_x are complex displacements. Equations (4.49)–(4.52) can be written as

$$\mathbf{M}\ddot{\mathbf{x}} + \mathbf{K}\mathbf{x} = \mathbf{f}(t) \tag{4.58}$$

$$\text{with } \mathbf{M} = \begin{bmatrix} m & 0 & 0 & 0 \\ 0 & m & 0 & 0 \\ 0 & 0 & I_d & 0 \\ 0 & 0 & 0 & I_d \end{bmatrix}; \mathbf{K} = \begin{bmatrix} 2k_{xx} & 2k_{xy} & 0 & 0 \\ 2k_{yx} & 2k_{yy} & 0 & 0 \\ 0 & 0 & 0.5l^2k_{xx} & 0.5l^2k_{xy} \\ 0 & 0 & 0.5l^2k_{yx} & 0.5l^2k_{yy} \end{bmatrix}; \mathbf{x} = \begin{Bmatrix} x \\ y \\ \varphi_y \\ \varphi_x \end{Bmatrix};$$

$$\mathbf{f} = \begin{Bmatrix} f_x(t) \\ f_y(t) \\ M_{xz}(t) \\ M_{yz}(t) \end{Bmatrix}$$

The response vector takes the following form

$$\mathbf{x} = \mathbf{X}e^{j\omega t} \quad \text{so that} \quad \ddot{\mathbf{x}} = -\omega^2\mathbf{X}e^{j\omega t} \tag{4.59}$$

On substituting Eqs. (4.53)–(4.56) and (4.59) into equations of motion (4.58), we get

$$\left(-\omega^2\mathbf{M} + \mathbf{K}\right)\mathbf{X} = \mathbf{F} \quad \text{with} \quad \mathbf{X} = \begin{Bmatrix} X \\ Y \\ \Phi_y \\ \Phi_x \end{Bmatrix} \quad \text{and} \quad \mathbf{F} = \begin{Bmatrix} F_x \\ F_y \\ \bar{M}_{xz} \\ \bar{M}_{yz} \end{Bmatrix} \tag{4.60}$$

which can be written as

$$\mathbf{DX} = \mathbf{F} \quad \text{with} \quad \mathbf{D} = \left(\mathbf{K} - \omega^2\mathbf{M}\right) \tag{4.61}$$

The unbalance response vector can be obtained as

$$\mathbf{X} = \mathbf{D}^{-1}\mathbf{F} \tag{4.62}$$

$$\text{with} \quad \mathbf{X} = \begin{Bmatrix} X \\ Y \\ \Phi_y \\ \Phi_x \end{Bmatrix} \equiv \begin{Bmatrix} X_r + \mathrm{j}X_i \\ Y_r + \mathrm{j}Y_i \\ \Phi_{yr} + \mathrm{j}\Phi_{yi} \\ \Phi_{xr} + \mathrm{j}\Phi_{xi} \end{Bmatrix}; \quad |X| = \sqrt{X_r^2 + X_i^2}; \quad \angle X = \tan^{-1}\frac{X_i}{X_r} \tag{4.63}$$

Putting the values in the mass and stiffness matrices from Exercise 4.3, we get

$$\mathbf{M}=\begin{bmatrix}1&0&0&0\\0&1&0&0\\0&0&0&0\\0&0&0&0\end{bmatrix};\quad \mathbf{K}=\begin{bmatrix}2,200&400&0&0\\200&3,600&0&0\\0&0&8.59&1.56\\0&0&0.78&14.06\end{bmatrix};\tag{4.64}$$

Also unbalance, me, of 25 gm-mm $=25\times10^{-6}$ kg-m with phase, $\phi=38°=38\pi/180$ rad is given so the force vector is given as

$$\mathbf{F}=\begin{Bmatrix}me\omega^2e^{j\phi}\\-jme\omega^2e^{j\phi}\\mee_z\omega^2e^{j\phi}\\-jmee_z\omega^2e^{j\phi}\end{Bmatrix}=\begin{Bmatrix}25\times10^{-6}\omega^2\times e^{j(38\pi/180)}\\-j25\times10^{-6}\omega^2e^{j(38\pi/180)}\\25\times10^{-6}\times0\times\omega^2e^{j(38\pi/180)}\\-j25\times10^{-6}\times0\times\omega^2e^{j(38\pi/180)}\end{Bmatrix}=\begin{Bmatrix}\left(1.9700\times10^{-5}+j1.5392\times10^{-5}\right)\omega^2\\\left(1.5392\times10^{-5}-j1.9700\times10^{-5}\right)\omega^2\\0\\0\end{Bmatrix}\tag{4.65}$$

Responses (amplitude and phase, refer Eq. (4.63)) are plotted in Figure 4.12 in the horizontal (x) and vertical (y) directions. It can be seen that only two critical speeds are visible in the form of sharp peaks at 46.31 and 60.46 rad/s in both direction response amplitudes, which is due to coupling of two orthogonal displacements. Interestingly, there is sharp change in phase at these speeds indicating critical speed location. However, due to cyclic nature of the phase whenever it

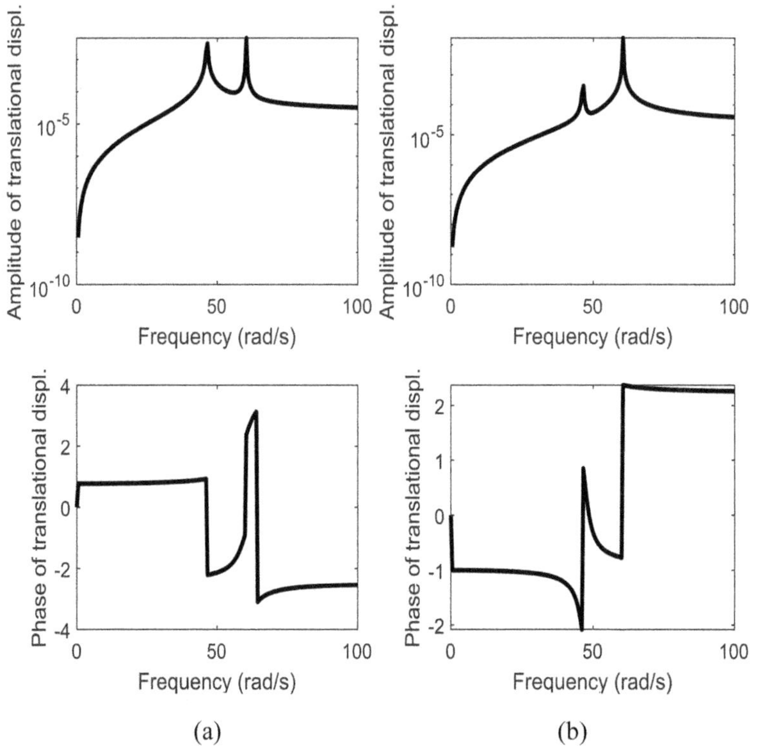

FIGURE 4.12 Shaft translational vibration amplitude and phase (a) horizontal direction (b) vertical direction.

reached phase of π^+ (slightly above π rad) or π^- (slightly below π rad) then it shows a shape change in the phase. This looks as if there is phase change but it is not a phase change, generally the phase change at resonance takes place of the order of π rad and not of the order of 2π rad, which is evident in this cyclic change in phase (refer left phase plot near second critical speed, the first due to the resonance and the second is due to cyclic change of the phase). The initial phase of unbalance is chosen as $38° = 0.66$ rad, in the phase plot at zero spin speed in x-direction phase plot this initial phase is visible. Whereas, in y-direction, which has 90° phase lag (i.e., $38-90=-52°=-0.91$ rad) the initial phase is −0.91 rad. Due to absence of the diametral mass moment of inertia of the rotor the critical speeds corresponding to tilting motion is not present the plot and that is why plot is shown up to 100 rad/s only.

In case the diametral mass moment of inertia and axial eccentricity is specified, we expect four critical speeds, two each for the translational and rotational motions. For translational motion Figure 4.12 is still valid, this remains the same due to uncoupled motion between translational and rotational motion. Figure 4.13 (left) and Figure 4.13 (right) are rotational response plots in horizontal and vertical planes, respectively, corresponding to $I_d = 0.01$ kg-m² (about centre of rotation) and $e_z = 0.001$ m. New two critical speeds can be seen in Figure 4.13 as 289.47 and 377.85 rad/s. Corresponding the mass matrix and force vector are (there is no change in **K** matrix)

$$\mathbf{M} = \begin{bmatrix} 1 & 0 & 0 & 0 \\ 0 & 1 & 0 & 0 \\ 0 & 0 & 0.01 & 0 \\ 0 & 0 & 0 & 0.01 \end{bmatrix} \tag{4.66}$$

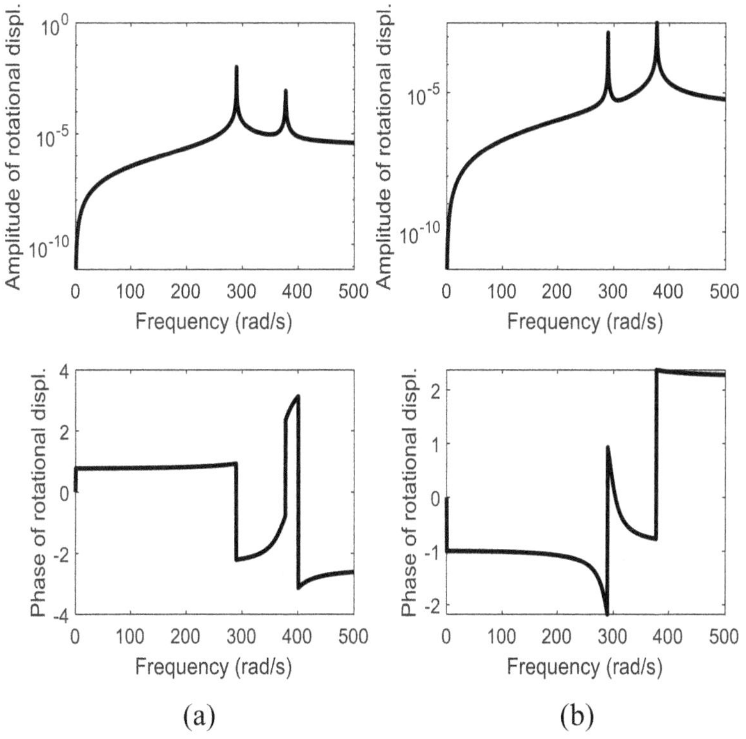

FIGURE 4.13 Shaft rotational (tilting) vibration amplitude and phase (a) horizontal plane (b) vertical plane.

$$\mathbf{F} = \begin{Bmatrix} me\omega^2 e^{j\phi} \\ -jme\omega^2 e^{j\phi} \\ mee_z\omega^2 e^{j\phi} \\ -jmee_z\omega^2 e^{j\phi} \end{Bmatrix} = \begin{Bmatrix} 25\times10^{-6}\omega^2 \times e^{j(38\pi/180)} \\ -j25\times10^{-6}\omega^2 e^{j(38\pi/180)} \\ 25\times10^{-6}\times0.001\times\omega^2 e^{j(38\pi/180)} \\ -j25\times10^{-6}\times0.001\times\omega^2 e^{j(38\pi/180)} \end{Bmatrix}$$

$$= \begin{Bmatrix} \left(1.9700\times10^{-5} + j1.5392\times10^{-5}\right)\omega^2 \\ \left(1.5392\times10^{-5} - j1.9700\times10^{-5}\right)\omega^2 \\ \left(1.9700\times10^{-8} + j1.5392\times10^{-8}\right)\omega^2 \\ \left(1.5392\times10^{-8} - j1.9700\times10^{-8}\right)\omega^2 \end{Bmatrix} \tag{4.67}$$

Exercise 4.5 For Example 4.4, the obtain critical speeds of the rotor-bearing-foundation system when the foundation (which is in series with bearing) has the following dynamic characteristics: $k_{f_x} = k_{f_y} = 100$ MN/m and $c_{f_x} = c_{f_y} = 50$ kN-s/m. Let the mass of each bearing be 2 kg. Plot the unbalance response amplitude and phase of the shaft ends and the bearing at B_1 with respect to the spin speed of shaft to show all critical speeds of the system. Take 25 g-mm of unbalance on the disc at a phase of 38° from a shaft reference point. Plot the variation of the bearing and foundation forces at left side (end A) with the spin speed. Choose a suitable range to cover all critical speeds.

***Solution*:** In the present section, a very simple model of the foundation is considered by ignoring cross-coupled terms of the stiffness and the damping (Figure 4.14).

The net displacement of the disc (when we consider shaft flexibility also but for the rigid shaft this can be ignored during formulation) is given by the vector sum of (i) the disc displacement relative to shaft ends (need to ignore for rigid shaft) (ii) that of shaft ends relative to the bearing and (iii) that of the bearing relative to the foundation. The theoretical analysis of the disc, shaft and bearing unbalanced responses, and that of the force transmissibility of such a system, can be carried out in a similar manner to that described in Exercise 4.4. Additional governing equations related to the foundation are derived, and how to relate them with governing equations of the disc and bearings are detailed here.

Bearings: The relationship between forces transmitted through bearings and displacements of shaft ends with respect to bearings are governed by the bearing stiffness and damping coefficients. The form of governing equation in frequency domain is given by Eq. (4.41) of Tiwari (2017), which is

$$\mathbf{f}_b = \mathbf{K}_b \mathbf{X}_b \tag{4.68}$$

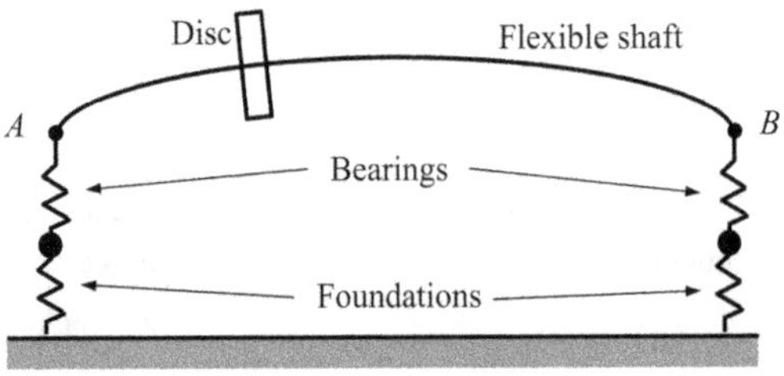

FIGURE 4.14 A flexible rotor–bearing–foundation system.

$$\text{with}\quad \mathbf{F}_b = \begin{Bmatrix} {}_AF_{bx} \\ {}_AF_{by} \\ {}_BF_{bx} \\ {}_BF_{by} \end{Bmatrix};\quad \mathbf{K}_b = \begin{bmatrix} {}_A(k_{xx}+j\omega c_{xx}) & {}_A(k_{xy}+j\omega c_{xy}) & 0 & 0 \\ {}_A(k_{yx}+j\omega c_{yx}) & {}_A(k_{yy}+j\omega c_{yy}) & 0 & 0 \\ 0 & 0 & {}_B(k_{xx}+j\omega c_{xx}) & {}_B(k_{xy}+j\omega c_{xy}) \\ 0 & 0 & {}_B(k_{yx}+j\omega c_{yx}) & {}_B(k_{yy}+j\omega c_{yy}) \end{bmatrix}$$

$$\text{and}\quad \mathbf{X}_b = \lfloor\ {}_AX_b \quad {}_AY_b \quad {}_BX_b \quad {}_BY_b\ \rfloor^T;$$

which gives shaft end displacements with respect to the bearing (*A*: left and *B*: right), as

$$\mathbf{X}_b = \mathbf{K}_b^{-1}\mathbf{F}_b \tag{4.69}$$

where $\mathbf{X}_b$ is the shaft end displacement relative to the bearing.

Foundation: Displacements of bearings with respect to foundations and the forces transmitted through bearings are shown in Figure 4.15.

The bearing will respond in the horizontal direction for an external force f_{bx}, which is governed by the following equation

$$f_{bx} - k_{fx}x_f - c_{fx}\dot{x}_f = m_b\ddot{x}_f \tag{4.70}$$

where x_f is the horizontal displacement of the bearing mass, m_b is the bearing mass of one bearing and k_{fx}, c_{fx}, k_{fy} and c_{fy} are the foundation stiffness and damping coefficients. Similarly, the response of the bearing mass in the vertical direction to a force f_{by} is given as

$$f_{by} - k_{fy}y_f - c_{fy}\dot{y}_f = m_b\ddot{y}_f \tag{4.71}$$

where y_f is the vertical displacement of bearing mass. The displacement of the bearing mass with respect to fixed foundation will take the form

$$x_f(t) = X_f e^{j\omega t} \quad \text{and} \quad y_f(t) = Y_f e^{j\omega t} \tag{4.72}$$

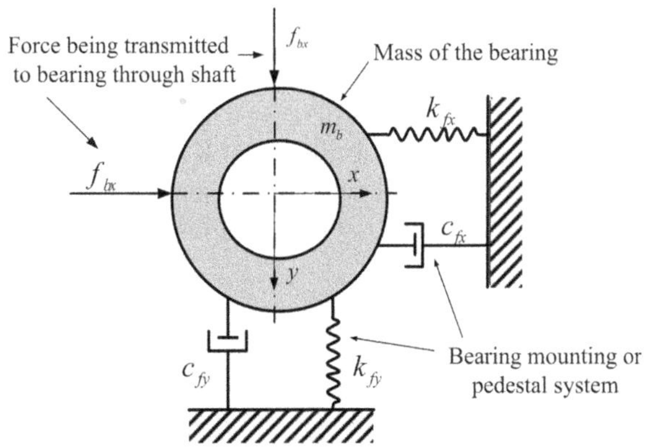

FIGURE 4.15 A bearing block (mass) mounted on a flexible foundation.

where X_f and Y_f are complex displacements in the x and y directions, respectively. On substituting Eq. (4.72) into equations of motion (4.70) and (4.71), and on combining in a matrix form (for bearing/foundation A), it gives

$$ {}_A\mathbf{K}_f \, {}_A\mathbf{X}_f = {}_A\mathbf{F}_b \tag{4.73}$$

with ${}_A\mathbf{K}_f = \left(\begin{bmatrix} k_{f_x} & 0 \\ 0 & k_{f_y} \end{bmatrix} - \omega^2 \begin{bmatrix} m_b & 0 \\ 0 & m_b \end{bmatrix} + j\omega \begin{bmatrix} c_{f_x} & 0 \\ 0 & c_{f_y} \end{bmatrix} \right)$; ${}_A\mathbf{X}_f = \left\{ \begin{matrix} X_f \\ Y_f \end{matrix} \right\}_A$;

${}_A\mathbf{F}_b = \left\{ \begin{matrix} F_{b_x} \\ F_{b_y} \end{matrix} \right\}_A$

For both bearings A and B, equations of form (f) can be combined as

$$ \mathbf{F}_b = \mathbf{K}_f \mathbf{X}_f \tag{4.74}$$

with $\mathbf{F}_b = \left\{ \begin{matrix} {}_A\mathbf{F}_b \\ {}_B\mathbf{F}_b \end{matrix} \right\}$; $\mathbf{K}_f = \begin{bmatrix} {}_A\mathbf{K}_f & \mathbf{0} \\ \mathbf{0} & {}_B\mathbf{K}_f \end{bmatrix}$; $\mathbf{X}_f = \left\{ \begin{matrix} {}_A\mathbf{X}_f \\ {}_B\mathbf{X}_f \end{matrix} \right\}$

which gives relative displacements between bearings and fixed foundations, as

$$ \mathbf{X}_f = \mathbf{K}_f^{-1}\mathbf{F}_b \equiv \boldsymbol{\alpha}_f \mathbf{F}_b \tag{4.75}$$

Bearing and Foundation: The total displacement of shaft ends under the action of an applied force $\mathbf{F}_b$ is given by the summation of individual displacements $\mathbf{X}_b$ (by Eq. (4.69)) and $\mathbf{X}_f$ (by Eq. (4.75)), i.e.

$$ \mathbf{X}_{bf} = \mathbf{X}_f + \mathbf{X}_b = \left(\mathbf{K}_f^{-1} + \mathbf{K}_b^{-1}\right)\mathbf{F}_b = \boldsymbol{\alpha}_{fb}\mathbf{F}_b \quad \text{or} \quad \mathbf{F}_b = \boldsymbol{\alpha}_{fb}^{-1}\mathbf{X}_{bf} = \mathbf{K}_{fb}\mathbf{X}_{bf} \tag{4.76}$$

$$ \text{with} \quad \boldsymbol{\alpha}_{fb} = \left(\mathbf{K}_f^{-1} + \mathbf{K}_b^{-1}\right) \quad \text{and} \quad \mathbf{K}_{fb} = \boldsymbol{\alpha}_{fb}^{-1} = \left(\mathbf{K}_f^{-1} + \mathbf{K}_b^{-1}\right)^{-1}$$

where $\boldsymbol{\alpha}_{fb}$ is a system equivalent dynamic receptance matrix describing the overall shaft support characteristics and allows for flexibilities of both bearings and foundations. The study of the disc motion may now proceed in the same manner as described in the previous section except the equivalent dynamic stiffness matrix $\mathbf{K}_{fb}$ should be substituted for $\mathbf{K}$. Once the disc displacement vector $\mathbf{U}_s$ is known, it is possible to substitute back and obtain $\mathbf{F}_s$, $\mathbf{X}_b$ and $\mathbf{F}_b$.

Foundation forces: Forces transmitted to foundations (A (left) or B (right) ends with appropriate back subscript) are given as

$$ f_{fx} = k_{fx}x_f + c_{fx}\dot{x}_f \quad \text{and} \quad f_{fy} = k_{fy}y_f + c_{fy}\dot{y}_f \tag{4.77}$$

For the unbalance excitation, we have

$$ f_{fx} = F_{fx}e^{j\omega t} \quad \text{and} \quad f_{fy} = F_{fy}e^{j\omega t} \tag{4.78}$$

On substituting Eq. (4.78) into Eq. (4.77), we get

$$\mathbf{F}_f = \mathbf{K}_p \mathbf{X}_f \quad \text{with} \quad \mathbf{F}_f = \begin{Bmatrix} F_{f_x} \\ F_{f_y} \end{Bmatrix}; \ \mathbf{K}_p = \begin{bmatrix} k_{f_x} & 0 \\ 0 & k_{f_y} \end{bmatrix} + j\omega \begin{bmatrix} c_{f_x} & 0 \\ 0 & c_{f_y} \end{bmatrix}; \ \mathbf{X}_f = \begin{Bmatrix} X_f \\ Y_f \end{Bmatrix}$$

For both pedestals at ends A and B, equations from above, can be combined as

$$\mathbf{F}_f = \mathbf{K}_p \mathbf{X}_f \tag{4.79}$$

$$\text{with} \quad \mathbf{F}_f = \begin{Bmatrix} {}_A\mathbf{F}_f \\ {}_B\mathbf{F}_f \end{Bmatrix}; \ \mathbf{K}_p = \begin{bmatrix} {}_A\mathbf{K}_p & \mathbf{0} \\ \mathbf{0} & {}_B\mathbf{K}_p \end{bmatrix}; \ \mathbf{X}_f = \begin{Bmatrix} {}_A\mathbf{X}_f \\ {}_B\mathbf{X}_f \end{Bmatrix}$$

which can be used to obtain forces transmitted to the foundation through pedestals, if we know the relative displacements between bearings and fixed foundations $\mathbf{X}_f$. Forces transmitted through foundations will not be the same as forces transmitted through bearings. Since bearing masses (i.e., inertia forces) will absorb some forces towards its acceleration. If bearing masses are negligible then bearings and foundations will transmit the same amount of forces, however, may be with some phase lag due to the presence of the damping. The amplitude and the phase of forces transmitted through foundations can be obtained from F_{fx1}, F_{fx2}, F_{fy1} and F_{fy2} as usual procedure from the complex form.

Shaft: Now we are bringing in the shaft and disc effect. The magnitude of reaction forces transmitted by bearings can also be evaluated in terms of forces applied to the shaft by the disc (refer Figure 4.16).

From Figure 4.16, the moment balance will be

$$\sum M_B = 0 \Rightarrow {}_A f_{by} l = f_y (l - a) + M_{yz} \quad \text{or} \quad {}_A f_{by} = f_y \left(1 - a/l\right) + M_{yz} \left(1/l\right) \tag{4.80}$$

$$\text{and} \quad \sum M_A = 0 \Rightarrow {}_B f_{by} l = f_y a - M_{yz} \quad \text{or} \quad {}_B f_{by} = f_y \left(a/l\right) - M_{yz} \left(1/l\right) \tag{4.81}$$

where f_{bx} and f_{by} are the bearing forces, f_x and f_y are the forces on the shaft from the disc (for example unbalance force for the rigid shaft and for the flexible shaft elastic forces), M_{zx} and M_{yz} are the moments on the shaft from the disc (for example unbalance moment for the rigid shaft and for the flexible shaft elastic moments), Similarly, forces in the horizontal direction may be written as

$${}_A f_{bx} = f_x \left(1 - a/l\right) - M_{zx} \left(1/l\right) \tag{4.82}$$

$$\text{and} \quad {}_B f_{bx} = f_x \left(a/l\right) + M_{zx} \left(1/l\right) \tag{4.83}$$

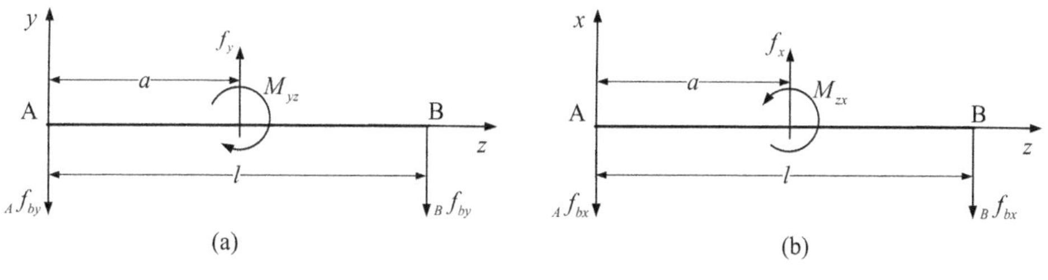

FIGURE 4.16 A free body diagram of the shaft (a) y–z plane (b) z–x plane.

Equations (4.80) through (4.83) can be combined in a matrix form as

$$\mathbf{f}_b = \mathbf{A}\mathbf{f}_s \tag{4.84}$$

$$\text{with}\quad \mathbf{f}_b = \begin{Bmatrix} {}_Af_{bx} \\ {}_Af_{by} \\ {}_Bf_{bx} \\ {}_Bf_{by} \end{Bmatrix};\quad \mathbf{f}_s = \begin{Bmatrix} f_x \\ f_y \\ M_{zx} \\ M_{yz} \end{Bmatrix};\quad \mathbf{A} = \begin{bmatrix} (1-a/l) & 0 & -1/l & 0 \\ 0 & (1-a/l) & 0 & 1/l \\ a/l & 0 & 1/l & 0 \\ 0 & a/l & 0 & -1/l \end{bmatrix}$$

For the unbalance excitation, we have

$$\mathbf{f}_b = \mathbf{F}_b e^{j\omega t} \quad \text{and} \quad \mathbf{f}_s = \mathbf{F}_s e^{j\omega t} \tag{4.85}$$

where subscript b refers to the bearing and s refers to the shaft. On substituting Eq. (4.85) into Eq. (4.84), we get

$$\mathbf{F}_b = \mathbf{A}\mathbf{F}_s \tag{4.86}$$

Shaft, Bearing and Foundation: In Eq. (4.86), bearing forces are related to the reaction forces and moments on the shaft by the disc. Now to connect the foundation flexibility and bearing flexibility, on equating Eqs. (4.76) and (4.86), we get

$$\mathbf{F}_b \equiv \mathbf{K}_{fb}\mathbf{X}_{bf} = \mathbf{A}\mathbf{F}_s \quad \text{or} \quad \mathbf{X}_{bf} = \mathbf{K}_{fb}^{-1}\mathbf{A}\mathbf{F}_s = \boldsymbol{\alpha}_{fb}\mathbf{A}\mathbf{F}_s \tag{4.87}$$

$$\text{with}\quad \mathbf{K}_{fb} = \boldsymbol{\alpha}_{fb}^{-1} = \left(\mathbf{K}_b^{-1} + \mathbf{K}_f^{-1}\right)^{-1}$$

Equation (4.87) relates the shaft end deflections to the reaction forces and moments on the shaft by the disc. It should be noted that relations between these forces ($\mathbf{F}_s$ and $\mathbf{F}_b$) have been derived considering simply-supported end conditions. For other support conditions (like over-hung and cantilever) suitably equations need to be changed.

Disc: The deflection at the location of the disc due to the movement of shaft ends can be obtained as follows. Consider the shaft to be rigid for some instant and let us denote the shaft end deflections in the horizontal direction to be ${}_Ax_{bf}$ and ${}_By_{bf}$ at ends A and B, respectively, as shown in Figure 4.17. It is to be noted that this displacement is deflection of shaft ends with respect to foundation (i.e., absolute displacements) and not with respect to bearing (i.e. relative displacements). These displacements are assumed to be small.

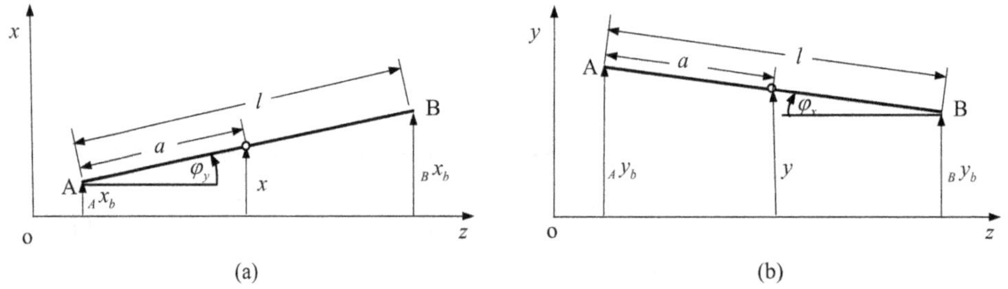

FIGURE 4.17 Rigid body movement of the shaft (a) z–x plane (b) y–z plane.

The translational displacement in the x-direction can be written as (refer Figure 4.17a)

$$x = {}_A x_{bf} + \frac{\left({}_B x_{bf} - {}_A x_{bf}\right)}{l} a = \left(1 - \frac{a}{l}\right){}_A x_{bf} + \left(\frac{a}{l}\right){}_B x_{bf} \tag{4.88}$$

The rotational displacement of the shaft in x–z plane will be (refer Figure 4.17a)

$$\varphi_y = \frac{\left({}_B x_{bf} - {}_A x_{bf}\right)}{l} = \left(-\frac{1}{l}\right){}_A x_{bf} + \left(\frac{1}{l}\right){}_B x_{bf} \tag{4.89}$$

Similarly, for the translational and rotational displacements in the y-direction and in the y–z plane, respectively; we have (refer Figure 4.17b)

$$y = {}_A y_{bf} - \frac{\left({}_A y_{bf} - {}_B y_{bf}\right)}{l}(a) = \left(\frac{l-a}{l}\right){}_A y_{bf} + \left(\frac{a}{l}\right){}_B y_{bf} \tag{4.90}$$

$$\text{and} \quad \varphi_x = \frac{\left({}_A y_{bf} - {}_B y_{bf}\right)}{l} = \left(\frac{1}{l}\right){}_A y_{bf} + \left(-\frac{1}{l}\right){}_B y_{bf} \tag{4.91}$$

Equations (4.88) through (4.91) can be combined in a matrix form as

$$\mathbf{u}_{s1} = \mathbf{B}\mathbf{x}_{bf} \tag{4.92}$$

$$\text{with} \quad \mathbf{u}_{s1} = \begin{Bmatrix} x \\ y \\ \varphi_y \\ \varphi_x \end{Bmatrix}_{s1}; \quad \mathbf{x}_b = \begin{Bmatrix} {}_A x_{bf} \\ {}_A y_{bf} \\ {}_B x_{bf} \\ {}_B y_{bf} \end{Bmatrix}; \quad \mathbf{B} = \begin{bmatrix} (1-a/l) & 0 & a/l & 0 \\ 0 & (1-a/l) & 0 & a/l \\ -1/l & 0 & 1/l & 0 \\ 0 & 1/l & 0 & -1/l \end{bmatrix}$$

where subscript s_1 represents that these displacements are due to the rigid body motion of the shaft. For the unbalance excitation (or for the free vibration analysis), shaft displacements at bearing locations and at the disc centre vary sinusoidally, such that

$$\mathbf{u}_{s1} = \mathbf{U}_{s1}e^{j\omega t} \quad \text{and} \quad \mathbf{x}_{bf} = \mathbf{X}_{bf}e^{j\omega t} \tag{4.93}$$

where ω is the spin speed (or the natural frequency in the case of free vibrations). On substituting Eq. (4.93) into Eq. (4.92), we have

$$\mathbf{U}_{s1} = \mathbf{B}\mathbf{X}_{bf} \tag{4.94}$$

On substituting Eq. (4.87) into Eq. (4.94), we get

$$\mathbf{U}_{s1} = \mathbf{B}\mathbf{K}_{fb}^{-1}\mathbf{A}\mathbf{F}_s = \boldsymbol{\alpha}_{fbs1}\mathbf{F}_s \quad \text{with} \quad \boldsymbol{\alpha}_{fbs1} = \mathbf{B}\mathbf{K}_{fb}^{-1}\mathbf{A} \tag{4.95}$$

Equation (4.95) will give deflections of the disc that is caused by only the movement of shaft ends (rigid body movement) on flexible bearings/foundations. To obtain the net disc deflection under a given load, we have to add the deflection due to the deformation of the shaft with respect to bearing/foundation locations also in Eq. (4.95). The deflection associated with flexure of the shaft alone has already been calculated in Chapter 2 (Tiwari, 2017) in terms of influence coefficients, which can be combined in a matrix form as

$$\mathbf{u}_{s2} = \boldsymbol{\alpha}_s\mathbf{f}_s \tag{4.96}$$

$$\text{with}\quad \mathbf{u}_{s2} = \begin{Bmatrix} x \\ y \\ \varphi_y \\ \varphi_x \end{Bmatrix}_{s2};\quad \mathbf{f}_s = \begin{Bmatrix} f_x \\ f_y \\ M_{yz} \\ M_{zx} \end{Bmatrix};\quad \boldsymbol{\alpha}_s = \begin{bmatrix} \alpha_{11} & 0 & \alpha_{12} & 0 \\ 0 & \alpha_{11} & 0 & \alpha_{12} \\ \alpha_{21} & 0 & \alpha_{22} & 0 \\ 0 & \alpha_{21} & 0 & \alpha_{22} \end{bmatrix}$$

where subscript s_2 represents that these displacements are due to the pure deformation of the shaft without any rigid body motion. It should be noted that these displacements have been derived considering simply-supported end conditions. For other support conditions (like, over-hung and cantilever) suitable equations need to be changed both for $\mathbf{u}_{s1}$ and $\mathbf{u}_{s2}$. For the unbalance excitation (or for the free vibration analysis), shaft reaction forces at the disc location and disc displacements vary sinusoidally and can be expressed as

$$\mathbf{u}_{s2} = \mathbf{U}_{s2} e^{j\omega t} \quad \text{and} \quad \mathbf{f}_s = \mathbf{F}_s e^{j\omega t} \tag{4.97}$$

On substituting Eq. (4.97) into Eq. (4.96), we get

$$\mathbf{U}_{s2} = \boldsymbol{\alpha}_{s2} \mathbf{F}_s \tag{4.98}$$

which is the deflection of disc due to the flexure of the shaft alone, without considering the bearing/foundation flexibility. In case the flexibility of the shaft needs to be ignored then this deflection part need not be considered. The net deflection of the disc caused by the deflection of bearings/foundation plus that due to the flexure of the shaft, is then given by

$$\mathbf{U}_s = \mathbf{U}_{s1} + \mathbf{U}_{s2} = \left(\boldsymbol{\alpha}_{fbs1} + \boldsymbol{\alpha}_{s2}\right)\mathbf{F}_s = \boldsymbol{\alpha}_{fbs}\mathbf{F}_s \tag{4.99}$$

$$\text{with}\quad \boldsymbol{\alpha}_{fbs1} = \mathbf{B}\mathbf{K}_{fb}^{-1}\mathbf{A} \quad \text{and} \quad \boldsymbol{\alpha}_{fbs} = \left(\boldsymbol{\alpha}_{fbs1} + \boldsymbol{\alpha}_{s2}\right)$$

where $\mathbf{U}_s$ contains absolute displacements of the shaft at the location of the disc. Equation (4.99) describes displacements of the shaft at the disc under the action of sinusoidal forces and moments applied at the disc (hence the matrix $\boldsymbol{\alpha}_{fbs}$ is similar to the effective influence coefficient matrix). Equation (4.99) can be written as

$$\mathbf{F}_s = \boldsymbol{\alpha}_{fbs}^{-1}\mathbf{U}_s = \mathbf{K}_{fbs}\mathbf{U}_s \tag{4.100}$$

$$\text{with}\quad \mathbf{K}_{fbs} = \alpha_{fbs}^{-1};\quad \boldsymbol{\alpha}_{fbs} = \left(\boldsymbol{\alpha}_{fbs1} + \boldsymbol{\alpha}_{s2}\right);\quad \boldsymbol{\alpha}_{fbs1} = \mathbf{B}\mathbf{K}_{fb}^{-1}\mathbf{A};\quad \mathbf{K}_{fb} = \boldsymbol{\alpha}_{fb}^{-1} = \left(\mathbf{K}_b^{-1} + \mathbf{K}_f^{-1}\right)^{-1}$$

where the matrix $\mathbf{K}_{fbs}$ is similar to the effective stiffness matrix (it is equivalent stiffness of the shaft and bearings/foundations experienced at the disc location). Equations of motion of the disc can be written in the x-direction and on the z–x plane (see Figure 4.18a), as

$$-f_x = m\frac{d^2}{dt^2}(x + e\cos\omega t) \quad \text{or} \quad me\omega^2\cos\omega t - f_x = m\ddot{x} \tag{4.101}$$

$$\text{and} \quad -M_{zx} = I_d\ddot{\varphi}_y \tag{4.102}$$

Similarly, equations of motion, in the y-direction and on the y–z plane (see Figure 4.18b), can be written as

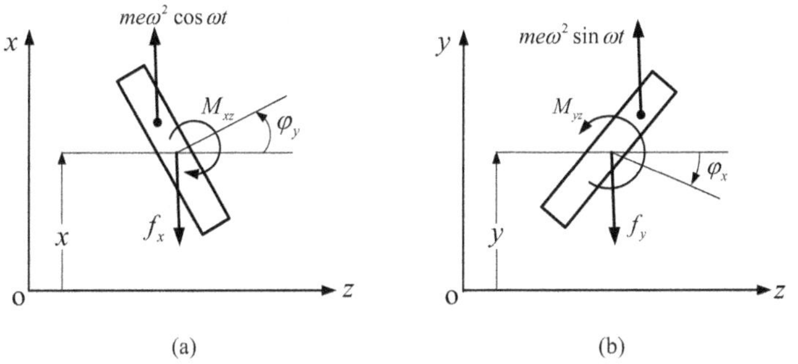

FIGURE 4.18 Free body diagram of the disc (a) in z–x plane (b) in y–z plane.

$$-f_y = m\frac{d^2}{dt^2}(y + e\sin\omega t) \quad \text{or} \quad me\omega^2 \sin\omega t - f_y = m\ddot{y} \tag{4.103}$$

$$\text{and} \quad -M_{yz} = I_d\ddot{\varphi}_x \tag{4.104}$$

Equations of motion (4.101) through (4.104) of the disc can be written in a matrix form as

$$\mathbf{M}\ddot{\mathbf{u}} + \mathbf{f}_s = \mathbf{f}_{unb} \tag{4.105}$$

$$\text{with} \quad \mathbf{M} = \begin{bmatrix} m & 0 & 0 & 0 \\ 0 & m & 0 & 0 \\ 0 & 0 & I_d & 0 \\ 0 & 0 & 0 & I_d \end{bmatrix}; \quad \mathbf{u} = \begin{Bmatrix} x \\ y \\ \varphi_y \\ \varphi_x \end{Bmatrix}; \quad \mathbf{f}_s = \begin{Bmatrix} f_x \\ f_y \\ M_{xz} \\ M_{yz} \end{Bmatrix}; \quad \mathbf{f}_{unb} = \begin{Bmatrix} me\omega^2 \\ -jme\omega^2 \\ 0 \\ 0 \end{Bmatrix} e^{j\omega t} = \mathbf{F}_{unb}e^{j\omega t}$$

Noting $\mathbf{f}_s = \mathbf{F}_s e^{j\omega t}$ and $\mathbf{u} = \mathbf{U}_s e^{j\omega t}$, equations of motion take the following form

$$-\omega^2\mathbf{M}\mathbf{U}_s + \mathbf{F}_s = \mathbf{F}_{unb} \tag{4.106}$$

Noting Eq. (4.100), Eq. (4.106) becomes

$$-\omega^2\mathbf{M}\mathbf{U}_s + \mathbf{K}_{fbs}\mathbf{U}_s = \mathbf{F}_{unb} \tag{4.107}$$

which gives

$$\mathbf{U}_s = \boldsymbol{\alpha}_{fbs}\mathbf{F}_{unb} \quad \text{with} \quad \boldsymbol{\alpha}_{fbs} = \left(-\omega^2\mathbf{M} + \mathbf{K}_{fbs}\right)^{-1} \tag{4.108}$$

where $\mathbf{K}_{fbs}$ and $\boldsymbol{\alpha}_{fbs}$ are the equivalent dynamic stiffness and flexibility matrices, respectively, as experienced by the disc, of the shaft and the bearing/foundation system. Once the response of the disc $\mathbf{U}_s$ has been obtained, from the aforementioned equation for a given unbalance force $\mathbf{F}_{unb}$, the loading applied to the shaft, $\mathbf{F}_s$, by the disc can be obtained by Eq. (4.100). Then from Eq. (4.87), we can get shaft end absolute displacements $\mathbf{X}_{fb}$ at each bearing, which is substituted in Eq. (4.76) to get bearing forces $\mathbf{F}_b$. It should be noted here that the dynamic stiffness matrix $\mathbf{K}_f$ includes not only stiffness and damping of foundation but the mass of bearing also, where $\mathbf{K}_p$ does not include mass of the bearing. Alternately, bearing forces can be used directly from Eq. (4.86) wherein it should be noted that as long as the shaft inertia is ignored and for small shaft deformation the forces and moments applied by disc on the shaft will be balance by bearing forces using the static equilibrium equation itself based on which Eq. (4.86) is derived. This bearing force can be used in Eq. (4.75) to get displacement between bearing and foundation $\mathbf{X}_f$, which can be used in Eq. (4.79) to get foundation force, $\mathbf{F}_f$. The difference between foundation force and bearing force is due to bearing mass considered, and if it is negligible, then these will be same. Displacements and forces have the complex form; the amplitude and the phase information can be extracted from the real and imaginary parts. Amplitudes will be the modulus of complex numbers, and phase angles of all these displacements can be evaluated by calculating arctangent of the ratio of the imaginary to real components.

The following rotor-bearing data are given from Exercise 4.3:

$$\text{Mass of the disc} = 1\ \text{kg}; \quad k_{xx} = 1.1\ \text{kN/m}; \quad k_{yy} = 1.8\ \text{kN/m}$$

$$k_{xy} = 0.20\ \text{kN/m}; \quad k_{yx} = 0.10\ \text{kN/m}$$

Also unbalance, *me*, of 25 gm-mm$=25\times10^{-6}$ kg-m with phase, $\phi=38°=38\pi/180$ is given from Exercise 4.4.

From the present case additionally, we have

Foundation dynamic characteristics are:	$k_{f_x} = k_{f_y} = 100$ MN/m; $c_{f_x} = c_{f_y} = 50$ kN-s/m.
The mass of each bearing:	2 kg

From Exercise 4.3 (rigid shaft and rigid foundation), we have

$$\omega_{cr1} = 46.3\ \text{rad/s} \quad \text{and} \quad \omega_{cr2} = 60.5\ \text{rad/s} \tag{4.109}$$

Since the overall procedure of obtaining unbalance responses of disc and bearings, and foundation forces have been given in algorithmic way so again various matrices are not shown here with given rotor parameters. It is left to the reader to work with computer code and get responses and critical speeds of the system. However, various critical speeds and plots are provided for reference. From the present data (rigid shaft and flexible foundation), the first two critical speeds (Refer Figure 4.19 through 4.23) are

$$\omega_{cr1} = 45.4\ \text{rad/s} \quad \text{and} \quad \omega_{cr2} = 59.3\ \text{rad/s} \tag{4.110}$$

Also, from the present data (flexible shaft and flexible foundation), the first two critical speeds (Refer Figure 4.24 through 4.28) are

$$\omega_{cr1} = 34.0\ \text{rad/s} \quad \text{and} \quad \omega_{cr2} = 61.2\ \text{rad/s} \tag{4.111}$$

For the present rotor data with the shaft and the foundation flexibility, the first natural frequency decreases by a consideration amount. The second natural frequency is not affected much. But regarding second natural frequency, a general conclusion cannot be made. Figures 4.19–4.23 show disc and bearing displacements, and foundation force variation with the spin speed of the shaft. Resonance condition can be seen in each of the plots. These plots are when the shaft has been considered as rigid. Similarly, Figures 4.24–4.28 show corresponding plots when the shaft flexibility is also considered. Since the resonance condition is same in all the plots, the displacement of shaft ends has not been plotted for brevity. In the present exercise, the diametral mass moment of inertia of the disc is ignored but formulation accounts for it so it can be easily added (for I_d=0.01 kg-m^2).

Without shaft flexibility: Refer Figures 4.19–4.23.

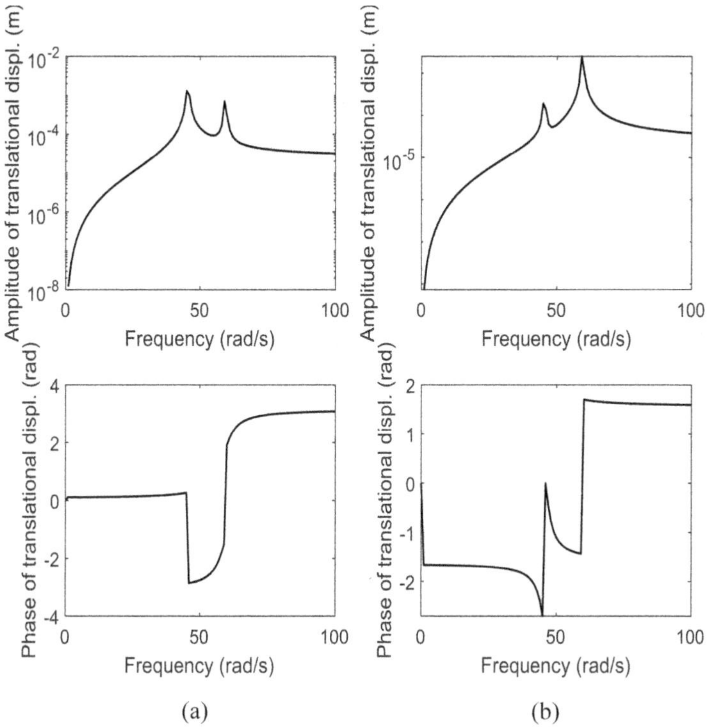

FIGURE 4.19 Amplitude and phase of translational displacements of disc in (a) x-direction (b) y-direction.

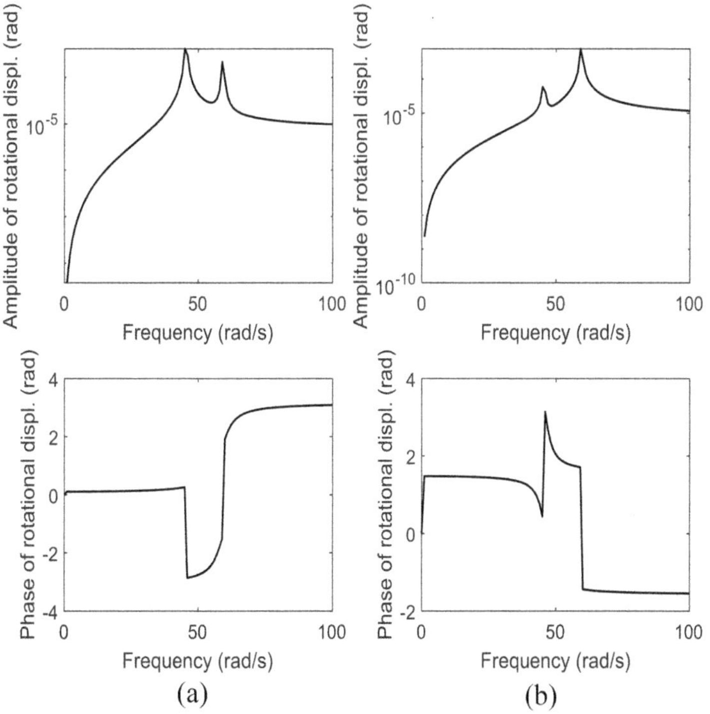

FIGURE 4.20 Amplitude and phase of rotational displacements of disc in (a) z–x plane (b) y–z plane.

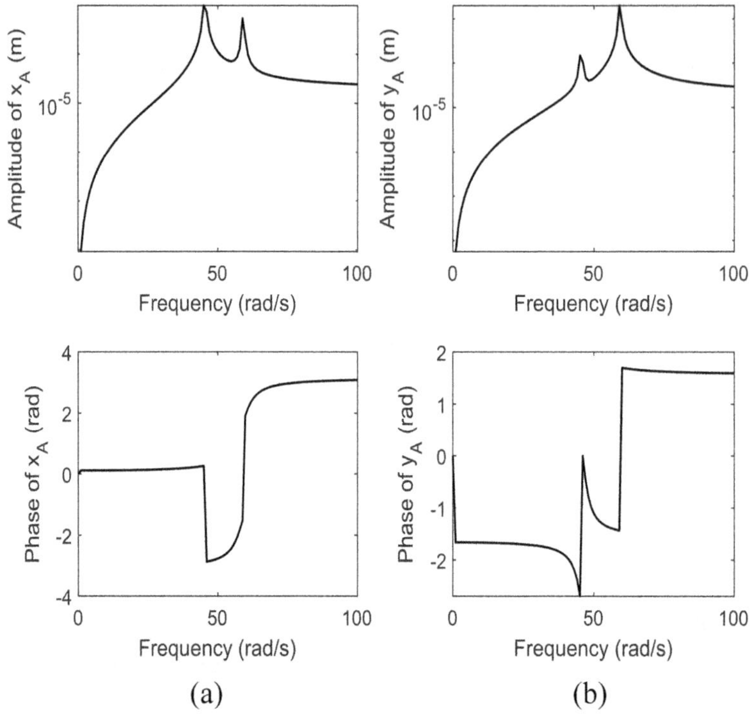

FIGURE 4.21 Amplitude and phase of translational displacements of bearing at end *A* (a) *x*-direction (b) *y*-direction.

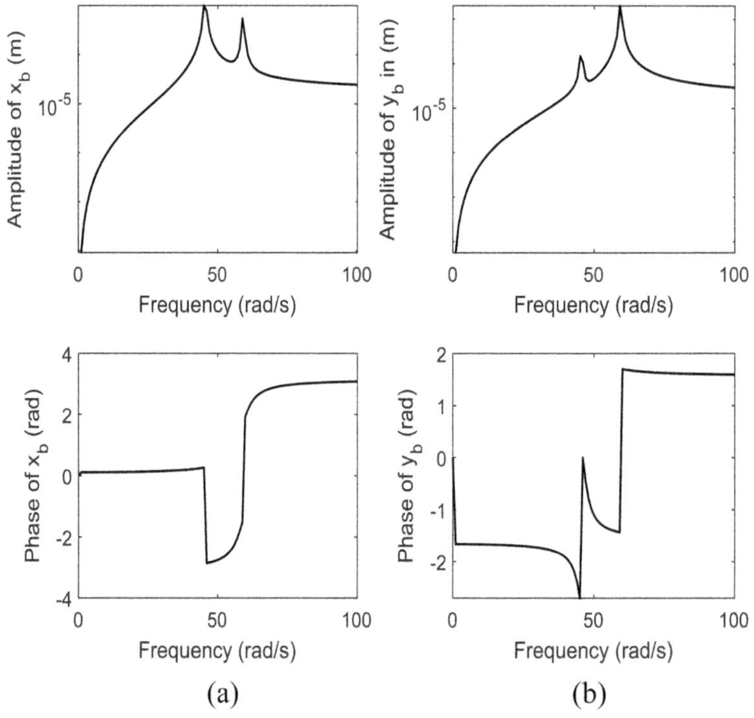

FIGURE 4.22 Amplitude and phase of translational displacements of bearing at end *B* (a) *x*-direction (b) *y*-direction.

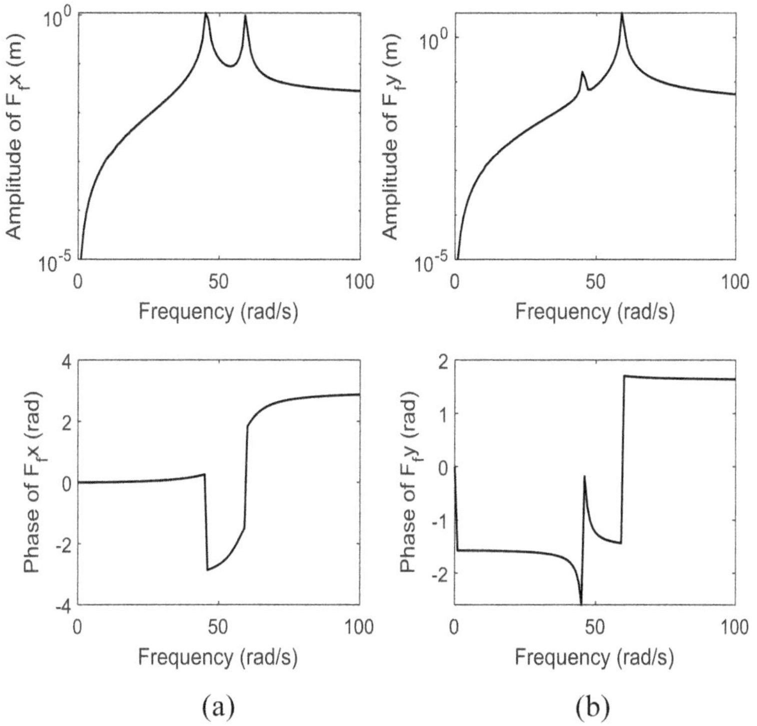

FIGURE 4.23 Amplitude and phase of foundation force at end *A* (a) *x*-direction (b) *y*-direction.

With shaft flexibility: Refer Figures 4.24–4.28.

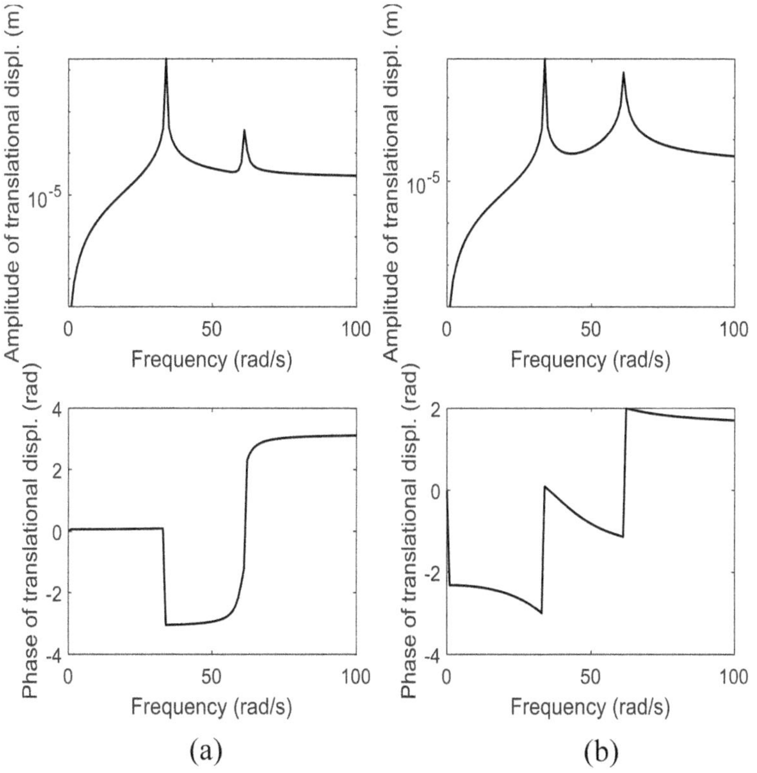

FIGURE 4.24 Amplitude and phase of translational displacements of disc in (a) *x*-direction (b) *y*-direction.

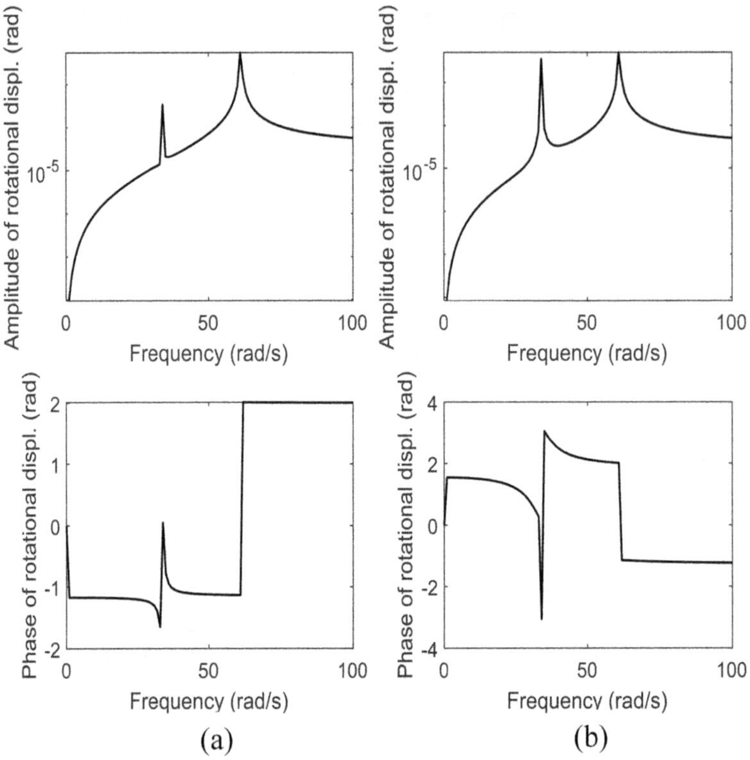

FIGURE 4.25 Amplitude and phase of rotational displacements of disc in (a) z–x plane (b) y–z plane.

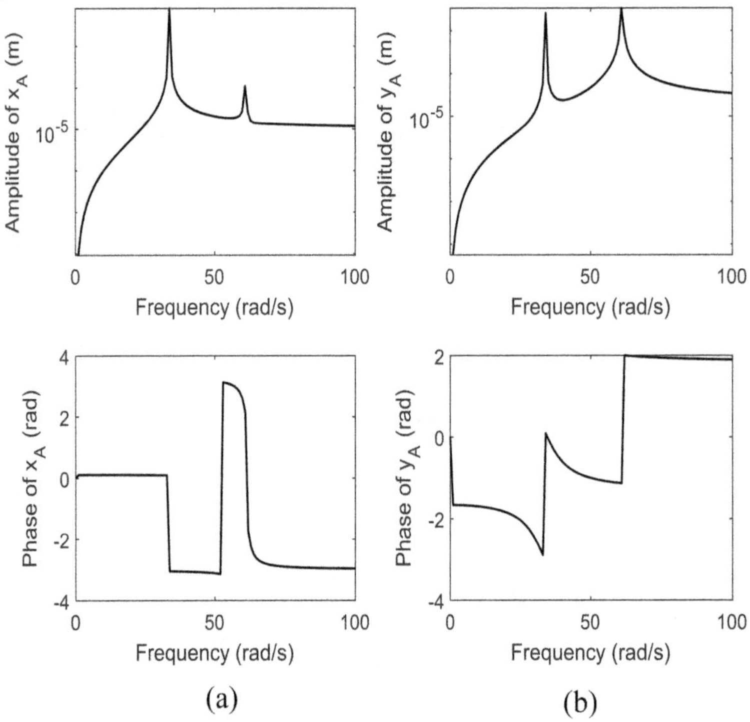

FIGURE 4.26 Amplitude and phase of translational displacements of bearing at end A (a) x-direction (b) y-direction.

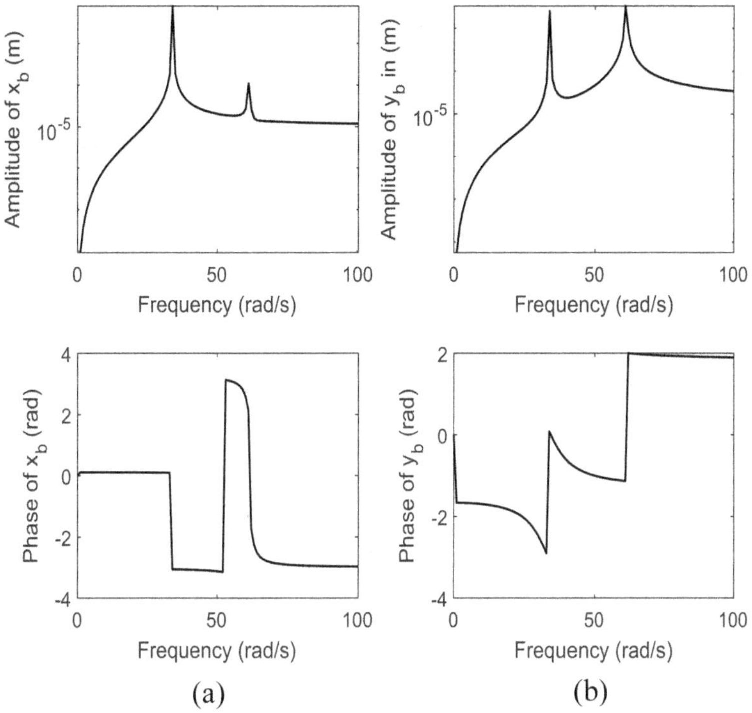

FIGURE 4.27 Amplitude and phase of translational displacements of bearing at end *B* (a) *x*-direction (b) *y*-direction.

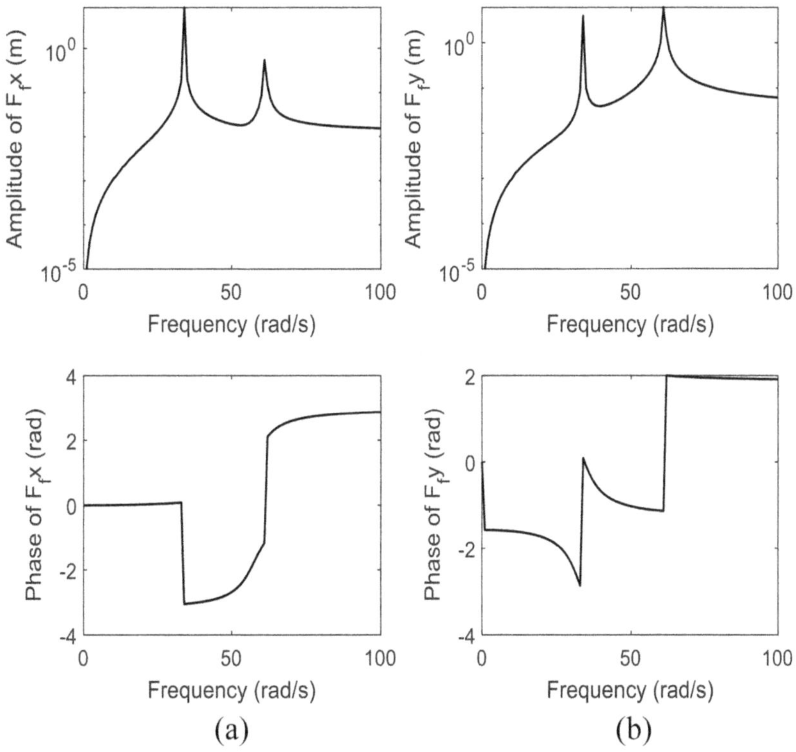

FIGURE 4.28 Amplitude and phase of foundation force at end *A* (a) *x*-direction (b) *y*-direction.

Exercise 4.6 Consider a simple rigid-rotor flexible-bearing system as shown in Figure 4.29. The rotor is supported on two different flexible bearings. In the figure, L_1 and L_2 are distances of bearings 1 and 2 from the centre of gravity of the rotor with $L = L_1 + L_2$, and R_1 and R_2 are distances of balancing planes (i.e., rigid discs) from the centre of gravity of the rotor. Consider linearised eight bearing dynamic parameters for each of the bearing based on the short bearing approximations (refer chapter 3).

Let m be the mass of the rotor, I_t is the transverse mass moment of inertia of the rotor about an axis passing through the centre of gravity, I_p is the polar mass moment of inertia of the rotor, k and c are, respectively, the stiffness and damping parameters, $f_x(t)$ and $f_y(t)$ are, respectively, the force in the horizontal and vertical directions, $u = m\,e$ is the unbalance, ϕ is the phase, x and y are translational displacements in the horizontal and vertical directions respectively, t is the time, and subscripts 1 and 2 represent the right and left-hand sides from the midspan of the rotor, respectively. Obtain equations of motion of the rotor-bearing system in terms of translational displacements (four in numbers, i.e., x_1, y_1, x_2, y_2) at two bearings. The motivation behind obtaining the equations of motion in terms of bearing response is that in real-life terms, often these responses can only be accessible to the practicing engineers.

***Solution*:** Refer Section 18.8 in book of Tiwari (2017) for more details, but in that case additionally active magnetic bearings also have been added, which can be omitted for the present formulation. Herein, a brief summary of main matrices and vectors are provided. The formulation is more general including effect of the gyroscopic effect (refer chapter 5 of Tiwari (2017)).

Equations of motion of the two offset discs in rigid rotor-flexible bearings (refer Figure 4.30) can be written as

$$\mathbf{M}\ddot{\mathbf{q}}(t) + (\mathbf{C} - \omega\mathbf{G})\dot{\mathbf{q}}(t) + \mathbf{K}\mathbf{q}(t) = \mathbf{f}_{unb} \tag{4.112}$$

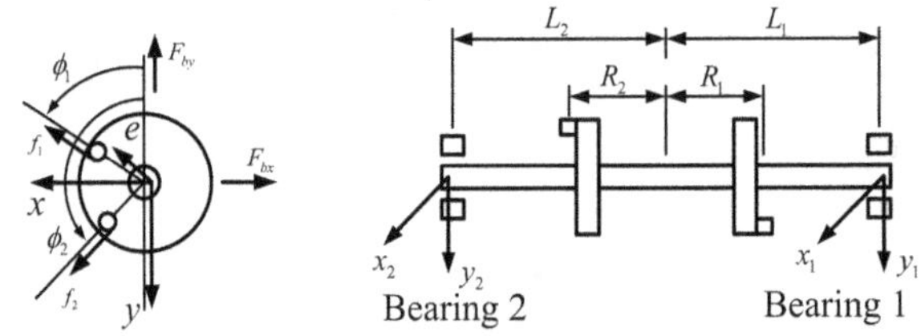

FIGURE 4.29 A rigid rotor on flexible bearings.

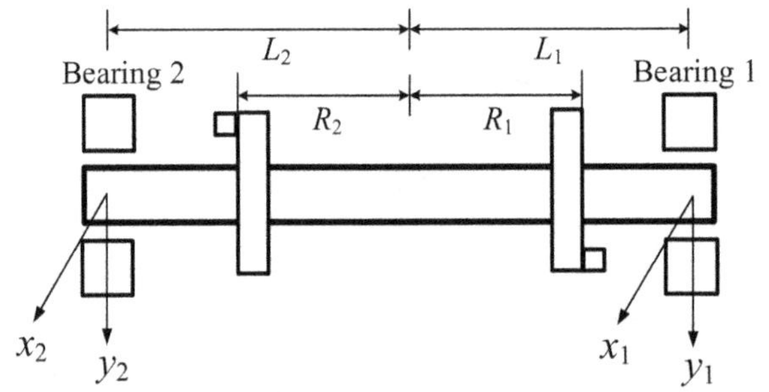

FIGURE 4.30 A rigid rotor on flexible bearings.

where the mass matrix **M** and the gyroscopic matrix **G** are

$$\mathbf{M}=\begin{bmatrix} \left(m\bar{l}_2^2+i_t\right) & 0 & \left(m\bar{l}_1\bar{l}_2-i_t\right) & 0 \\ 0 & \left(m\bar{l}_2^2+i_t\right) & 0 & \left(m\bar{l}_1\bar{l}_2-i_t\right) \\ \left(m\bar{l}_1\bar{l}_2-i_t\right) & 0 & \left(m\bar{l}_1^2+i_t\right) & 0 \\ 0 & \left(m\bar{l}_1\bar{l}_2-i_t\right) & 0 & \left(m\bar{l}_1^2+i_t\right) \end{bmatrix}; \quad \mathbf{G}=\begin{bmatrix} 0 & i_p & 0 & -i_p \\ -i_p & 0 & i_p & 0 \\ 0 & -i_p & 0 & i_p \\ i_p & 0 & -i_p & 0 \end{bmatrix} \tag{4.113}$$

$$\text{with} \quad i_t=\frac{I_t}{L^2}; \quad i_P=\left(\frac{I_P}{L^2}\right); \quad \bar{l}_1=\frac{L_1}{L}; \quad \bar{l}_2=\frac{L_2}{L}$$

The stiffness, **K**, and damping, **C**, matrices of bearings are

$$\mathbf{K}=\begin{bmatrix} k_{xx1} & k_{xy1} & 0 & 0 \\ k_{yx1} & k_{yy1} & 0 & 0 \\ 0 & 0 & k_{xx2} & k_{xy2} \\ 0 & 0 & k_{yx2} & k_{yy2} \end{bmatrix}; \quad \mathbf{C}=\begin{bmatrix} c_{xx1} & c_{xy1} & 0 & 0 \\ c_{yx1} & c_{yy1} & 0 & 0 \\ 0 & 0 & c_{xx2} & c_{xy2} \\ 0 & 0 & c_{yx2} & c_{yy2} \end{bmatrix} \tag{4.114}$$

The unbalance force vector is given as

$$\mathbf{f}_{unb}=\begin{Bmatrix} u_1\omega^2\cos(\omega t+\phi_1)(\bar{l}_2+\bar{r}_1)+u_2\omega^2\cos(\omega t+\phi_2)(\bar{l}_2-\bar{r}_2) \\ u_1\omega^2\sin(\omega t+\phi_1)(\bar{l}_2+\bar{r}_1)+u_2\omega^2\sin(\omega t+\phi_2)(\bar{l}_2-\bar{r}_2) \\ u_1\omega^2\cos(\omega t+\phi_1)(\bar{l}_1-\bar{r}_1)+u_2\omega^2\cos(\omega t+\phi_2)(\bar{l}_1+\bar{r}_2) \\ u_1\omega^2\sin(\omega t+\phi_1)(\bar{l}_1-\bar{r}_1)+u_2\omega^2\sin(\omega t+\phi_2)(\bar{l}_1+\bar{r}_2) \end{Bmatrix} \quad \text{with} \quad \bar{r}_1=\frac{R_1}{L}; \quad \bar{r}_2=\frac{R_2}{L} \tag{4.115}$$

The displacement vector is given as

$$\mathbf{q}(t)=\begin{Bmatrix} x_1 & y_1 & x_2 & y_2 \end{Bmatrix}^T$$

It is left to the reader to explore and re-derive EOMs by referring to chapters 5 and 18 of Tiwari (2017).

Exercise 4.7 Consider the equations of motion of Exercise 4.6 and the numerical data given in Table 4.1. Obtain the response (i.e., the amplitude and the phase) of the bearings with respect to the rotor speed and list down critical speeds of the rotor-bearing system.

***Solution*:** Equations of motion of the two offset discs in rigid rotor-flexible bearings as per Exercise 4.6 can be written as

$$\mathbf{M}\ddot{\mathbf{q}}(t)+(\mathbf{C}-\omega\mathbf{G})\dot{\mathbf{q}}(t)+\mathbf{K}\mathbf{q}(t)=\mathbf{f}_{unb} \tag{4.116}$$

TABLE 4.1
Details of the Rotor Model for the Numerical Simulation

Property	Numerical Value
Rotor	
Rotor shaft diameter	10 mm
Rotational speed, ω	100 Hz
Mass, m	4 kg
Length of rotor, L	0.425 m
Distance of bearings from centre of rotor	0.2125 m
Distance of discs from centre of rotor	0.130 m
Transverse mass moment of inertia, I_d	0.0786 kg-m^2
Rigid discs	
Inner diameter	10 mm
Outer diameter	74 mm
Thickness	25 mm
Bearings	
Diameter	25.4 mm
Length to diameter ratio	1
Radial clearance, c_r of bearing	0.075 mm
Kinetic viscosity	20.11 centi-Stokes
Temperature of lubricant	40°C
Specific gravity of lubricant	0.87

where all matrices are explicitly defined in Exercise 4.6. It should be noted the bearing stiffness and damping matrices have been obtained by the short bearing approximation of the fluid-film (or hydrodynamic bearing) given in chapter 3 of the present book. A flowchart showing overall solution procedure is provided in Figure 4.31. Also, a Simulink block diagram is shown in Figure 4.32. Figure 4.33 shows the time response of the bearings 1 and 2, whereas Figure 4.34 gives frequency domain amplitude and phase plot with the spin speed of the rotor. A ramp-up speed from 5 through 155 rad/s has been performed in 6 s such that transients are avoided.

Exercise 4.8 For a rigid rotor mounted on two isotropic bearings at ends, as shown in Figure 4.35, that has a varying cross-section along the longitudinal axis (e.g., a tapered rotor). For this case, the centre of gravity, G, of the rotor is axilly offset from the mid-span of the rotor, C. It is assumed that the rotor is perfectly balanced (i.e., it has no external radial force and corresponding external moment). Let m be the mass, I_d be the diametral mass moment of inertia of the rotor about centre of gravity, k_A and k_B be stiffness of bearings A and B, respectively, and l be the length of the rotor. Obtain governing equations of motion for the following sets of chosen generalised coordinates for a single plane motion of the rotor. Discuss forms of the mass and stiffness matrices in regards to the diagonal and off-diagonal terms (i.e., the mass and stiffness couplings) for a linearised system. (i) If we choose generalised coordinates as (x_G, φ_y), where the translational displacement of the centre of gravity is x_G, and tilting of the rotor about the vertical axis (i.e., y-axis) is φ_y. (ii) If we choose generalised coordinates as (x_E, φ_y), where the translational displacement of a point on the rotor where if a transverse force is applied then it produces pure translation of the rotor (i.e., $k_A l_{AE} = k_B l_{BE}$) is x_E, and tilting of the rotor remains same as for the first case. (iii) If we choose generalised coordinates as (x_A, φ_y), where the translational displacement of the extreme left end of the rotor is x_A, and tilting of the rotor remains same as for the first case. (iv) If we choose generalised coordinates as (x_A, x_B),

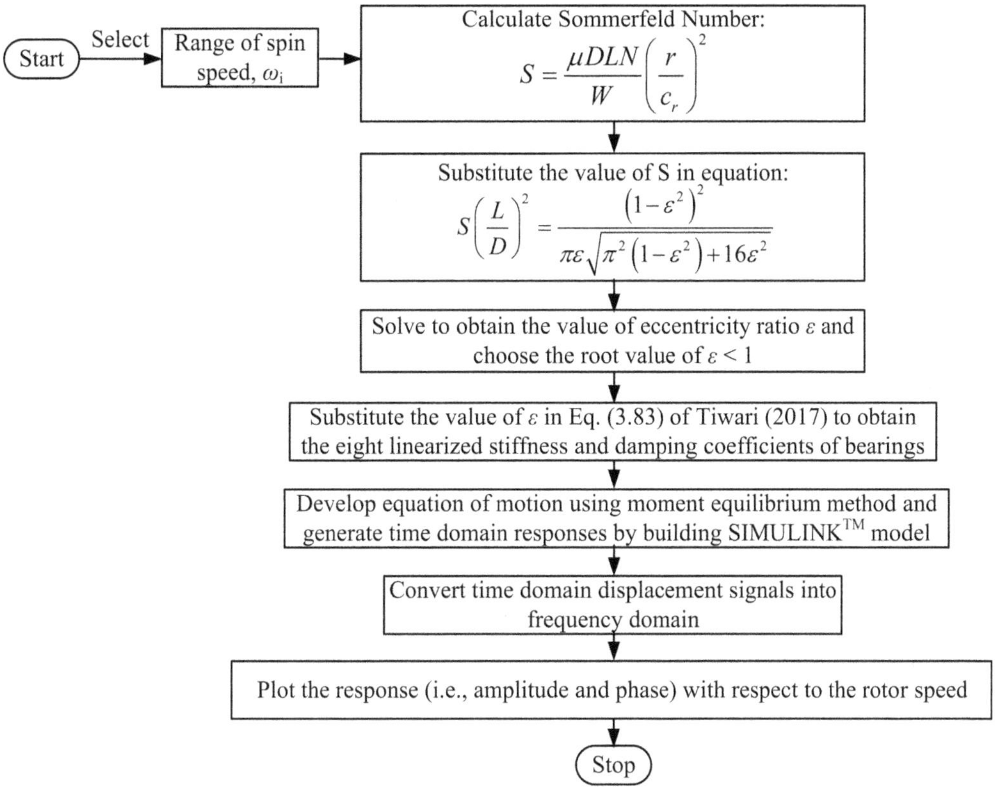

FIGURE 4.31 Flowchart of overall solution procedure to get unbalanced responses.

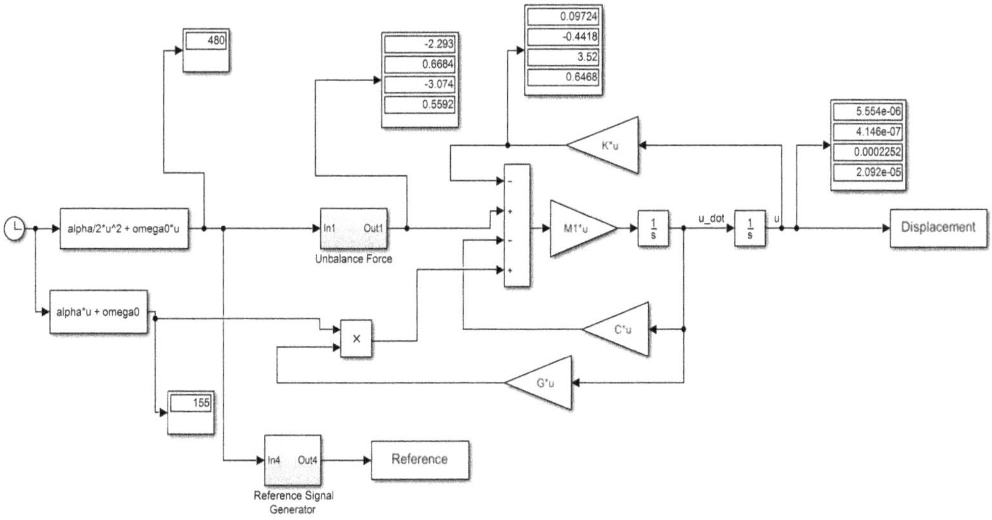

FIGURE 4.32 A Simulink block diagram of the present simulation.

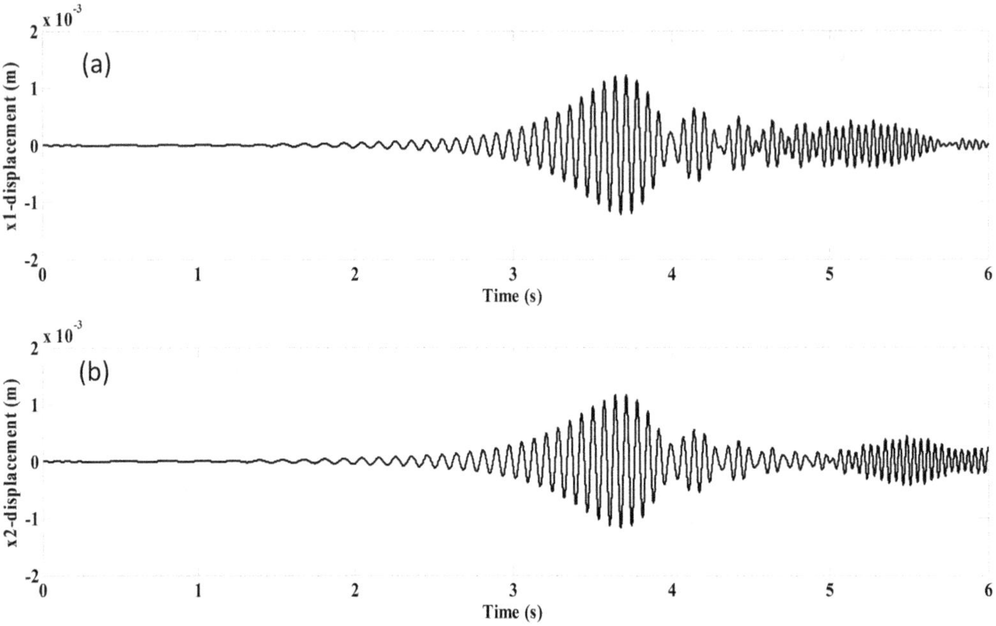

FIGURE 4.33 Displacement in x-direction at (a) bearing 1 (b) bearing 2.

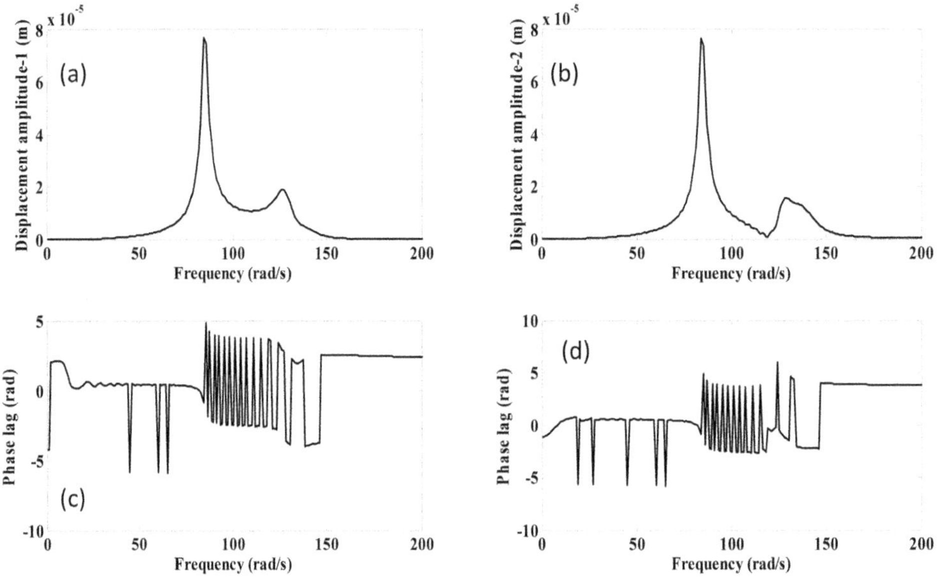

FIGURE 4.34 (a) Displacement amplitude versus rotor spin speeds at bearing-1 (b) displacement amplitude versus rotor spin speeds at bearing-2 (c) displacement phase versus rotor spin speeds at bearing-1 (d) displacement phase versus rotor spin speeds at bearing-2 (ramp-up speed from 5 through 155 rad/s).

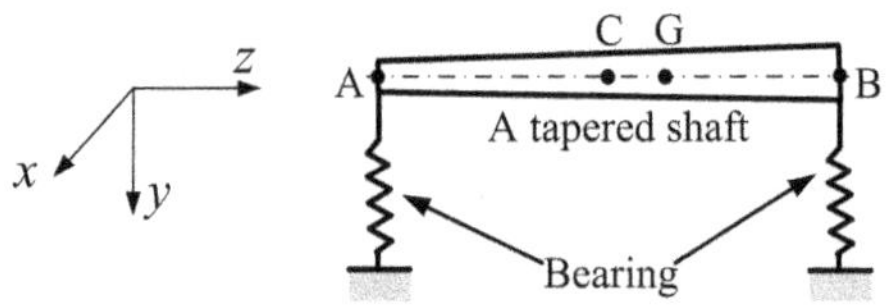

FIGURE 4.35 An axially asymmetric shaft mounted on flexible dissimilar bearings.

where the translational displacement of the extreme left and right ends of the rotor are x_A and x_B, respectively. (v) If we choose generalised coordinates as (x_C, φ_y), where the translational displacement of the mid-span is x_C, and tilting of the rotor about the vertical axis (i.e., y-axis) is φ_y. (vi) If we choose generalised coordinates as (x_E, x_G), where displacements have similar meanings as defined previously.

***Solution*:** The present exercise aims to show that for a system, if we choose different possible combinations of generalised coordinates the EOMs will be different and even the form of mass and stiffness matrices, and its characteristics will also be different. But if we solve each of them the natural frequencies obtain will remain the same. In fact, in Exercise 4.1, we have seen this feature to some extent (Exercise 4.9 dealts it in more detail numerically). In the present case, the energy form is used to derive EOMs from Lagrange's equation.

(i) Generalised co-ordinate (x_G, φ_y) (refer Figure 4.36):

The kinetic energy (KE) of the rotor system for the present case is

$$T = \tfrac{1}{2} m\dot{x}_G^2 + \tfrac{1}{2} I_{dG}\dot{\varphi}_y^2 \tag{4.117}$$

where m is the mass of the rotor and I_{dG} is the diametral mass moment of inertia of the rotor about its centre of gravity, G. The potential energy (PE) of the rotor system is

$$U = \tfrac{1}{2} k_A\left(x_G - l_{AG}\varphi_y\right)^2 + \tfrac{1}{2} k_B\left(x_G + l_{BG}\varphi_y\right)^2 \tag{4.118}$$

where k_A and k_B are the stiffness of support at bearings A and B, respectively, and l_{AG} and l_{BG} are the distances AG and BG, respectively. For getting equations of motion from Lagrange's equation for the generalised coordinate, x_G, we have,

$$\frac{d}{dt}\left(\frac{\partial T}{\partial \dot{x}_G}\right) - \frac{\partial T}{\partial x_G} + \frac{\partial U}{\partial x_G} = 0 \tag{4.119}$$

On substituting Eqs. (4.117) and (4.118) into parts of Eq. (4.119), we get

$$\frac{d}{dt}\left(\frac{\partial T}{\partial \dot{x}_G}\right) = \frac{d}{dt}\left\{\frac{\partial}{\partial \dot{x}_G}\left(\tfrac{1}{2} m\dot{x}_G^2 + \tfrac{1}{2} I_{dG}\dot{\varphi}_y^2\right)\right\} = m\ddot{x}_G \tag{4.120}$$

and
$$\begin{aligned}\frac{\partial U}{\partial x_G} &= \frac{\partial}{\partial x_G}\left\{\tfrac{1}{2} k_A\left(x_G - l_{AG}\varphi_y\right)^2 + \tfrac{1}{2} k_B\left(x_G + l_{BG}\varphi_y\right)^2\right\} = k_A\left(x_G - l_{AG}\varphi_y\right) + k_B\left(x_G + l_{BG}\varphi_y\right) \\ &= \left(k_A + k_B\right)x_G - \left(k_A l_{AG} - k_B l_{BG}\right)\varphi_y\end{aligned} \tag{4.121}$$

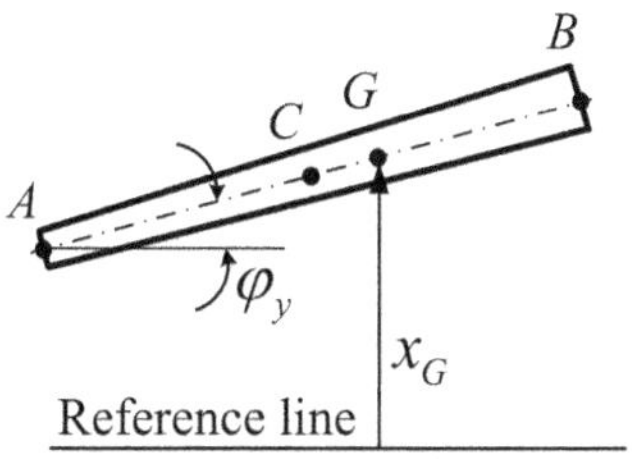

FIGURE 4.36 Deflected rigid shaft with generalized coordinates (x_G, φ_y).

On substituting Eqs. (4.120) and (4.121) into Eq. (4.119), we get

$$m\ddot{x}_G + (k_A + k_B)x_G - \{k_A l_{AG} - k_B l_{BG}\}\varphi_y = 0 \tag{4.122}$$

Similarly, for Lagrange's equation for the generalised coordinate, φ_y, we have

$$\frac{d}{dt}\left(\frac{\partial T}{\partial \dot{\varphi}_y}\right) - \frac{\partial T}{\partial \varphi_y} + \frac{\partial U}{\partial \varphi_y} = 0 \tag{4.123}$$

On substituting Eqs. (4.117) and (4.118) into parts of Eq. (4.123), we get

$$\frac{d}{dt}\left(\frac{\partial T}{\partial \dot{\varphi}_y}\right) = \frac{d}{dt}\left\{\frac{\partial}{\partial \dot{\varphi}_y}\left(\tfrac{1}{2}m\dot{x}_G^2 + \tfrac{1}{2}I_{dG}\dot{\varphi}_y^2\right)\right\} = I_{dG}\ddot{\varphi}_y \tag{4.124}$$

and

$$\begin{aligned}\frac{\partial U}{\partial \varphi_y} &= \frac{\partial}{\partial \varphi_y}\left\{\tfrac{1}{2}k_A\left(x_G - l_{AG}\varphi_y\right)^2 + \tfrac{1}{2}k_B\left(x_G + l_{BG}\varphi_y\right)^2\right\} \\ &= -(k_A l_{AG} - k_B l_{BG})x_G + \left(k_A l_{AG}^2 + k_B l_{BG}^2\right)\varphi_y\end{aligned} \tag{4.125}$$

On substituting Eqs. (4.124) and (4.125) into Eq. (4.123), we get

$$I_{dG}\ddot{\varphi}_y - (k_A l_{AG} - k_B l_{BG})x_G + \left(k_A l_{AG}^2 + k_B l_{BG}^2\right)\varphi_y = 0 \tag{4.126}$$

Equations (4.122) and (4.126), in the matrix form can be written as,

$$\begin{bmatrix} m & 0 \\ 0 & I_{dG} \end{bmatrix}\begin{Bmatrix} \ddot{x}_G \\ \ddot{\varphi}_y \end{Bmatrix} + \begin{bmatrix} (k_A + k_B) & -(k_A l_{AG} - k_B l_{BG}) \\ -(k_A l_{AG} - k_B l_{BG}) & \left(k_A l_{AG}^2 + k_B l_{BG}^2\right) \end{bmatrix}\begin{Bmatrix} x_G \\ \varphi_y \end{Bmatrix} = \begin{Bmatrix} 0 \\ 0 \end{Bmatrix} \tag{4.127}$$

(ii) Generalised coordinates $\left(x_E, \varphi_y\right)$ (refer Figure 4.37)

We have the following condition from the problem

$$k_A l_{AE} = k_B l_{BE}; \quad l_{AE} + l_{BE} = l_{AB}; \quad l_{AE} = \frac{k_B}{k_A + k_B} l_{AB}; \quad l_{EG} = l_{AG} - l_{AE} \tag{4.128}$$

From Figure 4.37, we have

$$x_G = \left(x_E + l_{EG}\varphi_y\right); \quad x_A = \left(x_E - l_{AE}\varphi_y\right); \quad x_B = \left(x_E + l_{BE}\varphi_y\right) \tag{4.129}$$

where l_{EG} is distance between points E and G locations on the shaft. The KE can be written as (as per Eq. (4.117)) and noting Eq. (4.129), we get

$$T = \tfrac{1}{2}m\dot{x}_G^2 + \tfrac{1}{2}I_{dG}\dot{\varphi}_y^2 = \tfrac{1}{2}m\left(\dot{x}_E + l_{EG}\dot{\varphi}_y\right)^2 + \tfrac{1}{2}I_{dG}\dot{\varphi}_y^2 \tag{4.130}$$

$$\text{or} \quad T = \tfrac{1}{2}m\left(\dot{x}_E^2 + l_{EG}^2\dot{\varphi}_y^2 + 2\dot{x}_E l_{EG}\dot{\varphi}_y\right) + \tfrac{1}{2}I_{dG}\dot{\varphi}_y^2$$

$$\text{or} \quad T = \tfrac{1}{2}m\dot{x}_E^2 + \tfrac{1}{2}\left(I_{dG} + ml_{EG}^2\right)\dot{\varphi}_y^2 + \tfrac{1}{2}m\left(2l_{EG}\dot{x}_E\dot{\varphi}_y\right) = \tfrac{1}{2}m\dot{x}_E^2 + \tfrac{1}{2}I_{dE}\dot{\varphi}_y^2 + \tfrac{1}{2}m\left(2l_{EG}\dot{x}_E\dot{\varphi}_y\right) \tag{4.131}$$

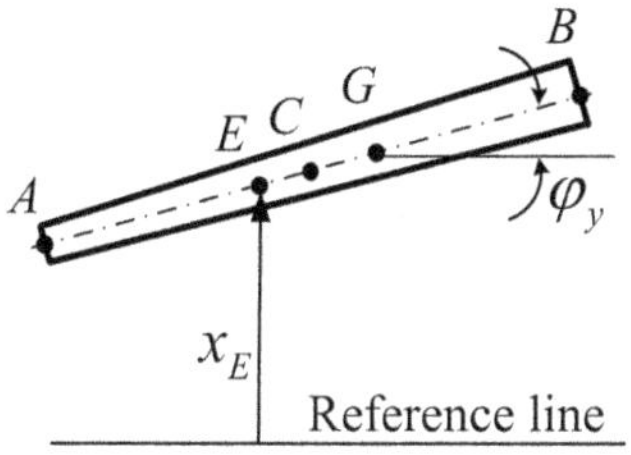

FIGURE 4.37 Deflected shaft with generalized coordinates (x_E, φ_y).

The PE of the rotor system from Eq. (4.118) and noting Eq. (4.129) is

$$U = \tfrac{1}{2}k_A x_A^2 + \tfrac{1}{2}k_B x_B^2 = \tfrac{1}{2}k_A\left(x_E - l_{AE}\varphi_y\right)^2 + \tfrac{1}{2}k_B\left(x_E + l_{BE}\varphi_y\right)^2 \tag{4.132}$$

Lagrange's equation for the generalised coordinate, x_E, is given as

$$\frac{d}{dt}\left(\frac{\partial T}{\partial \dot{x}_E}\right) - \frac{\partial T}{\partial x_E} + \frac{\partial U}{\partial x_E} = 0 \tag{4.133}$$

On substituting Eqs. (4.130) and (4.132) into parts of Eq. (4.133), we get

$$\frac{d}{dt}\left(\frac{\partial T}{\partial \dot{x}_E}\right) = \frac{d}{dt}\left\{\frac{\partial}{\partial \dot{x}_E}\left\{\tfrac{1}{2}m\left(\dot{x}_E + l_{EG}\dot{\varphi}_y\right)^2 + \tfrac{1}{2}I_{dG}\dot{\varphi}_y^2\right\}\right\} = \frac{d}{dt}\left\{m\left(\dot{x}_E + l_{EG}\dot{\varphi}_y\right)\right\} = m\left(\ddot{x}_E + l_{EG}\ddot{\varphi}_y\right) \tag{4.134}$$

and

$$\frac{\partial U}{\partial x_E} = \frac{\partial}{\partial x_E}\left\{\tfrac{1}{2}k_A\left(x_E - l_{AE}\varphi_y\right)^2 + \tfrac{1}{2}k_B\left(x_E + l_{BE}\varphi_y\right)^2\right\} = k_A\left(x_E - l_{AE}\varphi_y\right) + k_B\left(x_E + l_{BE}\varphi_y\right) \tag{4.135}$$

On substituting Eqs. (4.134) and (4.135) into Eq. (4.132), we get

$$m\left(\ddot{x}_E + l_{EG}\ddot{\varphi}_y\right) + k_A\left(x_E - l_{AE}\varphi_y\right) + k_B\left(x_E + l_{BE}\varphi_y\right) = 0$$

$$\text{or}\quad m\ddot{x}_E + m l_{EG}\ddot{\varphi}_y + \left(k_A + k_B\right)x_E - \left(k_A l_{AE} - k_B l_{BE}\right)\varphi_y = 0 \tag{4.136}$$

Lagrange's equation for the generalised coordinate φ_y is given as

$$\frac{d}{dt}\left(\frac{\partial T}{\partial \dot{\varphi}_y}\right) - \frac{\partial T}{\partial \varphi_y} + \frac{\partial U}{\partial \varphi_y} = 0 \tag{4.137}$$

On substituting Eqs. (4.130) and (4.132) into Eq. (4.130), we get

$$\frac{d}{dt}\left(\frac{\partial T}{\partial \dot{\varphi}_y}\right) = \frac{d}{dt}\left\{\frac{\partial}{\partial \dot{\varphi}_y}\left\{\tfrac{1}{2}m\left(\dot{x}_E + l_{EG}\dot{\varphi}_y\right)^2 + \tfrac{1}{2}I_{dG}\dot{\varphi}_y^2\right\}\right\} = \frac{d}{dt}\left\{m\left(\dot{x}_E + l_{EG}\dot{\varphi}_y\right)l_{EG} + I_{dG}\dot{\varphi}_y\right\}$$
$$= m l_{EG}\ddot{x}_E + \left(I_{dG} + m l_{EG}^2\right)\ddot{\varphi}_y \tag{4.138}$$

$$\text{and} \quad \frac{\partial U}{\partial \varphi_y} = \frac{\partial}{\partial \varphi_y}\left\{\frac{1}{2}k_A\left(x_E - l_{AE}\varphi_y\right)^2 + \frac{1}{2}k_B\left(x_E + l_{BE}\varphi_y\right)^2\right\}$$
$$= -\left(k_A l_{AE} - k_B l_{BE}\right)x_E + \left(k_A l_{AE}^2 + k_B l_{BE}^2\right)\varphi_y \tag{4.139}$$

On substituting Eqs. (4.138) and (4.139) into Eq. (4.137), we get

$$ml_{EG}\ddot{x}_E + \left(I_{dG} + ml_{EG}^2\right)\ddot{\varphi}_y - \left(k_A l_{AE} - k_B l_{BE}\right)x_E + \left(k_A l_{AE}^2 + k_B l_{BE}^2\right)\varphi_y = 0 \tag{4.140}$$

In matrix form, Eqs. (4.136) and (4.140) can be combined as

$$\begin{bmatrix} m & ml_{EG} \\ ml_{EG} & I_{dG} + ml_{EG}^2 \end{bmatrix}\begin{Bmatrix} \ddot{x}_E \\ \ddot{\varphi}_y \end{Bmatrix} + \begin{bmatrix} k_A + k_B & -\left(k_A l_{AE} - k_B l_{BE}\right) \\ -\left(k_A l_{AE} - k_B l_{BE}\right) & k_A l_{AE}^2 + k_B l_{BE}^2 \end{bmatrix}\begin{Bmatrix} x_E \\ \varphi_y \end{Bmatrix} = \begin{Bmatrix} 0 \\ 0 \end{Bmatrix}$$

$$\text{or} \quad \begin{bmatrix} m & ml_{EG} \\ ml_{EG} & I_{dG} + ml_{EG}^2 \end{bmatrix}\begin{Bmatrix} \ddot{x}_E \\ \ddot{\varphi}_y \end{Bmatrix} + \begin{bmatrix} k_A + k_B & 0 \\ 0 & k_A l_{AE}^2 + k_B l_{BE}^2 \end{bmatrix}\begin{Bmatrix} x_E \\ \varphi_y \end{Bmatrix} = \begin{Bmatrix} 0 \\ 0 \end{Bmatrix} \tag{4.141}$$

Noting Eq. (4.128), we observe from Eq. (4.141) that the stiffness matrix is uncoupled. So, the choice of location of generalised coordinate made the stiffness matrix uncoupled but the mass matrix now coupled.

Note: If the KE of the rotor system is written as (often the common mistake is done in this step)

$$T = \tfrac{1}{2}m\dot{x}_E^2 + \tfrac{1}{2}I_{dE}\dot{\varphi}_y^2 = \tfrac{1}{2}m\dot{x}_E^2 + \tfrac{1}{2}\left(I_{dG} + ml_{EG}^2\right)\dot{\varphi}_y^2 \tag{4.142}$$
$$\text{with} \quad I_{dE} = I_{dG} + ml_{EG}^2$$

It should be noted that since the translational generalised chosen is at point E, so rotational KE also needs to be taken about the same point (i.e., I_{dE}). But with this, we will not get coupling in mass matrix (compare Eq. (4.142) with Eq. (4.131)). So, we will have

$$\frac{d}{dt}\left(\frac{\partial T}{\partial \dot{x}_E}\right) = \frac{d}{dt}\left[\frac{\partial}{\partial \dot{x}_E}\left\{\tfrac{1}{2}m\dot{x}_E^2 + \tfrac{1}{2}\left(I_{dG} + ml_{EG}^2\right)\dot{\varphi}_y^2\right\}\right] = \frac{d}{dt}\left(m\dot{x}_E\right) = m\ddot{x}_E \tag{4.143}$$

and

$$\frac{d}{dt}\left(\frac{\partial T}{\partial \dot{\varphi}_y}\right) = \frac{d}{dt}\left[\frac{\partial}{\partial \dot{\varphi}_y}\left\{\tfrac{1}{2}m\dot{x}_E^2 + \tfrac{1}{2}\left(I_{dG} + ml_{EG}^2\right)\dot{\varphi}_y^2\right\}\right] = \frac{d}{dt}\left\{\left(I_{dG} + ml_{EG}^2\right)\dot{\varphi}_y\right\} = \left(I_{dG} + ml_{EG}^2\right)\ddot{\varphi}_y \tag{4.144}$$

The PE is given as

$$U = \tfrac{1}{2}k_A x_A^2 + \tfrac{1}{2}k_B x_B^2 = \tfrac{1}{2}k_A\left(x_E - l_{AE}\varphi_y\right)^2 + \tfrac{1}{2}k_B\left(x_E + l_{BE}\varphi_y\right)^2 \tag{4.145}$$

which is the same as Eq. (4.133) so the stiffness matrix terms we will remain the same. So, the erroneous equations of motion will be

$$\begin{bmatrix} m & 0 \\ 0 & I_{dG} + ml_{EG}^2 \end{bmatrix}\begin{Bmatrix} \ddot{x}_E \\ \ddot{\varphi}_y \end{Bmatrix} + \begin{bmatrix} k_A + k_B & -\left(k_A l_{AE} - k_B l_{BE}\right) \\ -\left(k_A l_{AE} - k_B l_{BE}\right) & k_A l_{AE}^2 + k_B l_{BE}^2 \end{bmatrix}\begin{Bmatrix} x_E \\ \varphi_y \end{Bmatrix} = \begin{Bmatrix} 0 \\ 0 \end{Bmatrix} \tag{4.146}$$

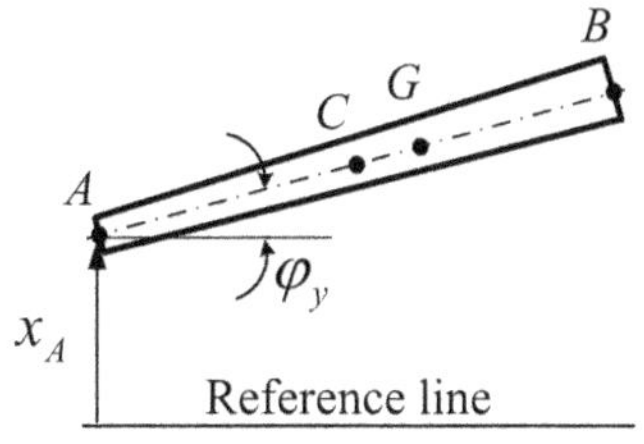

FIGURE 4.38 Deflected rigid shaft with generalized coordinates (x_A, φ_y).

So, on comparing Eq. (4.146) with Eq. (4.141), we observe that in the mass matrix the off-diagonal terms in Eq. (4.146) is missing. On comparing KE, we observe that the first two terms of Eq. (4.131) is same as Eq. (4.142). However, in Eq. (4.131) a third term is also appearing, which is due to Coriolis component of acceleration, which is a higher order term, which get omitted when we write KE as in Eq. (4.142). Moreover, on applying condition given in Eq. (4.128), then both mass and stiffness matrices will be coupled, which is not correct. So, we should write KE using generalised coordinates at centre of gravity and use relations to convert it to desired location generalised coordinates.

(iii) Generalised coordinates, (x_A, φ_y) (refer Figure 4.38)

From Figure 4.38, we have

$$x_G = (x_A + l_{AG}\varphi_y); \quad x_B = (x_A + l_{AB}\varphi_y) \tag{4.147}$$

where l_{AG} is distance between points A and G locations on the shaft, and l_{AG} is distance between points A and B. The KE can be written, as (as per Eq. (4.117)) and noting Eq. (4.147)

$$T = \tfrac{1}{2}m\dot{x}_G^2 + \tfrac{1}{2}I_{dG}\dot{\varphi}_y^2 = \tfrac{1}{2}m(\dot{x}_A + l_{AG}\dot{\varphi}_y)^2 + \tfrac{1}{2}I_{dG}\dot{\varphi}_y^2 \tag{4.148}$$

The potential energy (PE) of the rotor system is

$$U = \tfrac{1}{2}k_A x_A^2 + \tfrac{1}{2}k_B(x_A + l_{AB}\varphi_y)^2 \tag{4.149}$$

For getting the equation of motion from Lagrange's equation for generalised coordinate, x_A, we have,

$$\frac{d}{dt}\left(\frac{\partial T}{\partial \dot{x}_A}\right) - \frac{\partial T}{\partial x_A} + \frac{\partial U}{\partial x_A} = 0 \tag{4.150}$$

On substituting Eqs. (4.148) and (4.149) into parts of Eq. (4.150), we get

$$\frac{d}{dt}\left(\frac{\partial T}{\partial \dot{x}_A}\right) = \frac{d}{dt}\left[\frac{\partial}{\partial \dot{x}_A}\left\{\tfrac{1}{2}m(\dot{x}_A + l_{AG}\dot{\varphi}_y)^2 + \tfrac{1}{2}I_{dG}\dot{\varphi}_y^2\right\}\right] = \frac{d}{dt}\left\{m(\dot{x}_A + l_{AG}\dot{\varphi}_y)\right\} = m(\ddot{x}_A + l_{AG}\ddot{\varphi}_y) \tag{4.151}$$

and $$\frac{\partial U}{\partial x_A} = \frac{\partial}{\partial x_A}\left\{\tfrac{1}{2}k_A x_A^2 + \tfrac{1}{2}k_B(x_A + l_{AB}\varphi_y)^2\right\} = k_A x_A + k_B(x_A + l_{AB}\varphi_y) = (k_A + k_B)x_A + k_B l_{AB}\varphi_y \tag{4.152}$$

On substituting Eqs. (4.151) and (4.152) into Eq. (4.150), we get

$$m(\ddot{x}_A + l_{AG}\ddot{\varphi}_y) + (k_A + k_B)x_A + k_B l_{AB}\varphi_y = 0 \tag{4.153}$$

Similarly, for the generalised coordinate, φ_y, we have

$$\frac{d}{dt}\left(\frac{\partial T}{\partial \dot{\varphi}_y}\right)-\frac{\partial T}{\partial \varphi_y}+\frac{\partial U}{\partial \varphi_y}=0 \tag{4.154}$$

On substituting Eqs. (4.148) and (4.149) into parts of Eq. (4.154), we get

$$\frac{d}{dt}\left(\frac{\partial T}{\partial \dot{\varphi}_y}\right)=\frac{d}{dt}\left[\frac{\partial}{\partial \dot{\varphi}_y}\left\{\tfrac{1}{2}m\left(\dot{x}_A+l_{AG}\dot{\varphi}_y\right)^2+\tfrac{1}{2}I_{dG}\dot{\varphi}_y^2\right\}\right]=\frac{d}{dt}\left\{m\left(\dot{x}_A+l_{AG}\dot{\varphi}_y\right)l_{AG}+I_{dG}\dot{\varphi}_y\right\}$$
$$=ml_{AG}\ddot{x}_A+\left(I_{dG}+ml_{AG}^2\right)\ddot{\varphi}_y \tag{4.155}$$

and $$\frac{\partial U}{\partial \varphi_y}=\frac{\partial}{\partial \varphi_y}\left\{\tfrac{1}{2}k_A x_A^2+\tfrac{1}{2}k_B\left\{x_A+l_{AB}\varphi_y\right\}^2\right\}=k_B\left(x_A+l_{AB}\varphi_y\right)\left(l_{AB}\right)=k_B l_{AB}x_A+k_B l_{AB}^2\varphi_y \tag{4.156}$$

On substituting Eqs. (4.155) and (4.156) into Eq. (4.154), we get

$$ml_{AG}\ddot{x}_A+\left(I_{dG}+ml_{AG}^2\right)\ddot{\varphi}_y+k_B l_{AB}x_A+k_B l_{AB}^2\varphi_y=0 \tag{4.157}$$

Equations (4.153) and (4.157), in the matrix form can be written as,

$$\begin{bmatrix} m & ml_{AG} \\ ml_{AG} & I_{dG}+ml_{AG}^2 \end{bmatrix}\begin{Bmatrix} \ddot{x}_A \\ \ddot{\varphi}_y \end{Bmatrix}+\begin{bmatrix} (k_A+k_B) & k_B l_{AB} \\ k_B l_{AB} & k_B l_{AB}^2 \end{bmatrix}\begin{Bmatrix} x_A \\ \varphi_y \end{Bmatrix}=\begin{Bmatrix} 0 \\ 0 \end{Bmatrix} \tag{4.158}$$

(iv) Generalised coordinates, (x_A, x_B) (refer Figure 4.39)
From Figure 4.39, we have

$$\varphi_y=\frac{x_B-x_A}{l_{AB}}=\left(-\frac{1}{l_{AB}}\right)x_A+\left(\frac{1}{l_{AB}}\right)x_B \tag{4.159}$$

and $$x_G=\left(x_A+l_{AG}\varphi_y\right)=x_A+l_{AG}\left\{\left(-\frac{1}{l_{AB}}\right)x_A+\left(\frac{1}{l_{AB}}\right)x_B\right\}=\left(1-\frac{l_{AG}}{l_{AB}}\right)x_A+\left(\frac{l_{AG}}{l_{AB}}\right)x_B \tag{4.160}$$

The KE, as per Eq. (4.117) and noting Eqs. (4.159) and (4.160), we get

$$T=\tfrac{1}{2}m\dot{x}_G^2+\tfrac{1}{2}I_{dG}\dot{\varphi}_y^2=\tfrac{1}{2}m\left\{\left(1-\frac{l_{AG}}{l_{AB}}\right)\dot{x}_A+\left(\frac{l_{AG}}{l_{AB}}\right)\dot{x}_B\right\}^2+\tfrac{1}{2}I_{dG}\left\{\left(-\frac{1}{l_{AB}}\right)\dot{x}_A+\left(\frac{1}{l_{AB}}\right)\dot{x}_B\right\}^2 \tag{4.161}$$

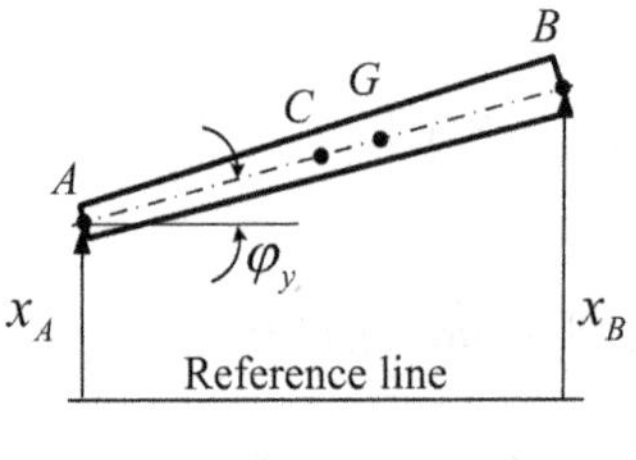

FIGURE 4.39 Deflected rigid shaft with generalized coordinates (x_A, x_B).

The PE of the rotor system from Eq. (4.118) is

$$U = \tfrac{1}{2}k_A x_A^2 + \tfrac{1}{2}k_B x_B^2 \tag{4.162}$$

Lagrange's equation for the generalised coordinate, x_A, is given as

$$\frac{d}{dt}\left(\frac{\partial T}{\partial \dot{x}_A}\right) - \frac{\partial T}{\partial x_A} + \frac{\partial U}{\partial x_A} = 0 \tag{4.163}$$

On substituting Eqs. (4.161) and (4.162) into parts of Eq. (4.163), we get

$$\frac{d}{dt}\left(\frac{\partial T}{\partial \dot{x}_A}\right) = \frac{d}{dt}\left\{\frac{\partial}{\partial \dot{x}_A}\left[\frac{1}{2}m\left\{\left(1-\frac{l_{AG}}{l_{AB}}\right)\dot{x}_A + \left(\frac{l_{AG}}{l_{AB}}\right)\dot{x}_B\right\}^2 + \frac{1}{2}I_{dG}\left\{\left(-\frac{1}{l_{AB}}\right)\dot{x}_A + \left(\frac{1}{l_{AB}}\right)\dot{x}_B\right\}^2\right]\right\}$$

$$= \frac{d}{dt}\left[m\left\{\left(1-\frac{l_{AG}}{l_{AB}}\right)\dot{x}_A + \left(\frac{l_{AG}}{l_{AB}}\right)\dot{x}_B\right\}\left(1-\frac{l_{AG}}{l_{AB}}\right) + I_{dG}\left\{\left(-\frac{1}{l_{AB}}\right)\dot{x}_A + \left(\frac{1}{l_{AB}}\right)\dot{x}_B\right\}\left(-\frac{1}{l_{AB}}\right)\right]$$

or

$$\frac{d}{dt}\left(\frac{\partial T}{\partial \dot{x}_A}\right) = \frac{d}{dt}\left\{m\left(1-\frac{l_{AG}}{l_{AB}}\right)^2\dot{x}_A + m\left(\frac{l_{AG}}{l_{AB}}\right)\left(1-\frac{l_{AG}}{l_{AB}}\right)\dot{x}_B + I_{dG}\left(\frac{1}{l_{AB}}\right)^2\dot{x}_A + I_{dG}\left(\frac{1}{l_{AB}}\right)\left(-\frac{1}{l_{AB}}\right)\dot{x}_B\right\}$$

$$= \left\{m\left(1-\frac{l_{AG}}{l_{AB}}\right)^2 + I_{dG}\left(\frac{1}{l_{AB}}\right)^2\right\}\ddot{x}_A + \left\{m\left(\frac{l_{AG}}{l_{AB}}\right)\left(1-\frac{l_{AG}}{l_{AB}}\right) - I_{dG}\left(\frac{1}{l_{AB}}\right)^2\right\}\ddot{x}_B \tag{4.164}$$

and
$$\frac{\partial U}{\partial x_A} = \frac{\partial}{\partial x_A}\left\{\frac{1}{2}k_A x_A^2 + \frac{1}{2}k_B x_B^2\right\} = k_A x_A \tag{4.165}$$

So that

$$\left\{m\left(1-\frac{l_{AG}}{l_{AB}}\right)^2 + I_{dG}\left(\frac{1}{l_{AB}}\right)^2\right\}\ddot{x}_A + \left\{m\left(\frac{l_{AG}}{l_{AB}}\right)\left(1-\frac{l_{AG}}{l_{AB}}\right) - I_{dG}\left(\frac{1}{l_{AB}}\right)^2\right\}\ddot{x}_B + k_A x_A = 0 \tag{4.166}$$

Lagrange's equation for the generalised coordinate, x_B, is given as

$$\frac{d}{dt}\left(\frac{\partial T}{\partial \dot{x}_B}\right) - \frac{\partial T}{\partial x_B} + \frac{\partial U}{\partial x_B} = 0 \tag{4.167}$$

On substituting Eqs. (4.161) and (4.162) into parts of Eq. (4.167), we get

$$\frac{d}{dt}\left(\frac{\partial T}{\partial \dot{x}_B}\right) = \frac{d}{dt}\left\{\frac{\partial}{\partial \dot{x}_B}\left[\tfrac{1}{2}m\left\{\left(1-\frac{l_{AG}}{l_{AB}}\right)\dot{x}_A + \left(\frac{l_{AG}}{l_{AB}}\right)\dot{x}_B\right\}^2 + \tfrac{1}{2}I_{dG}\left\{\left(-\frac{1}{l_{AB}}\right)\dot{x}_A + \left(\frac{1}{l_{AB}}\right)\dot{x}_B\right\}^2\right]\right\}$$

$$= \frac{d}{dt}\left[m\left\{\left(1-\frac{l_{AG}}{l_{AB}}\right)\dot{x}_A + \left(\frac{l_{AG}}{l_{AB}}\right)\dot{x}_B\right\}\left(\frac{l_{AG}}{l_{AB}}\right) + I_{dG}\left\{\left(-\frac{1}{l_{AB}}\right)\dot{x}_A + \left(\frac{1}{l_{AB}}\right)\dot{x}_B\right\}\left(\frac{1}{l_{AB}}\right)\right]$$

or

$$\frac{d}{dt}\left(\frac{\partial T}{\partial \dot{x}_A}\right)=\frac{d}{dt}\left\{m\left(1-\frac{l_{AG}}{l_{AB}}\right)\left(\frac{l_{AG}}{l_{AB}}\right)\dot{x}_A+m\left(\frac{l_{AG}}{l_{AB}}\right)^2\dot{x}_B+I_{dG}\left(-\frac{1}{l_{AB}}\right)\left(\frac{1}{l_{AB}}\right)\dot{x}_A+I_{dG}\left(\frac{1}{l_{AB}}\right)^2\dot{x}_B\right\}$$

$$=\left\{m\left(1-\frac{l_{AG}}{l_{AB}}\right)\left(\frac{l_{AG}}{l_{AB}}\right)-I_{dG}\left(\frac{1}{l_{AB}}\right)^2\right\}\ddot{x}_A+\left\{m\left(\frac{l_{AG}}{l_{AB}}\right)^2+I_{dG}\left(\frac{1}{l_{AB}}\right)^2\right\}\ddot{x}_B \tag{4.168}$$

and

$$\frac{\partial U}{\partial x_B}=\frac{\partial}{\partial x_B}\left\{\tfrac{1}{2}k_A x_A^2+\tfrac{1}{2}k_B x_B^2\right\}=k_B x_B \tag{4.169}$$

On substituting Eqs. (4.168) and (4.169) into Eq. (4.167), we get

$$\left\{m\left(1-\frac{l_{AG}}{l_{AB}}\right)\left(\frac{l_{AG}}{l_{AB}}\right)-I_{dG}\left(\frac{1}{l_{AB}}\right)^2\right\}\ddot{x}_A+\left\{m\left(\frac{l_{AG}}{l_{AB}}\right)^2+I_{dG}\left(\frac{1}{l_{AB}}\right)^2\right\}\ddot{x}_B+k_B x_B=0 \tag{4.170}$$

In matrix form, Eqs. (4.166) and (4.170) can be combined as

$$\begin{bmatrix} m\left(1-\frac{l_{AG}}{l_{AB}}\right)^2+I_{dG}\left(\frac{1}{l_{AB}}\right)^2 & m\left(\frac{l_{AG}}{l_{AB}}\right)\left(1-\frac{l_{AG}}{l_{AB}}\right)-I_{dG}\left(\frac{1}{l_{AB}}\right)^2 \\ m\left(1-\frac{l_{AG}}{l_{AB}}\right)\left(\frac{l_{AG}}{l_{AB}}\right)-I_{dG}\left(\frac{1}{l_{AB}}\right)^2 & m\left(\frac{l_{AG}}{l_{AB}}\right)^2+I_{dG}\left(\frac{1}{l_{AB}}\right)^2 \end{bmatrix}\begin{Bmatrix}\ddot{x}_A\\ \ddot{x}_B\end{Bmatrix}$$
$$+\begin{bmatrix} k_A & 0\\ 0 & k_B\end{bmatrix}\begin{Bmatrix}x_A\\ x_B\end{Bmatrix}=\begin{Bmatrix}0\\0\end{Bmatrix} \tag{4.171}$$

Here also the stiffness matrix is uncoupled since chosen generalized coordinates are at bearings itself.

(v) Generalised coordinates, (x_C, φ_y) (refer Figure 4.40)

It is similar to the second case but with some differences. We have the following condition from the problem

$$l_{AC}=l_{BC}=0.5l_{AB} \tag{4.172}$$

In Eq. (4.172), piont C is the mid-point of the shaft. From Figure 4.40, we have

$$x_G=\left(x_C+l_{CG}\varphi_y\right);\quad x_A=\left(x_C-l_{AC}\varphi_y\right);\quad x_B=\left(x_C+l_{BC}\varphi_y\right) \tag{4.173}$$

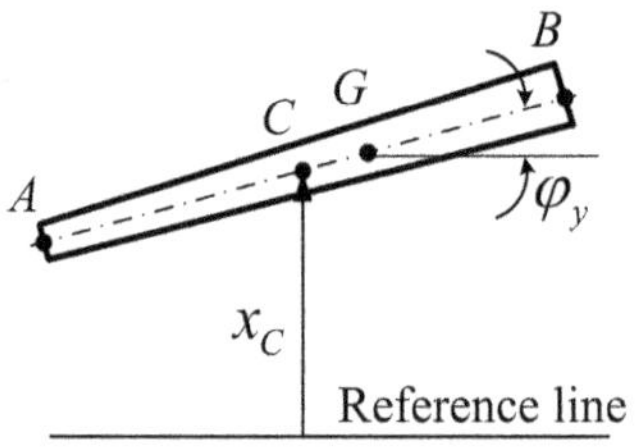

FIGURE 4.40 Deflected shaft with generalized coordinates (x_C, φ_y).

where l_{CG} is distance between points C and G locations on the shaft, l_{AC} is distance between points A and C, and l_{BC} is distance between points B and C. The KE can be written as (as per Eq. (4.117)) and noting Eq. (4.173), we get

$$T = \tfrac{1}{2} m\dot{x}_G^2 + \tfrac{1}{2} I_{dG}\dot{\varphi}_y^2 = \tfrac{1}{2} m\left(\dot{x}_C + l_{CG}\dot{\varphi}_y\right)^2 + \tfrac{1}{2} I_{dG}\dot{\varphi}_y^2 \tag{4.174}$$

The PE of the rotor system from Eq. (4.118) and noting Eq. (4.173) is

$$U = \tfrac{1}{2} k_A x_A^2 + \tfrac{1}{2} k_B x_B^2 = \tfrac{1}{2} k_A\left(x_C - l_{AC}\varphi_y\right)^2 + \tfrac{1}{2} k_B\left(x_C + l_{BC}\varphi_y\right)^2 \tag{4.175}$$

Lagrange's equation for the generalised coordinate, x_C, is given as

$$\frac{d}{dt}\left(\frac{\partial T}{\partial \dot{x}_C}\right) - \frac{\partial T}{\partial x_C} + \frac{\partial U}{\partial x_C} = 0 \tag{4.176}$$

On substituting Eqs. (4.174) and (4.175) into parts of Eq. (4.176), we get

$$\frac{d}{dt}\left(\frac{\partial T}{\partial \dot{x}_C}\right) = \frac{d}{dt}\left[\frac{\partial}{\partial \dot{x}_C}\left\{\tfrac{1}{2} m\left(\dot{x}_C + l_{CG}\dot{\varphi}_y\right)^2 + \tfrac{1}{2} I_{dG}\dot{\varphi}_y^2\right\}\right] = \frac{d}{dt}\left\{m\left(\dot{x}_C + l_{CG}\dot{\varphi}_y\right)\right\} = m\left(\ddot{x}_C + l_{CG}\ddot{\varphi}_y\right) \tag{4.177}$$

and $$\frac{\partial U}{\partial x_C} = \frac{\partial}{\partial x_C}\left\{\tfrac{1}{2} k_A\left(x_C - l_{AC}\varphi_y\right)^2 + \tfrac{1}{2} k_B\left(x_C + l_{BC}\varphi_y\right)^2\right\} = k_A\left(x_C - l_{AC}\varphi_y\right) + k_B\left(x_C + l_{BC}\varphi_y\right) \tag{4.178}$$

On substituting Eqs. (4.177) and (4.178) into Eq. (4.176), we get

$$m\left(\ddot{x}_C + l_{CG}\ddot{\varphi}_y\right) + k_A\left(x_C - l_{AC}\varphi_y\right) + k_B\left(x_C + l_{BC}\varphi_y\right) = 0$$

$$\text{or}\quad m\ddot{x}_C + ml_{CG}\ddot{\varphi}_y + (k_A + k_B)x_C - (k_A l_{AC} - k_B l_{BC})\varphi_y = 0 \tag{4.179}$$

Lagrange's equation for the generalised coordinate φ_y is given as

$$\frac{d}{dt}\left(\frac{\partial T}{\partial \dot{\varphi}_y}\right) - \frac{\partial T}{\partial \varphi_y} + \frac{\partial U}{\partial \varphi_y} = 0 \tag{4.180}$$

On substituting Eqs. (4.174) and (4.175) into Eq. (4.180), we get

$$\begin{aligned}\frac{d}{dt}\left(\frac{\partial T}{\partial \dot{\varphi}_y}\right) &= \frac{d}{dt}\left[\frac{\partial}{\partial \dot{\varphi}_y}\left\{\tfrac{1}{2} m\left(\dot{x}_C + l_{CG}\dot{\varphi}_y\right)^2 + \tfrac{1}{2} I_{dG}\dot{\varphi}_y^2\right\}\right] = \frac{d}{dt}\left\{m\left(\dot{x}_C + l_{CG}\dot{\varphi}_y\right)l_{CG} + I_{dG}\dot{\varphi}_y\right\} \\ &= ml_{CG}\ddot{x}_C + \left(I_{dG} + ml_{CG}^2\right)\ddot{\varphi}_y\end{aligned} \tag{4.181}$$

$$\text{and}\quad \begin{aligned}\frac{\partial U}{\partial \varphi_y} &= \frac{\partial}{\partial \varphi_y}\left\{\tfrac{1}{2} k_A\left(x_C - l_{AC}\varphi_y\right)^2 + \tfrac{1}{2} k_B\left(x_C + l_{BC}\varphi_y\right)^2\right\} \\ &= -(k_A l_{AC} - k_B l_{BC})x_C + \left(k_A l_{AC}^2 + k_B l_{BC}^2\right)\varphi_y\end{aligned} \tag{4.182}$$

On substituting Eqs. (4.181) and (4.182) into Eq. (4.180), we get

$$ml_{CG}\ddot{x}_C + \left(I_{dG} + ml_{CG}^2\right)\ddot{\varphi}_y - (k_A l_{AC} - k_B l_{BC})x_C + \left(k_A l_{AC}^2 + k_B l_{BC}^2\right)\varphi_y = 0 \tag{4.183}$$

In the matrix form, Eqs. (4.179) and (4.183) can be combined as

$$\begin{bmatrix} m & ml_{CG} \\ ml_{CG} & I_{dG}+ml_{CG}^2 \end{bmatrix}\begin{Bmatrix} \ddot{x}_C \\ \ddot{\varphi}_y \end{Bmatrix}+\begin{bmatrix} k_A+k_B & -(k_Al_{AC}-k_Bl_{BC}) \\ -(k_Al_{AC}-k_Bl_{BC}) & k_Al_{AC}^2+k_Bl_{BC}^2 \end{bmatrix}\begin{Bmatrix} x_C \\ \varphi_y \end{Bmatrix}=\begin{Bmatrix} 0 \\ 0 \end{Bmatrix} \tag{4.184}$$

Noting Eq. (4.172), Eq. (4.184) gives

$$\begin{bmatrix} m & ml_{CG} \\ ml_{CG} & I_{dG}+ml_{CG}^2 \end{bmatrix}\begin{Bmatrix} \ddot{x}_C \\ \ddot{\varphi}_y \end{Bmatrix}+\begin{bmatrix} k_A+k_B & -0.5(k_A-k_B)l_{AB} \\ -0.5(k_A-k_B)l_{AB} & 0.25(k_A+k_B)l_{AB}^2 \end{bmatrix}\begin{Bmatrix} x_C \\ \varphi_y \end{Bmatrix}=\begin{Bmatrix} 0 \\ 0 \end{Bmatrix} \tag{4.185}$$

It should be noted that, for the present case, if both bearings are identical then both mass and stiffness matrix will be uncoupled. This will lead to translational and rotational motion uncoupled and they can be analysed independent of each other.

Exercise 4.9 For a perfectly balanced rigid rotor mounted on flexible bearings, as shown in Figure 4.35, the following data are given: $m=10\,\text{kg}$, $I_d=0.015$ kg-m^2, $l=1$ m, $l_{AG}=0.6$m, $k_A=120$ kN/m, $k_B=140$ kN/m. Consider a single-plane motion with two DOFs and coupling in the generalised coordinates. Consider various cases of generalised coordinates as treated in Exericse 4.8. Obtain the transverse natural frequencies and mode shapes of the rotor-bearing system.

Solution: For the present rotor system, the following data are given

$$m=10\text{ kg};\quad I_d=0.015\text{ kg-m}^2;\quad l_{AB}=l=\ 1\text{ m};\quad l_{AG}=0.6\text{ m};$$

$$k_A=120\times10^3\text{ N/m};\quad k_B=140\times10^3\text{ N/m};$$

We additionally assumed the following rotor properties:

$$l_{BG}=l_{AB}-l_{AG}=1.0-0.6=0.4\text{ m};\qquad l_{AE}=l_{AB}k_B/(k_A+k_B)=0.5385\text{ m};$$
$$l_{BE}=l_{AB}-l_{AE}=0.4615\text{ m};\qquad l_{EG}=l_{AG}-l_{AE}=0.6-0.5385=0.0615\text{ m};$$
$$l_{AC}=l_{BC}=l_{BC}/2=0.5\text{ m}.$$

Equations of motion with various generalised coordinates are given (refer Exercise 4.8), as
Case I: (x_G,φ_y): (refer Figure 4.36 and Eq. (4.127))

$$\begin{bmatrix} m & 0 \\ 0 & I_{dG} \end{bmatrix}\begin{Bmatrix} \ddot{x}_G \\ \ddot{\varphi}_y \end{Bmatrix}+\begin{bmatrix} (k_A+k_B) & -(k_Al_{AG}-k_Bl_{BG}) \\ -(k_Al_{AG}-k_Bl_{BG}) & (k_Al_{AG}^2+k_Bl_{BG}^2) \end{bmatrix}\begin{Bmatrix} x_G \\ \varphi_y \end{Bmatrix}=\begin{Bmatrix} 0 \\ 0 \end{Bmatrix} \tag{4.186}$$

$$\text{or}\quad\begin{bmatrix} 10 & 0 \\ 0 & 0.015 \end{bmatrix}\begin{Bmatrix} \ddot{x}_G \\ \ddot{\varphi}_y \end{Bmatrix}+\begin{bmatrix} 260{,}000 & -16{,}000 \\ -16{,}000 & 65{,}600 \end{bmatrix}\begin{Bmatrix} x_G \\ \varphi_y \end{Bmatrix}=\begin{Bmatrix} 0 \\ 0 \end{Bmatrix} \tag{4.187}$$

which gives natural frequencies as

$$\omega_{nf1} = 160.02 \text{ rad/s} \quad \text{and} \quad \omega_{nf2} = 2091.3 \text{ rad/s}. \tag{4.188}$$

Case II: (x_E, φ_y): (refer Figure 4.37 and Eq. (4.141))

$$\begin{bmatrix} m & ml_{EG} \\ ml_{EG} & I_{dG} + ml_{EG}^2 \end{bmatrix} \begin{Bmatrix} \ddot{x}_E \\ \ddot{\varphi}_y \end{Bmatrix} + \begin{bmatrix} k_A + k_B & 0 \\ 0 & k_A l_{AE}^2 + k_B l_{BE}^2 \end{bmatrix} \begin{Bmatrix} x_E \\ \varphi_y \end{Bmatrix} = \begin{Bmatrix} 0 \\ 0 \end{Bmatrix} \tag{4.189}$$

$$\text{with} \quad k_A l_{AE} = k_B l_{BE}; \quad l_{AE} + l_{BE} = l_{AB}; \quad l_{AE} = \frac{k_B}{k_A + k_B} l_{AB}; \quad l_{EG} = l_{AG} - l_{AE} \tag{4.190}$$

$$\text{or} \begin{bmatrix} 10 & 0.6154 \\ 0.6154 & 0.0529 \end{bmatrix} \begin{Bmatrix} \ddot{x}_E \\ \ddot{\varphi}_y \end{Bmatrix} + 10^6 \begin{bmatrix} 2.6000 & 0 \\ 0 & 0.6462 \end{bmatrix} \begin{Bmatrix} x_E \\ \varphi_y \end{Bmatrix} = \begin{Bmatrix} 0 \\ 0 \end{Bmatrix} \tag{4.191}$$

which gives natural frequencies as

$$\omega_{nf1} = 160.02 \text{ rad/s} \quad \text{and} \quad \omega_{nf2} = 2091.3 \text{ rad/s} \tag{4.192}$$

Case III: (x_A, φ_y): (refer Figure 4.38 and Eq. (4.158))

$$\begin{bmatrix} m & ml_{AG} \\ ml_{AG} & I_{dG} + ml_{AG}^2 \end{bmatrix} \begin{Bmatrix} \ddot{x}_A \\ \ddot{\varphi}_y \end{Bmatrix} + \begin{bmatrix} (k_A + k_B) & k_B l_{AB} \\ k_B l_{AB} & k_B l_{AB}^2 \end{bmatrix} \begin{Bmatrix} x_A \\ \varphi_y \end{Bmatrix} = \begin{Bmatrix} 0 \\ 0 \end{Bmatrix} \tag{4.193}$$

$$\text{or} \begin{bmatrix} 10 & 6 \\ 6 & 3.6150 \end{bmatrix} \begin{Bmatrix} \ddot{x}_A \\ \ddot{\varphi}_y \end{Bmatrix} + \begin{bmatrix} 260{,}000 & 140{,}000 \\ 140{,}000 & 140{,}000 \end{bmatrix} \begin{Bmatrix} x_A \\ \varphi_y \end{Bmatrix} = \begin{Bmatrix} 0 \\ 0 \end{Bmatrix} \tag{4.194}$$

which gives natural frequencies as

$$\omega_{nf1} = 160.02 \text{ rad/s} \quad \text{and} \quad \omega_{nf2} = 2091.3 \text{ rad/s} \tag{4.195}$$

Case IV: (x_A, x_B): (refer Figure 4.39 and Eq. (4.171))

$$\begin{bmatrix} m\left(1 - \dfrac{l_{AG}}{l_{AB}}\right)^2 + I_{dG}\left(\dfrac{1}{l_{AB}}\right)^2 & m\left(\dfrac{l_{AG}}{l_{AB}}\right)\left(1 - \dfrac{l_{AG}}{l_{AB}}\right) - I_{dG}\left(\dfrac{1}{l_{AB}}\right)^2 \\ m\left(1 - \dfrac{l_{AG}}{l_{AB}}\right)\left(\dfrac{l_{AG}}{l_{AB}}\right) - I_{dG}\left(\dfrac{1}{l_{AB}}\right)^2 & m\left(\dfrac{l_{AG}}{l_{AB}}\right)^2 + I_{dG}\left(\dfrac{1}{l_{AB}}\right)^2 \end{bmatrix} \begin{Bmatrix} \ddot{x}_A \\ \ddot{x}_B \end{Bmatrix}$$

$$+ \begin{bmatrix} k_A & 0 \\ 0 & k_B \end{bmatrix} \begin{Bmatrix} x_A \\ x_B \end{Bmatrix} = \begin{Bmatrix} 0 \\ 0 \end{Bmatrix} \tag{4.196}$$

$$\text{or} \begin{bmatrix} 1.6150 & 2.3850 \\ 2.3850 & 3.6150 \end{bmatrix} \begin{Bmatrix} \ddot{x}_A \\ \ddot{x}_B \end{Bmatrix} + \begin{bmatrix} 120{,}000 & 0 \\ 0 & 140{,}000 \end{bmatrix} \begin{Bmatrix} x_A \\ x_B \end{Bmatrix} = \begin{Bmatrix} 0 \\ 0 \end{Bmatrix} \tag{4.197}$$

which gives natural frequencies as

$$\omega_{nf1} = 160.02 \text{ rad/s} \quad \text{and} \quad \omega_{nf2} = 2091.3 \text{ rad/s} \tag{4.198}$$

Case V: (x_C, φ_y): (x_A, x_B): (refer Figure 4.40 and Eq. (4.184))

$$\begin{bmatrix} m & ml_{CG} \\ ml_{CG} & I_{dG} + ml_{CG}^2 \end{bmatrix} \begin{Bmatrix} \ddot{x}_C \\ \ddot{\varphi}_y \end{Bmatrix} + \begin{bmatrix} k_A + k_B & -0.5(k_A - k_B)l_{AB} \\ -0.5(k_A - k_B)l_{AB} & 0.25(k_A + k_B)l_{AB}^2 \end{bmatrix} \begin{Bmatrix} x_C \\ \varphi_y \end{Bmatrix} = \begin{Bmatrix} 0 \\ 0 \end{Bmatrix} \tag{4.199}$$

$$\text{or} \quad \begin{bmatrix} 10 & 1 \\ 1 & 0.1150 \end{bmatrix} \begin{Bmatrix} \ddot{x}_C \\ \ddot{\varphi}_y \end{Bmatrix} + \begin{bmatrix} 260{,}000 & 10{,}000 \\ 10{,}000 & 65{,}000 \end{bmatrix} \begin{Bmatrix} x_C \\ \varphi_y \end{Bmatrix} = \begin{Bmatrix} 0 \\ 0 \end{Bmatrix} \tag{4.200}$$

which gives natural frequencies as

$$\omega_{nf1} = 160.02 \text{ rad/s} \quad \text{and} \quad \omega_{nf2} = 2091.3 \text{ rad/s} \tag{4.201}$$

It can be seen that as expected the natural frequencies obtained are same since the system is same. The mode shapes will change and it is left to readers to obtain the same.

Exercise 4.10 Obtain transverse natural frequencies of a rotor-bearing system, as shown in Figure 4.41, for a pure translatory motion of the shaft. Consider the shaft to be rigid, and the whole mass of the shaft is assumed to be concentrated at its mid-span. The shaft has a span of 1 m and the diameter is 0.05 m with a mass density of 7,800 kg/m^3. The shaft is supported at the ends by flexible bearings. Consider the motion in both the vertical and horizontal planes. Use the following bearing properties: for both bearings A and B: k_{xx} = 200 MN/m, k_{yy} = 150 MN/m, k_{xy} = 15 MN/m and k_{yx} = 10 MN/m.

***Solution*:** It is given that only a pure translational motion needs to consider. The following rotor data are given:

$$k_{xx} = 200 \text{ MN/m}; \quad k_{yy} = 150 \text{ MN/m}; \quad k_{xy} = 15 \text{ MN/m}; \quad k_{yx} = 10 \text{ MN/m}$$

$$\rho = 7{,}800 \text{ kg/m}^3; \quad d = 0.05 \text{ m}; \quad l = 1 \text{ m}; \quad m = \rho \frac{\pi d^2}{4} l = 15.315 \text{ kg}.$$

In y–z plane (refer Figure 4.42a): The equation of motion is given as

$$m\ddot{y} = -(2k_{yy}y + 2k_{yx}x) \quad \text{or} \quad m\ddot{y} + 2k_{yy}y + 2k_{yx}x = 0 \tag{4.202}$$

In z–x plane (refer Figure 4.42b): The equation of motion is given as

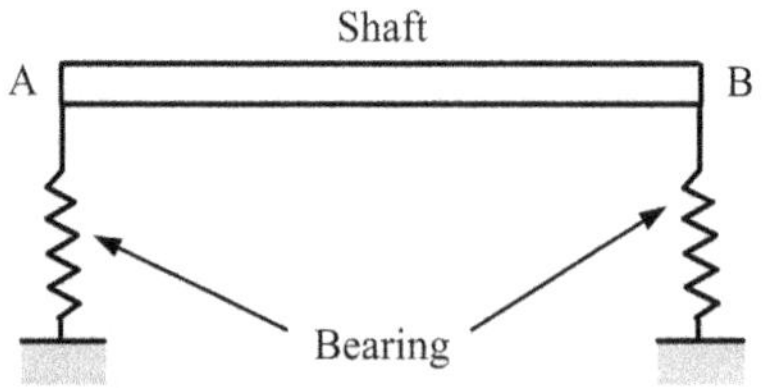

FIGURE 4.41 A rigid rotor mounted on two bearings.

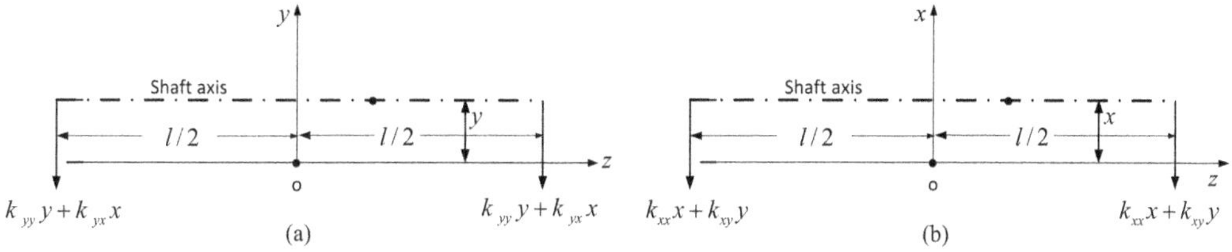

FIGURE 4.42 Free body diagram of shaft in (a) y–z plane and (b) z–x plane

$$m\ddot{x} = -(2k_{xx}x + 2k_{xy}y) \quad \text{or} \quad m\ddot{x} + 2k_{xx}x + 2k_{xy}y = 0 \tag{4.203}$$

On combining Eqs. (4.202) and (4.203) in a matrix form, we get

$$\begin{bmatrix} m & 0 \\ 0 & m \end{bmatrix} \begin{Bmatrix} \ddot{x} \\ \ddot{y} \end{Bmatrix} + \begin{bmatrix} 2k_{xx} & 2k_{xy} \\ 2k_{yx} & 2k_{yy} \end{bmatrix} \begin{Bmatrix} x \\ y \end{Bmatrix} = 0 \tag{4.204}$$

Herein, the stiffness matrix has elastic coupling in two orthogonal plane motions. On substituting given rotor-bearing parameters in Eq. (4.204), we get

$$\begin{bmatrix} 15.3153 & 0 \\ 0 & 15.3153 \end{bmatrix} \begin{Bmatrix} \ddot{x} \\ \ddot{y} \end{Bmatrix} + 10^6 \times \begin{bmatrix} 400 & 30 \\ 20 & 300 \end{bmatrix} \begin{Bmatrix} x \\ y \end{Bmatrix} = 0 \tag{4.205}$$

For free vibration, substituting $\ddot{x} = -\omega_{nf}^2 x$ and $\ddot{y} = -\omega_{nf}^2 y$, and for the non-trial solution, we get

$$\begin{vmatrix} 400\times10^6 - 15.3153\omega_{nf}^2 & 30\times10^6 \\ 20\times10^6 & 300\times10^6 - 15.3153\omega_{nf}^2 \end{vmatrix} = 0 \tag{4.206}$$

which gives the characteristic polynomial as

$$234.5573\omega_{nf}^4 - 1.0721\times10^{10}\omega_{nf}^2 + 1.1940\times10^{17} = 0 \tag{4.207}$$

On solving Eq. (4.207), we get

$$\omega_{nf1} = 4{,}383.8 \text{ rad/s} \quad \text{and} \quad \omega_{nf2} = 5{,}146.7 \text{ rad/s}$$

Readers can obtain corresponding mode shapes (i.e. relative displacements for each natural frequency).

Exercise 4.11 Obtain transverse critical speeds of a rotor-bearing system, as shown in Figure 4.43, for a pure tilting motion of the shaft. Consider the shaft to be rigid and the whole mass of the shaft is assumed to be concentrated at its mid-span. The shaft has a span of 1 m and the diameter is 0.05 m with a mass density of 7,800 kg/m³. The shaft is supported at the ends by flexible bearings. Consider the motion in both the vertical and horizontal planes. Use the following bearing properties: for both bearings A and B: k_{xx} = 200 MN/m, k_{yy} = 150 MN/m, k_{xy} = 1.5 MN/m and k_{yx} = 0.5 MN/m.

Solution: Figure 4.44a and b show free body diagram of the rotor in y–z and z–x planes, respectively, for pure tilting motion of the shaft.

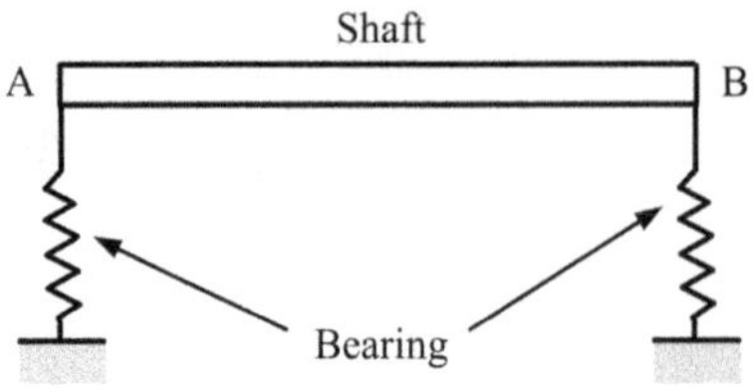

FIGURE 4.43 A rigid rotor mounted on two similar bearings.

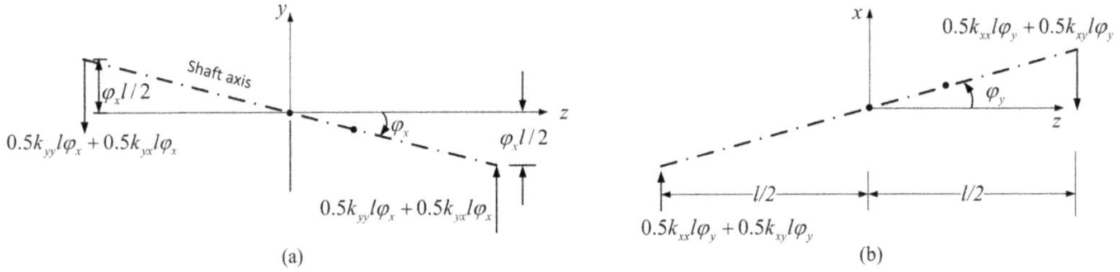

FIGURE 4.44 Free body diagram of rotor in (a) y–z plane (b) z–x plane.

Writing the moment balance equations from the free body diagrams (Figure 4.44), we get

$$I_d\ddot{\varphi}_x + \left(0.5k_{yy}l^2\right)\varphi_x + \left(0.5k_{yx}l^2\right)\varphi_y = 0 \tag{4.208}$$

$$\text{and} \quad I_d\ddot{\varphi}_y + \left(0.5k_{xy}l^2\right)\varphi_x + \left(0.5k_{xx}l^2\right)\varphi_y = 0 \quad \text{with} \quad I_d = \tfrac{1}{4}mr^2 + \tfrac{1}{12}ml^2 \tag{4.209}$$

Arranging the set of Eqs. (4.208) and (4.209) in a matrix form, as

$$\begin{bmatrix} I_d & 0 \\ 0 & I_d \end{bmatrix}\begin{Bmatrix} \ddot{\varphi}_x \\ \ddot{\varphi}_y \end{Bmatrix} + \begin{bmatrix} 0.5k_{yy}l^2 & 0.5k_{yx}l^2 \\ 0.5k_{xy}l^2 & 0.5k_{xx}l^2 \end{bmatrix}\begin{Bmatrix} \varphi_x \\ \varphi_y \end{Bmatrix} = 0 \tag{4.210}$$

The eigenvalue formulation for the aforementioned equation is

$$\left(-\omega_{nf}^2\mathbf{M} + \mathbf{K}\right)\eta = \mathbf{0} \quad \text{with} \quad \mathbf{M} = \begin{bmatrix} I_d & 0 \\ 0 & I_d \end{bmatrix}; \quad \mathbf{K} = \begin{bmatrix} 0.5k_{yy}l^2 & 0.5k_{yx}l^2 \\ 0.5k_{xy}l^2 & 0.5k_{xx}l^2 \end{bmatrix} \tag{4.211}$$

which gives

$$\left(0.5k_{yy}l^2 - \omega^2 I_d\right)\left(0.5k_{xx}l^2 - \omega^2 I_d\right) - 0.25k_{xy}k_{yx}l^4 = 0 \tag{4.212}$$

The following rotor data are given

$$k_{xx} = 200 \text{ MN/m}; \quad k_{yy} = 150 \text{ MN/m}; \quad k_{xy} = 15 \text{ MN/m}; \quad k_{yx} = 10 \text{ MN/m}$$

$$\rho = 7{,}800 \text{ kg/m}^3; \quad d = 0.05 \text{ m}; \quad l = 1 \text{ m}; \quad m = \rho\frac{\pi d^2}{4}l = 15.315 \text{ kg}.$$

On substituting given parameter values in Eq. (4.212), we get

$$1.6350\omega_{nf}^4 - 2.2377\times 10^8\omega_{nf}^2 + 7.4998\times 10^{15} = 0 \tag{4.213}$$

On solving the aforementioned characteristic equation, we get

$$\omega_{nf1} = 7{,}658.3 \text{ rad/s} \quad \text{and} \quad \omega_{nf2} = 8{,}843.8 \text{ rad/s}.$$

For the present case, a pure tilting motion of the shaft is considered. In case the rotor has centre of gravity offset from the mid-span, the readers can try out for the coupled translational and tilt motion to get corresponding four natural frequencies and mode shapes.

Exercise 4.12 Obtain equations of motion (put them in an expanded matrix form) for the transverse vibration of a rotor-bearing-coupling system as shown in Figure 4.45. Two identical rigid rotors are connected by a coupling (a pin joint or a universal joint) and are supported on four bearings (modelled as springs, each of which has a stiffness of k) as shown in the figure. Choose appropriately generalised coordinates to define the motion of the system. In the figure, l is the distance between the bearings and $a = a_1 = a_2$ is the overhang portion of each rotor towards the coupling. Consider a single-plane motion of the system.

Solution: The rotor system with generalised coordinates (y, φ_{x1} and φ_{x2}) are shown in Figure 4.46. At coupling since both the rotors have common transverse displacement so it is represented as y. Tilting angles of the left and right-side rotors have been designated by φ_{x1} and φ_{x2}.

In the present case, a single plane motion needs to be considered but translational and rotational motions are coupled. Since both the shaft are rigid and have no fixed point about which they have rotation so it is better to write KE in terms of displacements of individual rotors about its centre of gravity. Initially, the KE and PE will be written in a more general form and then it will be simplified

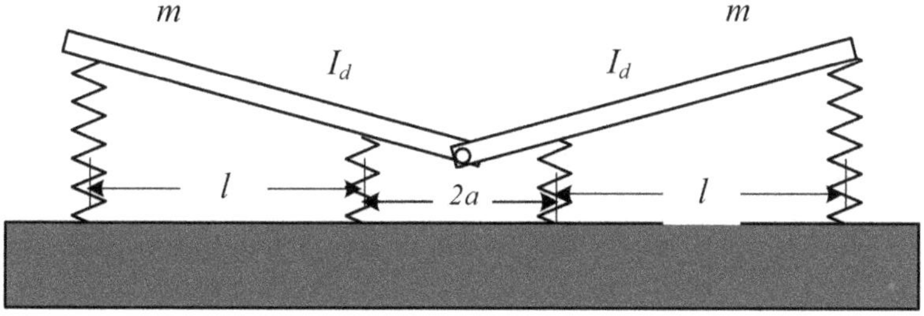

FIGURE 4.45 A rotor-bearing-coupling system model.

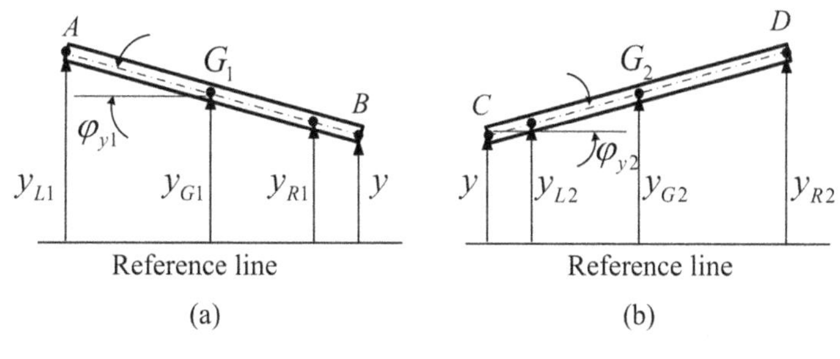

FIGURE 4.46 A rotor system with generalised coordinates.

as per the present problem so that expressions are relatively simpler. The kinetic energy of the rotor system is given as

$$T = \tfrac{1}{2} m_1 \dot{y}_{G1}^2 + \tfrac{1}{2} m_2 \dot{y}_{G2}^2 + \tfrac{1}{2} I_{dG1} \dot{\varphi}_{x1}^2 + \tfrac{1}{2} I_{dG2} \dot{\varphi}_{x2}^2 \tag{4.214}$$

$$\text{with} \quad y_{G1} = y + 0.5(a_1 + l_1)\varphi_{x1}; \quad y_{G2} = y + 0.5(a_2 + l_2)\varphi_{x2};$$

where l_1 and l_2 are the distance between two bearings in the left and right rotors, respectively; a_1 and a_2 are the distance between the coupling joint and the nearest bearing in the left and right rotors, respectively. In general, subscripts 1 and 2 are for the left and right rotors, respectively; and subscripts L and R for the left and right bearing of a particular rotor. The potential energy of the rotor-bearing is written as

$$U = \frac{1}{2} k_{L1} y_{L1}^2 + \frac{1}{2} k_{R1} y_{R1}^2 + \frac{1}{2} k_{L2} y_{L2}^2 + \frac{1}{2} k_{R2} y_{R2}^2 \tag{4.215}$$

$$\text{with} \quad y_{L1} = y + (a_1 + l_1)\varphi_{x1}; \quad y_{R1} = y + a_1\varphi_{x1}; \quad y_{L2} = y + a_2\varphi_{x2}; \quad y_{R2} = y + (a_2 + l_2)\varphi_{x2}.$$

One can derive EOMs from the KE and PE given in the form of Eqs. (4.214) and (4.215), respectively. However, these equations are now simplified as per the present problem, as

$$T = \tfrac{1}{2} m \dot{y}_{G1}^2 + \tfrac{1}{2} m \dot{y}_{G2}^2 + \tfrac{1}{2} I_{dG} \dot{\varphi}_{x1}^2 + \tfrac{1}{2} I_{dG} \dot{\varphi}_{x2}^2 \tag{4.216}$$

$$\text{with} \quad y_{G1} = y + 0.5(a + L)\varphi_{x1}; \quad y_{G2} = y + 0.5(a + L)\varphi_{x2}$$

$$\text{or} \quad T = \tfrac{1}{2} m\{\dot{y} + 0.5(a+L)\dot{\varphi}_{x1}\}^2 + \tfrac{1}{2} m\{\dot{y} + 0.5(a+L)\dot{\varphi}_{x2}\}^2 + \tfrac{1}{2} I_{dG}\dot{\varphi}_{x1}^2 + \tfrac{1}{2} I_{dG}\dot{\varphi}_{x2}^2 \tag{4.217}$$

and

$$U = \tfrac{1}{2} k y_{L1}^2 + \tfrac{1}{2} k y_{R1}^2 + \tfrac{1}{2} k y_{L2}^2 + \tfrac{1}{2} k y_{R2}^2 \tag{4.218}$$

$$\text{with} \quad y_{L1} = y + (a + L)\varphi_{x1}; \quad y_{R1} = y + a\varphi_{x1}; \quad y_{L2} = y + a\varphi_{x2}; \quad y_{R2} = y + (a + L)\varphi_{x2}.$$

$$\text{or} \quad U = \tfrac{1}{2} k\{y + (a+L)\varphi_{x1}\}^2 + \tfrac{1}{2} k(y + a\varphi_{x1})^2 + \tfrac{1}{2} k(y + a\varphi_{x2})^2 + \tfrac{1}{2} k\{y + (a+L)\varphi_{x2}\}^2 \tag{4.219}$$

Lagrange's equation for the generalised coordinate y is given as

$$\frac{d}{dt}\left(\frac{\partial T}{\partial \dot{y}}\right) + \left(\frac{\partial U}{\partial y}\right) = 0 \tag{4.220}$$

We can obtain individual component of Eq. (4.220), as

$$\frac{d}{dt}\left(\frac{\partial T}{\partial \dot{y}}\right) = \frac{d}{dt}\left(\frac{\partial}{\partial \dot{y}}\left[\tfrac{1}{2} m\{\dot{y} + 0.5(a+L)\dot{\varphi}_{x1}\}^2 + \tfrac{1}{2} m\{\dot{y} + 0.5(a+L)\dot{\varphi}_{x2}\}^2 + \tfrac{1}{2} I_{dG}\dot{\varphi}_{x1}^2 + \tfrac{1}{2} I_{dG}\dot{\varphi}_{x2}^2\right]\right)$$

$$\text{or} \quad \frac{d}{dt}\left(\frac{\partial T}{\partial \dot{y}}\right) = 2m\ddot{y} + \{0.5m(a+L)\}\ddot{\varphi}_{x1} + \{0.5m(a+L)\}\ddot{\varphi}_{x2} \tag{4.221}$$

and

$$\left(\frac{\partial U}{\partial y}\right)=\frac{\partial}{\partial y}\left[\tfrac{1}{2}k\{y+(a+L)\varphi_{x1}\}^2+\tfrac{1}{2}k(y+a\varphi_{x1})^2+\tfrac{1}{2}k(y+a\varphi_{x2})^2+\tfrac{1}{2}k\{y+(a+L)\varphi_{x2}\}^2\right]$$

$$\text{or}\quad\left(\frac{\partial U}{\partial y}\right)=4ky+k(2a+L)\varphi_{x1}+k(2a+L)\varphi_{x2}\tag{4.222}$$

Hence, on substituting Eqs. (4.221) and (4.222) into Eq. (4.220), the EOM becomes

$$2m\ddot{y}+\{0.5m(a+L)\}\ddot{\varphi}_{x1}+\{0.5m(a+L)\}\ddot{\varphi}_{x2}+4ky+k(2a+L)\varphi_{x1}+k(2a+L)\varphi_{x2}=0\tag{4.223}$$

Lagrange's equation for the generalised coordinate φ_{x1} is given as

$$\frac{d}{dt}\left(\frac{\partial T}{\partial\dot{\varphi}_{x1}}\right)+\left(\frac{\partial U}{\partial\varphi_{x1}}\right)=0\tag{4.224}$$

We can obtain individual component of Eq. (4.224), as

$$\frac{d}{dt}\left(\frac{\partial T}{\partial\dot{\varphi}_{x1}}\right)=\frac{d}{dt}\left(\frac{\partial}{\partial\dot{\varphi}_{x1}}\left[\tfrac{1}{2}m\{\dot{y}+0.5(a+L)\dot{\varphi}_{x1}\}^2+\tfrac{1}{2}m\{\dot{y}+0.5(a+L)\dot{\varphi}_{x2}\}^2+\tfrac{1}{2}I_{dG}\dot{\varphi}_{x1}^2+\tfrac{1}{2}I_{dG}\dot{\varphi}_{x2}^2\right]\right)$$

$$\text{or}\quad\frac{d}{dt}\left(\frac{\partial T}{\partial\dot{\varphi}_{x1}}\right)=0.5m(a+L)\ddot{y}+\{0.25m(a+L)^2+I_{dG}\}\ddot{\varphi}_{x1}\tag{4.225}$$

and

$$\left(\frac{\partial U}{\partial\varphi_{x1}}\right)=\frac{\partial}{\partial\varphi_{x1}}\left[\frac{1}{2}k\{y+(a+L)\varphi_{x1}\}^2+\frac{1}{2}k(y+a\varphi_{x1})^2+\frac{1}{2}k(y+a\varphi_{x2})^2+\frac{1}{2}k\{y+(a+L)\varphi_{x2}\}^2\right]$$

$$\text{or}\quad\left(\frac{\partial U}{\partial\varphi_{x1}}\right)=\left[k(2a+L)y+k\{(a+L)^2+a^2\}\varphi_{x1}\right]\tag{4.226}$$

Hence, on substituting Eqs. (4.225) and (4.226) into Eq. (4.224), the EOM becomes

$$0.5m(a+L)\ddot{y}+\{0.25m(a+L)^2+I_{dG}\}\ddot{\varphi}_{x1}+k(2a+L)y+k\{(a+L)^2+a^2\}\varphi_{x1}=0\tag{4.227}$$

Lagrange's equation for the generalised coordinate φ_{x2} is given as

$$\frac{d}{dt}\left(\frac{\partial T}{\partial\dot{\varphi}_{x2}}\right)+\left(\frac{\partial U}{\partial\varphi_{x2}}\right)=0\tag{4.228}$$

We can obtain individual component of Eq. (4.228), as

$$\frac{d}{dt}\left(\frac{\partial T}{\partial\dot{\varphi}_{x2}}\right)=\frac{d}{dt}\left(\frac{\partial}{\partial\dot{\varphi}_{x2}}\left[\tfrac{1}{2}m\{\dot{y}+0.5(a+L)\dot{\varphi}_{x1}\}^2+\tfrac{1}{2}m\{\dot{y}+0.5(a+L)\dot{\varphi}_{x2}\}^2+\tfrac{1}{2}I_{dG}\dot{\varphi}_{x1}^2+\tfrac{1}{2}I_{dG}\dot{\varphi}_{x2}^2\right]\right)$$

$$\text{or} \quad \frac{d}{dt}\left(\frac{\partial T}{\partial \dot{\varphi}_{x2}}\right)=\frac{d}{dt}\left[m\{\dot{y}+0.5(a+L)\dot{\varphi}_{x2}\}0.5(a+L)+I_{dG}\dot{\varphi}_{x2}\right]$$

$$\text{or} \quad \frac{d}{dt}\left(\frac{\partial T}{\partial \dot{\varphi}_{x2}}\right)=0.5m(a+L)\ddot{y}+\left\{0.25m(a+L)^2+I_{dG}\right\}\ddot{\varphi}_{x2} \tag{4.229}$$

and

$$\left(\frac{\partial U}{\partial \varphi_{x2}}\right)=\frac{\partial}{\partial \varphi_{x2}}\left[\tfrac{1}{2}k\{y+(a+L)\varphi_{x1}\}^2+\tfrac{1}{2}k(y+a\varphi_{x1})^2+\tfrac{1}{2}k(y+a\varphi_{x2})^2+\tfrac{1}{2}k\{y+(a+L)\varphi_{x2}\}^2\right]$$

$$\text{or} \quad \left(\frac{\partial U}{\partial \varphi_{x2}}\right)=\left[k\{y+(a+L)\varphi_{x2}\}(a+L)+k(y+a\varphi_{x2})a\right]$$

$$\text{or} \quad \left(\frac{\partial U}{\partial \varphi_{x2}}\right)=\left[k(2a+L)y+k\left\{(a+L)^2+a^2\right\}\varphi_{x2}\right] \tag{4.230}$$

Hence, on substituting Eqs. (4.229) and (4.230) into Eq. (4.228), the EOM becomes

$$0.5m(a+L)\ddot{y}+\left\{0.25m(a+L)^2+I_{dG}\right\}\ddot{\varphi}_{x2}+k(2a+L)y+k\left\{(a+L)^2+a^2\right\}\varphi_{x2}=0 \tag{4.231}$$

On combining Eqs. (4.223), (4.227) and (4.231) in a matrix form, we get

$$\begin{bmatrix} 2m & 0.5m(a+L) & 0.5m(a+L) \\ 0.5m(a+L) & 0.25m(a+L)^2+I_{dG} & 0 \\ 0.5m(a+L) & 0 & 0.25m(a+L)^2+I_{dG} \end{bmatrix}\begin{Bmatrix} \ddot{y} \\ \ddot{\varphi}_{x1} \\ \ddot{\varphi}_{x2} \end{Bmatrix}$$

$$+\begin{bmatrix} 4k & k(2a+L) & k(2a+L) \\ k(2a+L) & k\left\{(a+L)^2+a^2\right\} & 0 \\ k(2a+L) & 0 & k\left\{(a+L)^2+a^2\right\} \end{bmatrix}\begin{Bmatrix} y \\ \varphi_{x1} \\ \varphi_{x2} \end{Bmatrix}=\begin{Bmatrix} 0 \\ 0 \\ 0 \end{Bmatrix} \tag{4.232}$$

It should be noted that both mass and stiffness matrices are coupled and symmetric. The angular displacements of both rotors are coupled by common translational displacement at coupling point. Herein, also we started KE expression with respect to respective generalised coordinates at centre of gravity (for rotor 1: (y_{G1}, φ_{x1}) and for rotor 2: (y_{G2}, φ_{x2})) and then transformed into required common generalised coordinates (y, φ_{x1} and φ_{x2}). Writing KE directly in common generalised coordinates (y, φ_{x1} and φ_{x2}) may lead to erroneous EOMs, especially in mass matrices, the coupling will be absent. Readers are encouraged to derive more general EOMs using Eqs. (4.214) and (4.215) having both rotors and respective bearings dissimilar properties. The form of EOMs will be same as Eq. (4.232) but will have bigger expressions.

Exercise 4.13 For a symmetrical long rigid shaft on flexible anisotropic bearings, the following energy expressions are given:

$$T=\tfrac{1}{2}m\dot{x}^2+\tfrac{1}{2}m\dot{y}^2+\tfrac{1}{2}I_d\dot{\varphi}_x^2+\tfrac{1}{2}I_d\dot{\varphi}_y^2; \quad U=\tfrac{1}{2}(2k_x)x^2+\tfrac{1}{2}(2k_y)y^2+\tfrac{1}{2}\left(k_xl^2/2\right)\varphi_x^2+\tfrac{1}{2}\left(k_yl^2/2\right)\varphi_y^2;$$

$$\delta W_{nc}=\left(me\omega^2\cos\omega t\right)\delta x+\left(me\omega^2\sin\omega t\right)\delta y+\left(me\omega^2e_z\cos\omega t\right)\delta\varphi_y+\left(me\omega^2e_z\sin\omega t\right)\delta\varphi_x$$

where T is the kinetic energy, U is the potential energy, δW_{nc} is the non-conservative virtual work done, x and y are the coordinates of rotor geometrical centre (i.e., generalised coordinates), φ_x and φ_y are rotational coordinates of the rotor (i.e., generalised coordinates), e is the radial eccentricity, e_z is the axial eccentricity, m is the mass of the rotor, k_x and k_y are the stiffness of the each bearing, l is the length of the rotor, I_d is the diametral mass moment of inertia of the rotor, and ω is the spin speed of the rotor. Using Lagrange's equation (refer chapter 7 of Tiwari (2017)) obtain equations of motion of the rotor system.

***Solution*:** From the question, the given energies, we have

$$T = \tfrac{1}{2}m\dot{x}^2 + \tfrac{1}{2}m\dot{y}^2 + \tfrac{1}{2}I_d\dot{\varphi}_x^2 + \tfrac{1}{2}I_d\dot{\varphi}_y^2 \tag{4.233}$$

$$U = \tfrac{1}{2}(2k_x)x^2 + \tfrac{1}{2}(2k_y)y^2 + \tfrac{1}{2}(k_x l^2/2)\varphi_x^2 + \tfrac{1}{2}(k_y l^2/2)\varphi_y^2 \tag{4.234}$$

$$\delta W_{nc} = (me\omega^2\cos\omega t)\delta x + (me\omega^2\sin\omega t)\delta y + (me\omega^2 e_z\cos\omega t)\delta\varphi_y + (me\omega^2 e_z\sin\omega t)\delta\varphi_x \tag{4.235}$$

$$= Q_{ncx}\delta x + Q_{ncy}\delta y + Q_{nc\varphi_y}\delta\varphi_y + +Q_{nc\varphi_x}\delta\varphi_x$$

where T is the kinetic energy, U is the potential energy, δW_{nc} is the non-conservative virtual work done. We have the following Lagrange's equations corresponding to four generalised coordinates (x, y, φ_x, φ_y):

$$\frac{d}{dt}\left(\frac{\partial T}{\partial \dot{x}}\right) + \frac{\partial U}{\partial x} = Q_{ncx}; \quad \frac{d}{dt}\left(\frac{\partial T}{\partial \dot{y}}\right) + \frac{\partial U}{\partial y} = Q_{ncy}; \tag{4.236}$$

$$\frac{d}{dt}\left(\frac{\partial T}{\partial \dot{\varphi}_y}\right) + \frac{\partial U}{\partial \varphi_y} = Q_{nc\varphi_y}; \quad \frac{d}{dt}\left(\frac{\partial T}{\partial \dot{\varphi}_x}\right) + \frac{\partial U}{\partial \varphi_x} = Q_{nc\varphi_x} \tag{4.237}$$

On taking derivatives of components of Eqs. (4.236) and (4.237), we get

$$\frac{\partial T}{\partial \dot{x}} = m\dot{x}; \quad \frac{\partial U}{\partial x} = 2k_x x; \quad \frac{\partial T}{\partial \dot{y}} = m\dot{y}; \quad \frac{\partial U}{\partial y} = 2k_y y \tag{4.238}$$

$$\frac{\partial T}{\partial \dot{\varphi}_y} = I_d\dot{\varphi}_y; \quad \frac{\partial U}{\partial \varphi_y} = -0.5k_y l^2\varphi_y; \quad \frac{\partial T}{\partial \dot{\varphi}_x} = I_d\dot{\varphi}_x; \quad \frac{\partial L}{\partial \varphi_x} = -0.5k_x l^2\varphi_x \tag{4.239}$$

Substituting Eqs. (4.238) and (4.239) into the Lagrange's equations (Eqs. (4.236) and (4.237)), we get the equations of motion as

$$m\ddot{x} + 2k_x x = me\omega^2\cos\omega t; \quad m\ddot{y} + 2k_y y = me\omega^2\sin\omega t \tag{4.240}$$

$$I_d\ddot{\varphi}_y + 0.5k_y l^2\varphi_y = me\omega^2 e_z\cos\omega t; \quad I_d\ddot{\varphi}_x + 0.5k_x l^2\varphi_x = me\omega^2 e_z\sin\omega t \tag{4.241}$$

On combining Eqs. (4.240) through (4.241) into a matrix form, we get

$$\begin{bmatrix} m & 0 & 0 & 0 \\ 0 & m & 0 & 0 \\ 0 & 0 & I_d & 0 \\ 0 & 0 & 0 & I_d \end{bmatrix} \begin{Bmatrix} \ddot{x} \\ \ddot{y} \\ \ddot{\varphi}_y \\ \ddot{\varphi}_x \end{Bmatrix} + \begin{bmatrix} 2k_x & 0 & 0 & 0 \\ 0 & 2k_y & 0 & 0 \\ 0 & 0 & 0.5k_y l^2 & 0 \\ 0 & 0 & 0 & 0.5k_x l^2 \end{bmatrix} \begin{Bmatrix} x \\ y \\ \varphi_y \\ \varphi_x \end{Bmatrix} = \begin{Bmatrix} me\omega^2 \cos \omega t \\ me\omega^2 \sin \omega t \\ me\omega^2 e_z \cos \omega t \\ me\omega^2 e_z \sin \omega t \end{Bmatrix} \tag{4.242}$$

Both the mass and stiffness matrices are uncoupled and in fact all of them can be solved independently of each other just like single DOF system.

Exercise 4.14 For a symmetrical long rigid shaft on anisotropic bearings, the following energy expressions are given:

$$T = \tfrac{1}{2} m\dot{x}^2 + \tfrac{1}{2} m\dot{y}^2 + \tfrac{1}{2} I_d \dot{\varphi}_x^2 + \tfrac{1}{2} I_d \dot{\varphi}_y^2;$$

$$U = \tfrac{1}{2}(2k_{xx})x^2 + \tfrac{1}{2}(2k_{yy})y^2 + \tfrac{1}{2}\left(k_{xx}l^2 / 2\right)\varphi_x^2 + \tfrac{1}{2}\left(k_{yy}l^2 / 2\right)\varphi_y^2;$$

$$\begin{aligned} \delta W_{nc} = &\left(me\omega^2 \cos \omega t - 2k_{xy}y - 2c_{xx}\dot{x} - 2c_{xy}\dot{y}\right)\delta x + \left(me\omega^2 \sin \omega t - 2k_{yx}x - 2c_{yy}\dot{y} - 2c_{yx}\dot{x}\right)\delta y \\ &+ \left(me\omega^2 e_z \cos \omega t - 0.5l^2 k_{xy}\varphi_x - 0.5l^2 c_{xx}\dot{\varphi}_y - 0.5l^2 c_{xy}\dot{\varphi}_x\right)\delta\varphi_y \\ &+ \left(me\omega^2 e_z \sin \omega t - 0.5l^2 k_{yx}\varphi_y - 0.5l^2 c_{yy}\dot{\varphi}_x - 0.5l^2 c_{yx}\dot{\varphi}_y\right)\delta\varphi_x \end{aligned}$$

where T is the kinetic energy, U is the potential energy, δW_{nc} is the non-conservative virtual work done, x and y are the coordinates of rotor geometrical centre (i.e., generalised coordinates), φ_x and φ_y are rotational coordinates of the rotor (i.e., generalised coordinates), e is the radial eccentricity, e_z is the axial eccentricity, m is the mass of the rotor, k_{xx} and k_{yy} are the direct stiffness of each bearing, k_{xy} and k_{yx} are the cross-coupled stiffness of each bearing ($k_{xy} \neq k_{yx}$), c_{xx} and c_{yy} are the direct damping of each bearing, c_{xy} and c_{yx} are the cross-coupled damping of each bearing ($c_{xy} \neq c_{yx}$), l is the length of the rotor, I_d is the diametral mass moment of inertia of the rotor, and ω is the spin speed of the rotor. Using Lagrange's equation (refer chapter 7 of Tiwari (2017)) obtain equations of motion of the rotor system.

***Solution*:** The following energy terms are given

$$T = \tfrac{1}{2} m\dot{x}^2 + \tfrac{1}{2} m\dot{y}^2 + \tfrac{1}{2} I_d \dot{\varphi}_x^2 + \tfrac{1}{2} I_d \dot{\varphi}_y^2 \tag{4.243}$$

$$U = \tfrac{1}{2}(2k_{xx})x^2 + \tfrac{1}{2}(2k_{yy})y^2 + \tfrac{1}{2}\left(k_{xx}l^2 / 2\right)\varphi_x^2 + \tfrac{1}{2}\left(k_{yy}l^2 / 2\right)\varphi_y^2 \tag{4.244}$$

$$\begin{aligned} \delta W_{nc} = &\left(me\omega^2 \cos \omega t - 2k_{xy}y - 2c_{xx}\dot{x} - 2c_{xy}\dot{y}\right)\delta x + \left(me\omega^2 \sin \omega t - 2k_{yx}x - 2c_{yy}\dot{y} - 2c_{yx}\dot{x}\right)\delta y \\ &+ \left(me\omega^2 e_z \cos \omega t - 0.5l^2 k_{xy}\varphi_x - 0.5l^2 c_{xx}\dot{\varphi}_y - 0.5l^2 c_{xy}\dot{\varphi}_x\right)\delta\varphi_y \\ &+ \left(me\omega^2 e_z \sin \omega t - 0.5l^2 k_{yx}\varphi_y - 0.5l^2 c_{yy}\dot{\varphi}_x - 0.5l^2 c_{yx}\dot{\varphi}_y\right)\delta\varphi_x \\ &= Q_{ncx}\delta x + Q_{ncy}\delta y + Q_{nc\varphi_y}\delta\varphi_y + +Q_{nc\varphi_x}\delta\varphi_x \end{aligned} \tag{4.245}$$

where T is the kinetic energy, U is the potential energy, δW_{nc} is the non-conservative virtual work done. From Lagrange's equation corresponding to four generalised coordinates (x, y, φ_x, φ_y):

$$\frac{d}{dt}\left(\frac{\partial T}{\partial \dot{x}}\right)+\frac{\partial U}{\partial x}=Q_{ncx};\quad \frac{d}{dt}\left(\frac{\partial T}{\partial \dot{y}}\right)+\frac{\partial U}{\partial y}=Q_{ncy}; \tag{4.246}$$

$$\frac{d}{dt}\left(\frac{\partial T}{\partial \dot{\varphi}_y}\right)+\frac{\partial U}{\partial \varphi_y}=Q_{nc\varphi_y};\quad \frac{d}{dt}\left(\frac{\partial T}{\partial \dot{\varphi}_x}\right)+\frac{\partial U}{\partial \varphi_x}=Q_{nc\varphi_x} \tag{4.247}$$

On taking derivatives of components of Eqs. (4.246) and (4.247), we get

$$\frac{\partial T}{\partial \dot{x}}=m\dot{x};\quad \frac{\partial U}{\partial x}=2k_{xx}x;\quad \frac{\partial T}{\partial \dot{y}}=m\dot{y};\quad \frac{\partial U}{\partial y}=2k_{yy}y;\quad \frac{\partial T}{\partial \dot{\varphi}_y}=I_d\dot{\varphi}_y \tag{4.248}$$

$$\frac{\partial U}{\partial x}=-0.5k_{xx}l^2\varphi;\quad \frac{\partial T}{\partial \dot{y}}=I_d\dot{\varphi}_y;\quad \frac{\partial L}{\partial \varphi}=-0.5k_{xx}l^2\varphi \tag{4.249}$$

On substituting Eqs. (4.248) and (4.249) into the Lagrange's equations (Eqs. (4.246) and (4.247)), we get the equations of motion as

$$m\ddot{x}+2c_{xx}\dot{x}+2c_{xy}\dot{y}+2k_{xx}x+2k_{xy}y=me\omega^2\cos\omega t \tag{4.250}$$

$$m\ddot{y}+2c_{yy}\dot{y}+2c_{yx}\dot{x}+2k_{y}y+2k_{yx}x=me\omega^2\sin\omega t \tag{4.251}$$

$$I_d\ddot{\varphi}_y+0.5c_{yy}l^2\dot{\varphi}_x+0.5c_{yx}l^2\dot{\varphi}_y+0.5k_{yy}l^2\varphi_y+0.5k_{yx}l^2\varphi_y=me\omega^2e_z\cos\omega t \tag{4.252}$$

$$I_d\ddot{\varphi}_x+0.5c_{xx}l^2\dot{\varphi}_y+0.5c_{xy}l^2\dot{\varphi}_x+0.5k_{xx}l^2\varphi_x+0.5k_{xy}l^2\varphi_x=me\omega^2e_z\sin\omega t \tag{4.253}$$

On combining Eqs. (4.250) through (4.253) into a matrix form, we get

$$\begin{bmatrix} m & 0 & 0 & 0\\ 0 & m & 0 & 0\\ 0 & 0 & I_d & 0\\ 0 & 0 & 0 & I_d \end{bmatrix}\begin{Bmatrix}\ddot{x}\\ \ddot{y}\\ \ddot{\varphi}_y\\ \ddot{\varphi}_x\end{Bmatrix}+\begin{bmatrix} 2c_{xx} & 2c_{xy} & 0 & 0\\ 2c_{yx} & 2c_{yy} & 0 & 0\\ 0 & 0 & 0.5c_{yy}l^2 & 0.5c_{yx}l^2\\ 0 & 0 & 0.5c_{xy}l^2 & 0.5c_{xx}l^2 \end{bmatrix}\begin{Bmatrix}\dot{x}\\ \dot{y}\\ \dot{\varphi}_y\\ \dot{\varphi}_x\end{Bmatrix}$$
$$+\begin{bmatrix} 2k_{xx} & 2k_{xy} & 0 & 0\\ 2k_{yx} & 2k_{yy} & 0 & 0\\ 0 & 0 & 0.5k_{yy}l^2 & 0.5k_{yx}l^2\\ 0 & 0 & 0.5k_{xy}l^2 & 0.5k_{xx}l^2 \end{bmatrix}\begin{Bmatrix}x\\ y\\ \varphi_y\\ \varphi_x\end{Bmatrix}=\begin{Bmatrix}me\omega^2\cos\omega t\\ me\omega^2\sin\omega t\\ me\omega^2e_z\cos\omega t\\ me\omega^2e_z\sin\omega t\end{Bmatrix} \tag{4.254}$$

For the present case, the stiffness matrice is uncoupled, and in fact, all of them cannot be solved independently of each other. And it can be solved for translational motion separately and rotational motion separately. There is no coupling between the translational motion and the rotational motion; however, two orthogonal translational motions are coupled, and similarly, two orthogonal rotational motions are coupled.

Exercise 4.15 Let $x(t)=\sin(mt)$ and $y(t)=\sin(nt+\phi)$, where $m=1, 2, 3, \ldots, n=1, 2, 3, 4, \ldots,$ phase $\phi=(0-360)$. $x(t)$ and $y(t)$ could be thought as transverse displacements of a rotor. Plot (i) $x(t)$ versus t, (ii) $y(t)$ versus t and (iii) $x(t)$ versus $y(t)$. Vary various variables and see the plots (Lissajous figures) and interpret them. For example For $\phi=0$ (i) $m=1$, $n=1, 2, \ldots, 5$ (ii) $m=2$, $n=1, 2, \ldots, 5$ (iii) $m=3$, $n=1, 2, \ldots, 5$ (iv) $m=4$, $n=1, 2, \ldots, 5$ (v) $m=5$, $n=1, 2, \ldots, 5$. Similarly for $\phi=0, \pi/4, \pi/2, 3\pi/4, \pi, 5\pi/4, 3\pi/2, 7\pi/4, 2\pi$.

Solution: Plots for various parameters using the given equations and for various cases are shown in Figure 4.47. It be drawn in any convenient software.

Exercise 4.16 Obtain the transverse natural frequencies of a long, rigid, tapered rotor mounted on dissimilar springs. The length of the rotor is 1 m, and diameters of two ends are 0.10 m (LHS) and 0.15 m (RHS), respectively. The rotor is supported on springs at ends (L: left and R: right) of the rotor with $k_L=1$ kN/m and $k_R=1.2$ kN/m. Consider a single-plane motion only. The density of the rotor material is 7,800 kg/m^3. Give the steps for formulations and state the assumptions that are made.

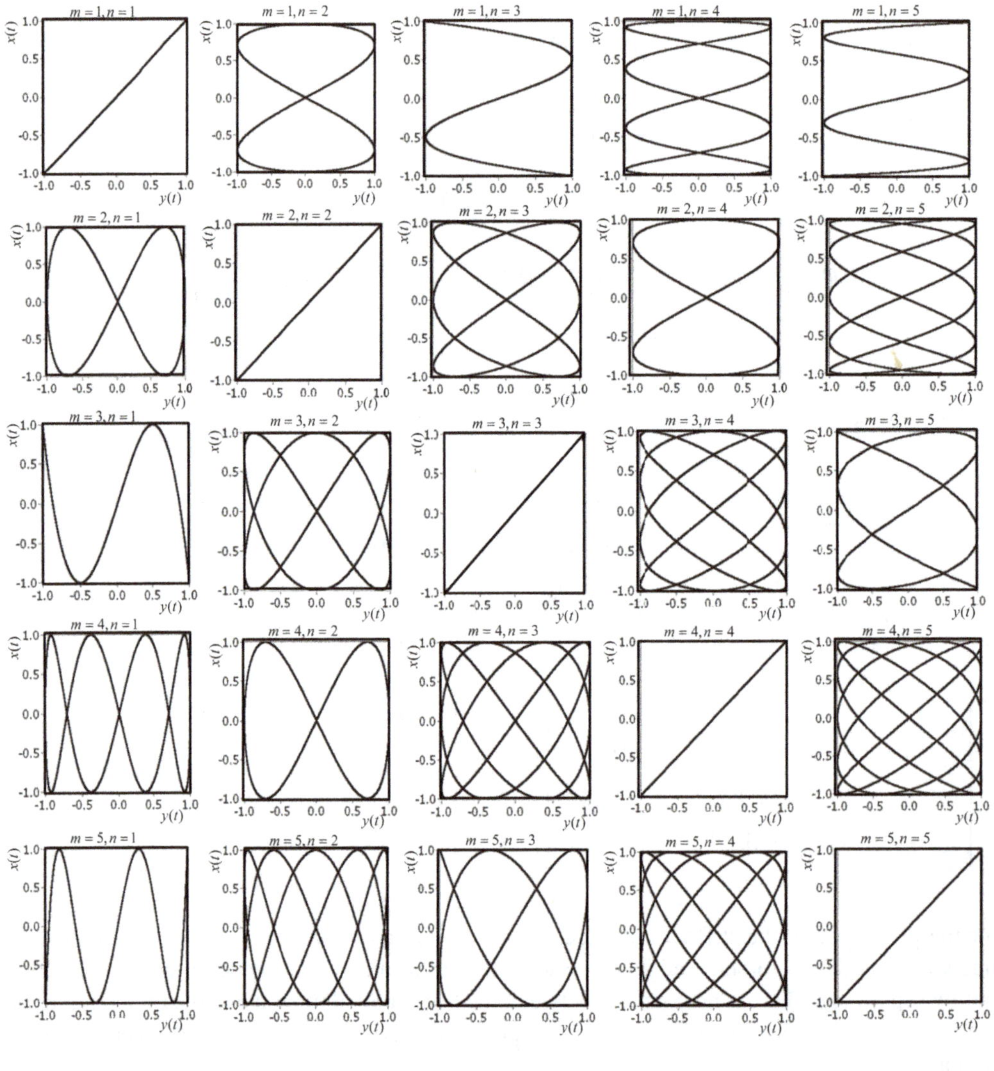

FIGURE 4.47 Lissajous plots for various parameters.

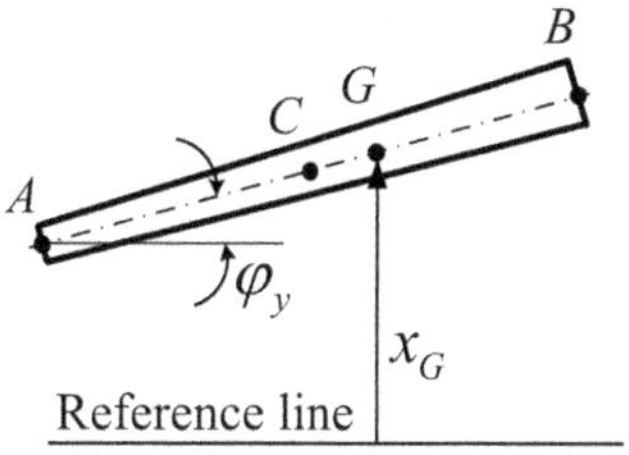

FIGURE 4.48 Deflected rigid shaft with generalized coordinates (x_G, φ_y).

***Solution*:** Generalised co-ordinates (x_G, φ_y) are chosen as shown in Figure 4.48.
The kinetic energy (KE) of the rotor system for the present case is

$$T = \tfrac{1}{2} m\dot{x}_G^2 + \tfrac{1}{2} I_{dG}\dot{\varphi}_y^2 \tag{4.255}$$

where m is the mass of the rotor and I_{dG} is the diametral mass moment of inertia of the rotor about its centre of gravity, G. The potential energy (PE) of the rotor system is

$$U = \tfrac{1}{2} k_L \left(x_G - l_{AG}\varphi_y\right)^2 + \tfrac{1}{2} k_R \left(x_G + l_{BG}\varphi_y\right)^2 \tag{4.256}$$

where k_L and k_R are the stiffness of support at the left and right bearings, respectively, and l_{AG} and l_{BG} are the distances AG and BG, respectively. For getting equations of motion from Lagrange's equation for the generalised coordinate, x_G, we have,

$$\frac{d}{dt}\left(\frac{\partial T}{\partial \dot{x}_G}\right) - \frac{\partial T}{\partial x_G} + \frac{\partial U}{\partial x_G} = 0 \tag{4.257}$$

On substituting Eqs. (4.255) and (4.256) into parts of Eq. (4.257), we get

$$\frac{d}{dt}\left(\frac{\partial T}{\partial \dot{x}_G}\right) = \frac{d}{dt}\left\{\frac{\partial}{\partial \dot{x}_G}\left(\frac{1}{2} m\dot{x}_G^2 + \frac{1}{2} I_{dG}\dot{\varphi}_y^2\right)\right\} = m\ddot{x}_G \tag{4.258}$$

and

$$\begin{aligned}\frac{\partial U}{\partial x_G} &= \frac{\partial}{\partial x_G}\left\{\frac{1}{2} k_L \left(x_G - l_{AG}\varphi_y\right)^2 + \frac{1}{2} k_R \left(x_G + l_{BG}\varphi_y\right)^2\right\} = k_L\left(x_G - l_{AG}\varphi_y\right) + k_R\left(x_G + l_{BG}\varphi_y\right) \\ &= \left(k_L + k_R\right)x_G - \left\{k_L l_{AG} - k_R l_{BG}\right\}\varphi_y\end{aligned} \tag{4.259}$$

On substituting Eqs. (4.258) and (4.259) into Eq. (4.257), we get

$$m\ddot{x}_G + \left(k_L + k_R\right)x_G - \left(k_L l_{AG} - k_R l_{BG}\right)\varphi_y = 0 \tag{4.260}$$

Similarly, for the generalised coordinate, φ_y, we have

$$\frac{d}{dt}\left(\frac{\partial T}{\partial \dot{\varphi}_y}\right) - \frac{\partial T}{\partial \varphi_y} + \frac{\partial U}{\partial \varphi_y} = 0 \tag{4.261}$$

On substituting Eqs. (4.255) and (4.256) into parts of Eq. (4.261), we get

$$\frac{d}{dt}\left(\frac{\partial T}{\partial \dot{\varphi}_y}\right)=\frac{d}{dt}\left[\frac{\partial}{\partial \dot{\varphi}_y}\left(\frac{1}{2}m\dot{x}_G^2+\frac{1}{2}I_{dG}\dot{\varphi}_y^2\right)\right]=I_{dG}\ddot{\varphi}_y \tag{4.262}$$

and

$$\begin{aligned}\frac{\partial U}{\partial \varphi_y}&=\frac{\partial}{\partial \varphi_y}\left\{\frac{1}{2}k_L\left(x_G-l_{AG}\varphi_y\right)^2+\frac{1}{2}k_R\left(x_G+l_{BG}\varphi_y\right)^2\right\}\\&=-\left(k_L l_{AG}-k_R l_{BG}\right)x_G+\left(k_L l_{AG}^2+k_R l_{BG}^2\right)\varphi_y\end{aligned} \tag{4.263}$$

On substituting Eqs. (4.262) and (4.263) into Eq. (4.261), we get

$$I_{dG}\ddot{\varphi}_y-\left(k_L l_{AG}-k_R l_{BG}\right)x_G+\left(k_L l_{AG}^2+k_R l_{BG}^2\right)\varphi_y=0 \tag{4.264}$$

Equations (4.260) and (4.264), in the matrix form can be written as,

$$\begin{bmatrix} m & 0\\ 0 & I_{dG}\end{bmatrix}\begin{Bmatrix}\ddot{x}_G\\ \ddot{\varphi}_y\end{Bmatrix}+\begin{bmatrix}(k_L+k_R) & -\left(k_L l_{AG}-k_R l_{BG}\right)\\ -\left(k_L l_{AG}-k_R l_{BG}\right) & \left(k_L l_{AG}^2+k_R l_{BG}^2\right)\end{bmatrix}\begin{Bmatrix}x_G\\ \varphi_y\end{Bmatrix}=\begin{Bmatrix}0\\0\end{Bmatrix} \tag{4.265}$$

For a tapered shaft, the inertia properties are obtained now. In Figure 4.49, point G is the centre of gravity of the tapered shaft and origin is taken of axis x–y at G. The origin of the shaft from the right end face is given as

$$l_2=\frac{r_1^2+2r_1r_2+3r_2^2}{4\left(r_1^2+r_1r_2+r_2^2\right)}l \quad \text{and} \quad l_1=l-l_2 \tag{4.266}$$

Now since the tapered line can be represented as

$$y=az+b \tag{4.267}$$

where a and b are constants and can be obtained by following boundary conditions:

$$\text{at} \quad z=l_2;\quad y=r_2;\quad \text{and} \quad \text{at} \quad z=-l_1;\quad y=r_1 \tag{4.268}$$

so that

$$r_2=al_2+b \quad \text{and} \quad r_1=-al_1+b \tag{4.269}$$

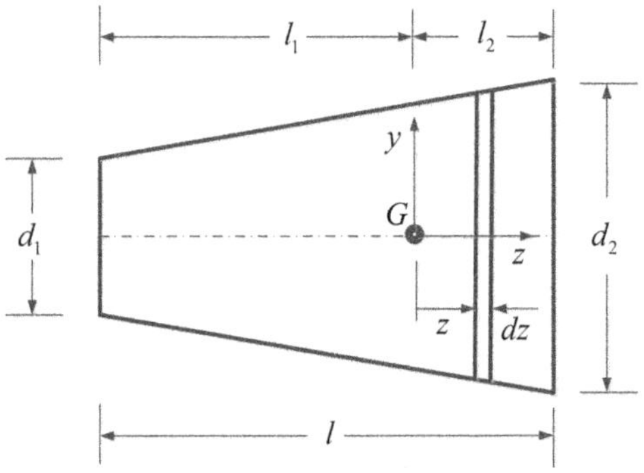

FIGURE 4.49 Tapered shaft geometry.

which gives

$$a = \frac{r_2 - r_1}{l} \quad \text{and} \quad b = \frac{l_1 r_2 + l_2 r_1}{l} \tag{4.270}$$

The diametral mass moment of inertia of a small section of the tapered shaft is given as

$$dI_{dx} = dI_{dy} = \tfrac{1}{4}\rho\pi y^4 dz + \rho\pi y^2 z^2 dz \tag{4.271}$$

So on integrating over the domain of the tapered shaft, we get

$$I_{dx} = I_{dy} = \int_{-l_1}^{l_2} \tfrac{1}{4}\rho\pi y^4\, dz + \int_{-l_1}^{l_2} \rho\pi y^2 z^2\, dz \tag{4.272}$$

Herein, y is substituted from Eq. (4.267) in terms of x. The closed form of the diametral mass moment of inertia is very cumbersome so not present here. For the present rotor system, the following data are given:

$$l_{AB} = l = 1\text{ m};\quad d_L = 0.10\text{ m};\quad d_R = 0.15\text{ m};\quad k_L = 1.0\times10^3\text{ N/m};\quad k_R = 1.2\times10^3\text{ N/m}.$$

We can have the following properties from the given data

$$V = \tfrac{\pi}{3} l\left(r_L^2 + r_R^2 + r_L r_R\right) = 0.0124\text{ m}^3;\quad m = \rho V = 96.9967\text{ kg};\quad I_d = 7.8493\text{ kg-m}^2;$$

$$l_{BG} = l_2 = 0.4342\text{ m};\quad l_{AG} = l - l_{BG} = 1 - 0.4342 = 0.5658\text{ m}.$$

For given values, equations of motion (4.265) are given as

$$\begin{bmatrix} 96.9967 & 0 \\ 0 & 7.8493 \end{bmatrix}\begin{Bmatrix} \ddot{x}_G \\ \ddot{\varphi}_y \end{Bmatrix} + 10^3\begin{bmatrix} 2.2000 & 0.0447 \\ 0.0447 & 0.5464 \end{bmatrix}\begin{Bmatrix} x_G \\ \varphi_y \end{Bmatrix} = \begin{Bmatrix} 0 \\ 0 \end{Bmatrix} \tag{4.273}$$

For the free vibration, we have

$$\left(-\omega_{nf}^2\begin{bmatrix} 96.9967 & 0 \\ 0 & 7.8493 \end{bmatrix} + 10^3\begin{bmatrix} 2.2000 & 0.0447 \\ 0.0447 & 0.5464 \end{bmatrix}\right)\begin{Bmatrix} x_G \\ \varphi_y \end{Bmatrix} = \begin{Bmatrix} 0 \\ 0 \end{Bmatrix} \tag{4.274}$$

which gives natural frequencies as

$$\omega_{nf1} = 4.76\text{ rad/s} \quad \text{and} \quad \omega_{nf2} = 8.35\text{ rad/s} \tag{4.275}$$

Exercise 4.17 Briefly distinguish (i) whirl frequency and critical speed and (ii) translatory and conical whirls.

***Solution*:** **(i)** Whirl Frequency: The rotatory motion of bent shaft (or rigid rotor on flexible bearings) about its bearing axis is called the whirling motion. The time rate of whirling is called whirling frequency. Depending on the whirling frequency, the whirling is further categorised as the synchronous whirling (refer Figure 4.50a) or anti-synchronous (refer Figure 4.50b) and asynchronous whirling.

Critical Speed: The spinning speed of the rotor at which it is in resonance condition is called the critical speed. At this speed, the whirl amplitude is at its peak value (refer Figure 4.51). A rotor shaft can have many critical speeds as the shaft-bearing system is a multi-DOF system. A rotor system operating speed should be designed in such a way that it avoids the critical speed. The critical speed of the rotor system depends on mass distribution, and its stiffness and the damping.

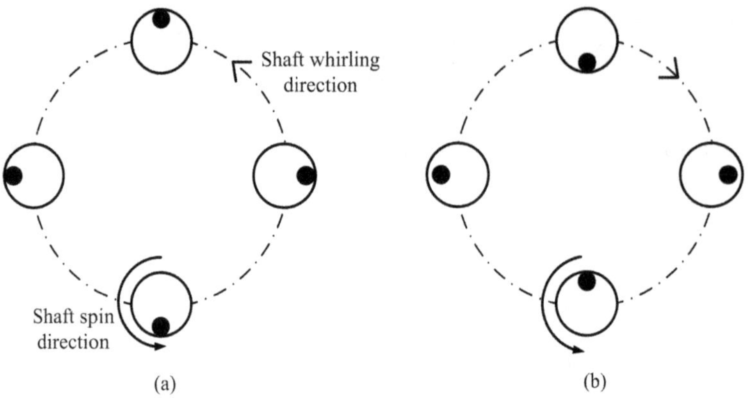

FIGURE 4.50 Whirling of shaft (a) synchronous whirl (b) anti-synchronous whirl.

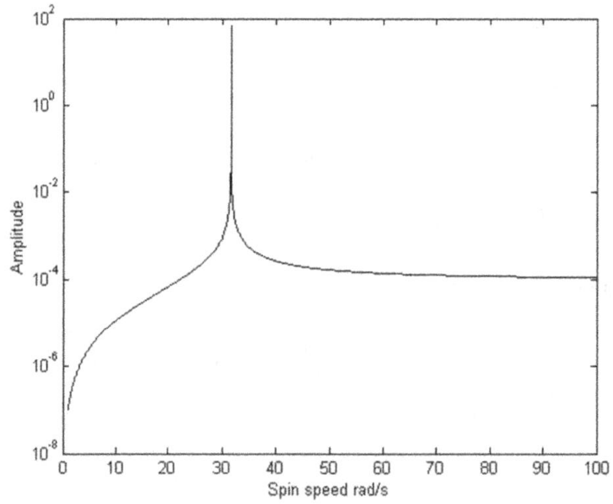

FIGURE 4.51 Plot of rotor amplitude with spin speed showing critical speed.

(ii) Translatory whirl: The whirling motion without any tilting of the rotor is called transverse translatory whirling.

$$\frac{x^2}{X^2} + \frac{y^2}{Y^2} = 1 \tag{4.276}$$

In the pure transverse translational motion, each particle of the rotor will have an elliptical path or orbit or trajectory (refer Figure 4.52). Such motion would be possible with the radial eccentricity, $e \neq 0$, and the axial eccentricity, $e_z = 0$. Additionally, the rotor has to operate near corresponding critical speed.

Conical Whirling: The whirling motion without any transverse translatory motion of the rotor is called conical whirling.

$$\frac{\varphi_x^{\,2}}{\phi_x^{\,2}} + \frac{\varphi_y^{\,2}}{\phi_y^{\,2}} = 1 \tag{4.277}$$

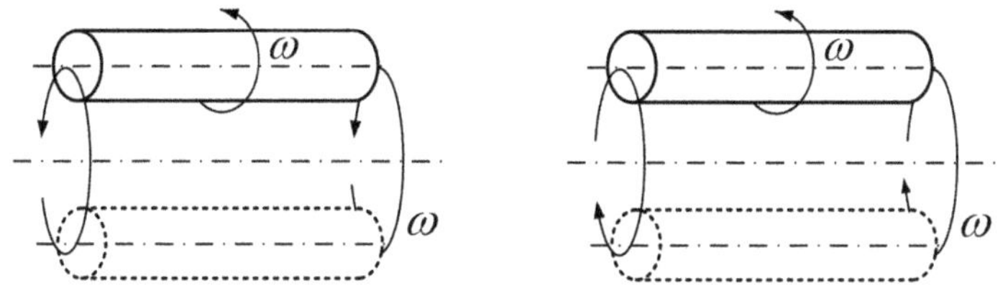

FIGURE 4.52 A pure transverse translational motion (or cylindrical whirl) (a) synchronous whirl (b) anti-synchronous whirl.

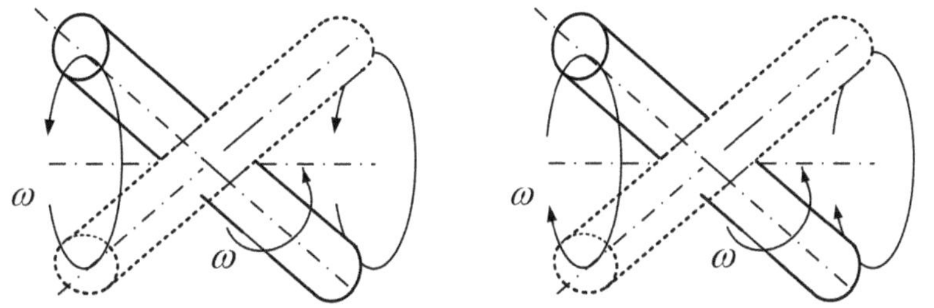

FIGURE 4.53 A pure transverse rotational motion (or conical whirling) (a) synchronous whirl (b) anti-synchronous whirl.

For the pure rotational (tilting) motion in the transverse planes, each particle of the rotor will have an elliptical path (refer Figure 4.53). Such motion would be possible with $e_z \neq 0$ and $e = 0$ (additionally, the rotor has to operate near corresponding critical speed).

Exercise 4.18 Obtain equations of motion of a rotor-bearing system shown in Figure 4.54 in explicit form. Assume single-plane motion with both translational and tilting motions of the disc, however, ignore the gyroscopic effect. Consider the shaft to be massless and flexible with EI as the flexural rigidity. Let m be the mass of the disc and I_d be the diametral mass moment of inertia of the disc.

***Solution*:** The present problem is solved herein by considering (i) only the translational motion of the disc and (ii) by considering both translational and rotational (tilting) motions.
(i) Only translational motion of the disc: We consider only vertical plane motion. The flexible shaft on flexible bearings (Figure 4.55a) have been considered in two cases, *Case I*: rigid shaft on flexible bearings (Figure 4.55b) and *Case II*: Flexible shaft on *rigid* bearings (Figure 4.55c).

Case-1: Rigid shaft on flexible bearings (refer Figure 4.56):

A force P is applied on the shaft at the disc location and that will be resisted by the springs (i.e. bearings), as

$$P = k_{b1} y_{b1} + k_{b2} y_{b2} \tag{4.278}$$

where k_{b1} and k_{b2} are the stiffness of the left and right bearings, respectively, and y_{b1} and y_{b2} are displacement of the shaft at the left and right bearing locations, respectively. Also, the moment balance about the location of the applied force P (i.e. the disc location) will be given as

$$k_{b1} y_{b1} a = k_{b2} y_{b2} b \tag{4.279}$$

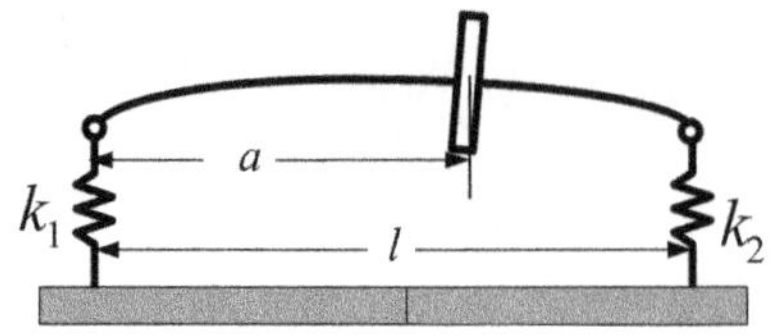

FIGURE 4.54 A flexible rotor on flexible bearings.

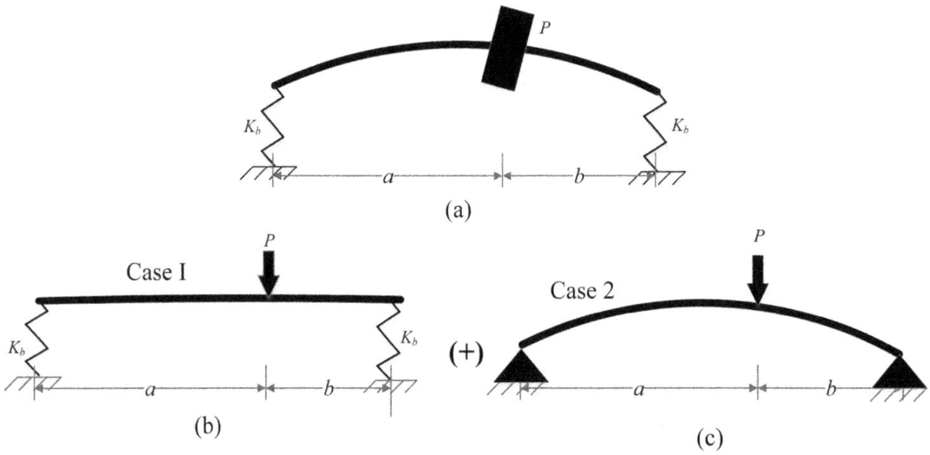

FIGURE 4.55 (a) Flexible shaft on flexible bearings (b) rigid shaft on flexible bearings and (c) flexible shaft on rigid bearings.

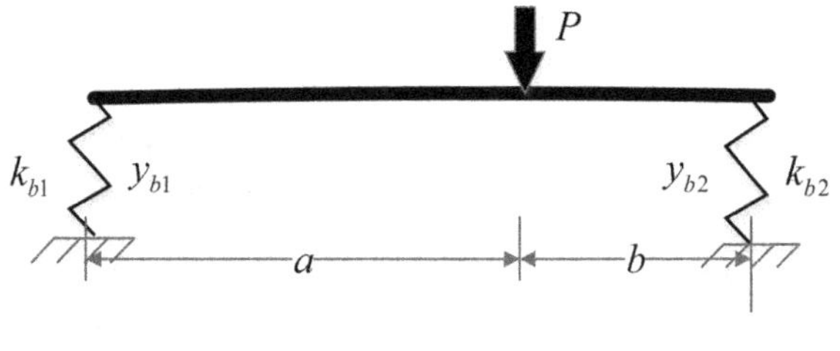

FIGURE 4.56 Rigid shaft on flexible bearings.

where a and b are distances of disc from the left and right bearings, respectively. Equation (4.279) can be written, as

$$y_{b1} = \frac{k_{b2} b y_{b2}}{k_{b1} a} \tag{4.280}$$

On substituting Eq. (4.280) into Eq. (4.278), we get

$$P = k_{b1} \frac{k_{b2} b y_{b2}}{k_{b1} a} + k_{b2} y_{b2} \Rightarrow P = k_{b2} y_{b2} \frac{l}{a} \Rightarrow y_{b2} = \frac{Pa}{k_{b2} l} \tag{4.281}$$

On substituting Eq. (4.281) into Eq. (4.279), we get

$$k_{b1} y_{b1} = P - \frac{Pa}{k_{b2} l} k_{b2} = P \frac{b}{l} \Rightarrow y_{b1} = \frac{Pb}{k_{b1} l} \tag{4.282}$$

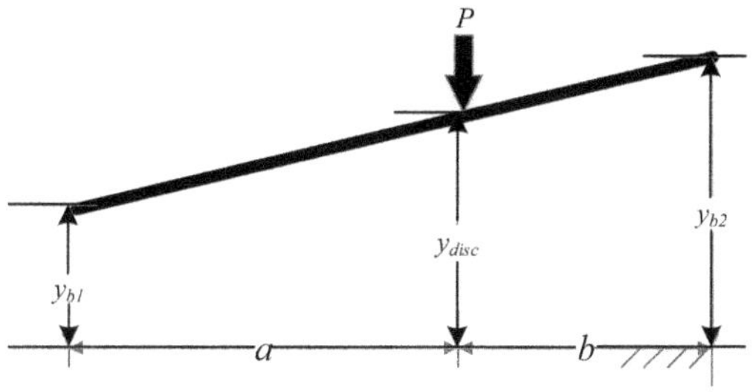

FIGURE 4.57 Displacement pattern of the rigid shaft on flexible supports.

To derive displacement at disc location y_{disc1} in terms of y_{b1} and y_{b2}, while still considering shaft as rigid (consider Figure 4.57).

For the case of a deflected rigid shaft slope is same along the length, so we have

$$\frac{y_{b2}-y_{disc1}}{b}=\frac{y_{disc1}-y_{b1}}{a}\Rightarrow ay_{b2}-ay_{disc1}=by_{disc1}-by_{b1}\Rightarrow y_{disc1}=\frac{ay_{b2}+by_{b1}}{l} \quad (4.283)$$

On substituting Eqs. (4.281) and (4.282), we get

$$y_{disc1}=\frac{P(k_{b1}a^2+k_{b2}b^2)}{k_{b1}k_{b2}l^2} \quad (4.284)$$

So the aforementioned equation gives the shaft displacement at the disc location due to deflections at bearings while considering the shaft as rigid. Now we will consider the flexible shaft and rigid bearings.

Case 2: Flexible shaft and rigid bearings (consider Figure 4.58)

For a flexible shaft on rigid bearings, the deflection at any point along the length, z, is given by (Appendix 2.1)

$$y(z)=\frac{Pbz}{6lEI}(l^2-z^2-b^2) \quad \text{for} \quad 0<z<a \quad (4.285)$$

The deflection at offset disc location (Figure 4.58) is given as (for $z=a$, displacement at the location of the force)

$$y_{disc2}=\frac{Pbz}{6lEI}(l^2-z^2-b^2)=\frac{Pba}{6lEI}(l^2-a^2-b^2)=\frac{Pab}{6lEI}(ab)=\frac{Pa^2b^2}{3lEI} \quad \text{with} \quad l=a+b \quad (4.286)$$

The total deflection at the disc due to bearing as well as the shaft flexibility, noting Eqs. (4.284) and (4.286), we get

$$y_{disc}=y_{disc}=y_{disc1}+y_{disc2}=\frac{P(k_{b1}a^2+k_{b2}b^2)}{k_{b1}k_{b2}l^2}+\frac{Pa^2b^2}{3lEI} \quad (4.287)$$

So, the equivalent stiffness at the disc location is given as

$$k_{eq}=\frac{P}{y_{disc}}=\frac{1}{\dfrac{(k_{b1}a^2+k_{b2}b^2)}{k_{b1}k_{b2}l^2}+\dfrac{a^2b^2}{3lEI}}=\frac{3k_{b1}k_{b2}EIl^3}{3EIl(k_{b1}a^2+k_{b2}b^2)+k_{b1}k_{b2}l^2a^2b^2} \quad (4.288)$$

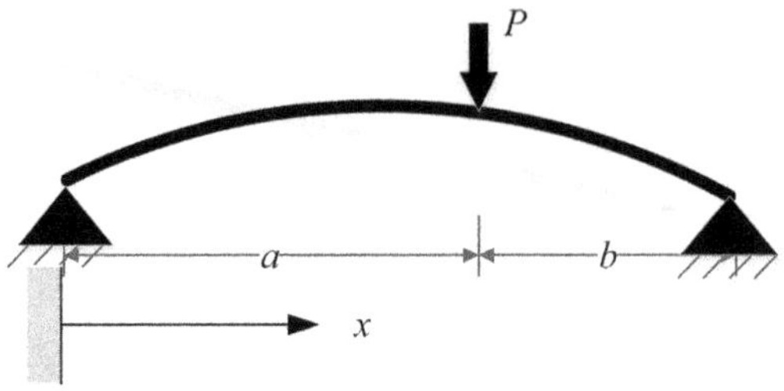

FIGURE 4.58 Defection pattern of flexible shaft on rigid supports.

Hence, the natural frequency is given by

$$\omega_{nf} = \sqrt{\frac{k_{eq}}{m}} = \sqrt{\frac{3k_{b1}k_{b2}EIl^3}{3mEIl(k_{b1}a^2 + k_{b2}b^2) + k_{b1}k_{b2}l^2a^2b^2}} \tag{4.289}$$

For following rotor parameters, let us obtain the natural frequency of the system (without considering the tilting motion of the disc)

$$m = 14\text{ kg};\quad P = 140\text{ N};\quad l = 0.4\text{ m};\quad d = 0.025\text{ m};\quad I = \frac{\pi}{64}d^4 = 1.91\times10^{-8}\text{ m}^4;$$

$$E = 2.1\times10^{11}\text{ N/m}^2;\quad k_{b1} = 5\times10^4\text{ N/m};\quad k_{b1} = 6\times10^4\text{ N/m};\quad a = 0.2\text{ m};\quad b = 0.2\text{ m};$$

$$k_{eq} = 1.0529\times10^5\text{ N/m}.\quad \omega_{nf} = \sqrt{\frac{1.0529\times10^5}{14}} = 86.72\text{ rad/s}.$$

Case-2: A Symmetrical Flexible Shaft on Different Bearings

For the present case, both the shaft and bearings are flexible as shown in Figure 4.55a. All generalised coordinates are with respect to a fixed frame of reference and they are absolute displacements. The motion in a single plane will be considered. The analysis allows finding different instantaneous displacements of the shaft at the disc (offset from the mid–span of shaft) and at bearings. The system will behave in a similar manner to that described in previous method, except that now the tilting of the disc is also considered. An equivalent set of system stiffness coefficients (or influence coefficients) is first evaluated, which allows for the flexibility of the shaft in addition to that of bearings, and is used in place of influence coefficients of the rigid bearing analysis. The total deflection of the disc is the vector sum of the deflection of the disc relative to the shaft ends, plus that of shaft ends in bearings. For the disc, we observe the displacement of its geometrical centre. The deflection of the shaft ends in bearings is related to the force transmitted through bearings by the bearing stiffness coefficients as (for one of the bearing)

$$f_{by} = k_{yy}y_b \tag{4.290}$$

where y_b is the instantaneous displacement of shaft ends relative to bearings in the vertical direction, and for a synchronous vibration it take the following form

$$y_b = Y_b e^{j\omega_{nf}t} \tag{4.291}$$

where Y_b is the complex displacement in y direction and ω_{nf} is the whirl natural frequency. It should be noted that bearings are modelled as a point connection with the shaft and only translational displacements of them are considered, since they support mainly radial (transverse) loads. Bearing forces will have the following form

$$f_{by} = F_{by}e^{j\omega_{nf}t} \tag{4.292}$$

where F_{by} is the complex force in y direction. On substituting in Eqs. (4.291) and (4.292) into Eq. (4.290), we get

$$F_{by} = k_{yy}Y_b \tag{4.293}$$

which is in the frequency domain and can be written in a matrix form for both bearings A (left bearing) and B (right bearing) in single plane motion in vertical y-direction, as

$$\mathbf{F}_b = \mathbf{K}_b\mathbf{Y}_b \quad \text{with} \quad \mathbf{F}_b = \begin{Bmatrix} {}_AF_{by} \\ {}_BF_{by} \end{Bmatrix}; \; \mathbf{K}_b = \begin{bmatrix} {}_Ak_{yy} & 0 \\ 0 & {}_Bk_{yy} \end{bmatrix}; \; \mathbf{Y}_b = \begin{Bmatrix} {}_AY_b \\ {}_BY_b \end{Bmatrix} \tag{4.294}$$

Two bearing motions are not coupled by stiffness coefficients. The magnitude of reaction forces transmitted by bearings can also be evaluated in terms of forces applied to the shaft by the disc (refer Figure 4.59).

From Figure 4.59, the moment balance in the vertical plane will be

$$\sum M_B = 0 \Rightarrow {}_Af_{by}l = f_y(l-a) + M_{yz} \quad \text{or} \quad {}_Af_{by} = f_y\left(1 - a/l\right) + M_{yz}\left(1/l\right) \tag{4.295}$$

$$\text{and} \quad \sum M_A = 0 \Rightarrow {}_Bf_{by}l = f_ya - M_{yz} \quad \text{or} \quad {}_Bf_{by} = f_y\left(a/l\right) - M_{yz}\left(1/l\right) \tag{4.296}$$

Equations (4.295) and (4.296) can be combined in a matrix form, for a single plane motion, as

$$\mathbf{f}_b = \mathbf{A}\mathbf{f}_s \quad \text{with} \quad \mathbf{f}_b = \begin{Bmatrix} {}_Af_{by} \\ {}_Bf_{by} \end{Bmatrix}; \; \mathbf{f}_s = \begin{Bmatrix} f_y \\ M_{yz} \end{Bmatrix}; \; \mathbf{A} = \begin{bmatrix} (1-a/l) & 1/l \\ a/l & -1/l \end{bmatrix} \tag{4.297}$$

For the motion under free vibration, we have

$$\mathbf{f}_b = \mathbf{F}_be^{j\omega_{nf}t} \quad \text{and} \quad \mathbf{f}_s = \mathbf{F}_se^{j\omega_{nf}t} \quad \text{with} \quad \mathbf{F}_b = \begin{Bmatrix} {}_AF_{by} \\ {}_BF_{by} \end{Bmatrix}; \; \mathbf{F}_s = \begin{Bmatrix} F_y \\ \bar{M}_{yz} \end{Bmatrix} \tag{4.298}$$

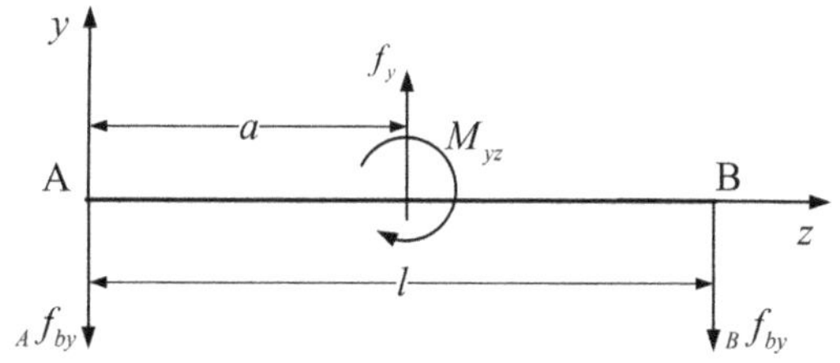

FIGURE 4.59 A free body diagram of the shaft y–z plane.

where subscript b refers to the bearing and s refers to the shaft, and vectors $\mathbf{F}_b$ and $\mathbf{F}_s$ are the complex force vector on the shaft at the bearing and disc locations, respectively. On substituting Eq. (4.298) into Eq. (4.297), we get

$$\mathbf{F}_b = \mathbf{A}\mathbf{F}_s \tag{4.299}$$

which is the frequency domain now (i.e., no time dependency). In Eq. (4.297), bearing forces are related to the reaction forces and moments on the shaft by the disc. On equating Eqs. (4.294) and (4.299), we get

$$\mathbf{F}_b \equiv \mathbf{K}_b\mathbf{Y}_b = \mathbf{A}\mathbf{F}_s \quad \text{or} \quad \mathbf{Y}_b = \mathbf{K}_b^{-1}\mathbf{A}\mathbf{F}_s \tag{4.300}$$

Equation (4.300) relates the shaft end deflections to the reaction forces and moments on the shaft by the disc. The deflection at the location of the disc due to movement of the shaft ends can be obtained as follows. Consider the shaft to be rigid for time being and let us denote the shaft end deflections in the vertical direction to be ${}_Ay_b$ and ${}_By_b$ at ends A and B, respectively, as shown in Figure 4.60. These displacements are assumed to be small.

The translational displacement in the y-direction can be written as (refer Figure 4.60)

$$y = {}_Ay_b - \frac{\left({}_Ay_b - {}_By_b\right)}{l}a = \left(\frac{l-a}{l}\right){}_Ay_b + \left(\frac{a}{l}\right){}_By_b \tag{4.301}$$

And similarly, for the rotational displacements in the y–z plane, we have (refer Figure 4.60)

$$\varphi_x = \frac{\left({}_Ay_b - {}_By_b\right)}{l} = \left(\frac{1}{l}\right){}_Ay_b + \left(-\frac{1}{l}\right){}_By_b \tag{4.302}$$

Equations (4.301) and (4.302) can be combined in a matrix form as

$$\mathbf{u}_{s_1} = \mathbf{B}\mathbf{y}_b \quad \text{with} \quad \mathbf{u}_{s_1} = \begin{Bmatrix} y \\ \varphi_x \end{Bmatrix}_{s_1}; \quad \mathbf{y}_b = \begin{Bmatrix} {}_Ay_b \\ {}_By_b \end{Bmatrix}; \quad \mathbf{B} = \begin{bmatrix} (1-a/l) & a/l \\ 1/l & -1/l \end{bmatrix} \tag{4.303}$$

where subscript s_1 represents that these displacements are due to the rigid body motion of the shaft. For the free vibration analysis, shaft displacements at bearing locations and at the disc centre vary in a sinusoidal motion, such that

$$\mathbf{u}_{s1} = \mathbf{U}_{s1}e^{j\omega_{nf}t} \quad \text{and} \quad \mathbf{y}_b = \mathbf{Y}_b e^{j\omega_{nf}t} \tag{4.304}$$

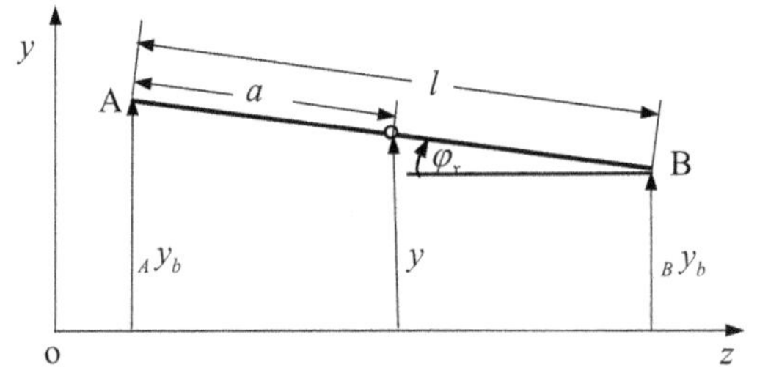

FIGURE 4.60 Rigid body movement of the shaft in y–z plane.

where ω_{nf} is the whirl natural frequency in the case of free vibration. On substituting Eq. (4.304) into Eq. (4.303), we have

$$\mathbf{U}_{s1} = \mathbf{B}\mathbf{Y}_b \tag{4.305}$$

which is in frequency domain. On substituting Eq. (4.300) into Eq. (4.305), we get

$$\mathbf{U}_{s1} \equiv \mathbf{B}\mathbf{Y}_b = \mathbf{B}\mathbf{K}_b^{-1}\mathbf{A}\mathbf{F}_s = \mathbf{C}\mathbf{F}_s \quad \text{with} \quad \mathbf{C} = \mathbf{B}\mathbf{K}_b^{-1}\mathbf{A} \tag{4.306}$$

which gives the deflection of the disc due to the shaft elastic force and moment, when the shaft is rigid. Equation (4.306) will give deflections of the disc that is caused by only the movement of shaft ends (rigid body movement) on flexible bearings. To obtain the net rotor deflection under a given load, we have to add the deflection due to the deformation of the shaft with respect to bearing locations also in Eq. (4.306). The deflection associated with flexure of the shaft alone has already dealt in Chapter 2, which can be combined in a matrix form as

$$\mathbf{u}_{s2} = \boldsymbol{\alpha}\mathbf{f}_s \quad \text{with} \quad \mathbf{u}_{s2} = \begin{Bmatrix} y \\ \varphi_x \end{Bmatrix}_{s2}; \quad \mathbf{f}_s = \begin{Bmatrix} f_y \\ M_{yz} \end{Bmatrix}; \quad \boldsymbol{\alpha} = \begin{bmatrix} \alpha_{11} & \alpha_{12} \\ \alpha_{21} & \alpha_{22} \end{bmatrix} \tag{4.307}$$

where subscript s_2 represents that these displacements are due to the pure deformation of the shaft without any rigid body motion and α_{ij} $(i, j=1, 2)$ is the influence coefficient. For the undamped free vibration analysis, the shaft reaction forces at the disc location and disc displacements vary in a sinusoidal motion and can be expressed as

$$\mathbf{u}_{s2} = \mathbf{U}_{s2}e^{j\omega_{nf}t} \quad \text{and} \quad \mathbf{f}_s = \mathbf{F}_s e^{j\omega_{nf}t} \tag{4.308}$$

On substituting Eq. (4.308) into Eq. (4.307), to get equation in frequency domain, as

$$\mathbf{U}_{s2} = \boldsymbol{\alpha}\mathbf{F}_s \tag{4.309}$$

which is the deflection of disc due to the flexure of the shaft alone, without considering the bearing flexibility. The net deflection of the disc caused by the deflection of bearings plus that due to the flexure of the shaft is then given by (refer Eqs. (4.306) and (4.309))

$$\mathbf{U}_s = \mathbf{U}_{s1} + \mathbf{U}_{s2} = (\mathbf{C}+\boldsymbol{\alpha})\mathbf{F}_s = \mathbf{D}\mathbf{F}_s \quad \text{with} \quad \mathbf{D} = (\mathbf{C}+\boldsymbol{\alpha}) \equiv \begin{bmatrix} d_{11} & d_{12} \\ d_{21} & d_{22} \end{bmatrix} \quad \text{and} \quad d_{12} \neq d_{21} \tag{4.310}$$

where $\mathbf{U}_s$ contains absolute displacements of the shaft at the location of the disc. Equation (4.310) describes the displacements of the shaft at the disc under the action of forces and moments applied at the disc (hence the matrix **D** is effective influence coefficient matrix due to bearing **C**, and shaft $\boldsymbol{\alpha}$). On substituting matrcies **B**, $\mathbf{K}_b$ and **A** into expression of matrix **C**, we get

$$\mathbf{C} = \mathbf{B}\mathbf{K}_b^{-1}\mathbf{A} = \begin{bmatrix} \left(1-\frac{a}{l}\right) & \frac{a}{l} \\ \frac{1}{l} & -\frac{1}{l} \end{bmatrix} \begin{bmatrix} \frac{1}{k_1} & 0 \\ 0 & \frac{1}{k_2} \end{bmatrix} \begin{bmatrix} \left(1-\frac{a}{l}\right) & \frac{1}{l} \\ \frac{a}{l} & -\frac{1}{l} \end{bmatrix} = \begin{bmatrix} \left(1-\frac{a}{l}\right) & \frac{a}{l} \\ \frac{1}{l} & -\frac{1}{l} \end{bmatrix}$$

$$\begin{bmatrix} \frac{1}{k_1}\left(1-\frac{a}{l}\right) & \frac{1}{k_1}\left(\frac{1}{l}\right) \\ \frac{1}{k_2}\left(\frac{a}{l}\right) & \frac{1}{k_2}\left(-\frac{1}{l}\right) \end{bmatrix}$$

$$\text{or} \quad \mathbf{C} = \begin{bmatrix} \frac{1}{k_1}\left(1-\frac{a}{l}\right)^2 + \frac{1}{k_2}\left(\frac{a}{l}\right)^2 & \frac{1}{k_1}\left(1-\frac{a}{l}\right)\left(\frac{1}{l}\right) + \frac{1}{k_2}\frac{a}{l}\left(-\frac{1}{l}\right) \\ \frac{1}{k_1}\frac{1}{l}\left(1-\frac{a}{l}\right) + \frac{1}{k_2}\left(-\frac{1}{l}\right)\left(\frac{a}{l}\right) & \frac{1}{k_1}\left(\frac{1}{l}\right)^2 + \frac{1}{k_2}\left(-\frac{1}{l}\right)^2 \end{bmatrix} \equiv \begin{bmatrix} \beta_{11} & \beta_{12} \\ \beta_{21} & \beta_{22} \end{bmatrix} \tag{4.311}$$

$$\text{with} \quad \mathbf{A} = \begin{bmatrix} (1-a/l) & 1/l \\ a/l & -1/l \end{bmatrix}; \quad \mathbf{B} = \begin{bmatrix} (1-a/l) & a/l \\ 1/l & -1/l \end{bmatrix}; \quad \mathbf{K}_b^{-1} = \begin{bmatrix} 1/k_1 & 0 \\ 0 & 1/k_2 \end{bmatrix}$$

Since in general $\beta_{12} \neq \beta_{21}$ (and so $d_{12} \neq d_{21}$). So the effective influence coefficient matrix, which the thin disc will experience, will be

$$\mathbf{D} = \boldsymbol{\alpha} + \mathbf{C} = \begin{bmatrix} \alpha_{11} & \alpha_{12} \\ \alpha_{21} & \alpha_{22} \end{bmatrix} + \begin{bmatrix} \beta_{11} & \beta_{12} \\ \beta_{21} & \beta_{22} \end{bmatrix} \quad \text{with} \quad \mathbf{D} \equiv \begin{bmatrix} d_{11} & d_{12} \\ d_{21} & d_{22} \end{bmatrix}$$

$$= \begin{bmatrix} \alpha_{11}+\beta_{11} & \alpha_{12}+\beta_{12} \\ \alpha_{21}+\beta_{21} & \alpha_{22}+\beta_{22} \end{bmatrix} \tag{4.312}$$

In dynamic approach while considering the gyroscopic effect, we need to consider both orthogonal planes. So now how to obtain unbalance response in a non-gyroscopic effect case is illustrated. The advantage of the present formulation is that equations are in closed form and easy to analyse. Equation (4.310) can be written as

$$\mathbf{D}\mathbf{F}_s = \mathbf{U}_s \Rightarrow \mathbf{F}_s = \mathbf{D}^{-1}\mathbf{U}_s = \mathbf{K}_{bs}\mathbf{U}_s \quad \text{with} \quad \mathbf{K}_{bs} = \mathbf{D}^{-1} \tag{4.313}$$

where the matrix $\mathbf{K}_{bs}$ is similar to the stiffness matrix (it is equivalent stiffness of the shaft and bearings experienced at the disc location). Equations of motion, in the y-direction and on the y–z plane (see Figure 4.61), can be written as

$$-f_y = m\frac{d^2}{dt^2}(y + e\sin\omega t) \quad \text{or} \quad me\omega^2 \sin\omega t - f_y = m\ddot{y} \tag{4.314}$$

$$\text{and} \quad -M_{yz} = I_d\ddot{\varphi}_x \tag{4.315}$$

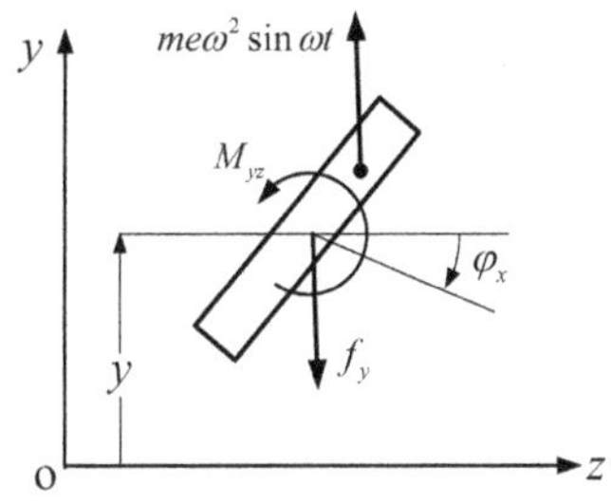

FIGURE 4.61 Free body diagram of the disc in y–z plane.

Equations of motion (4.314) and (4.315) of the disc can be written in a matrix form as

$$\mathbf{M\ddot{u}} + \mathbf{f}_s = \mathbf{f}_{unb} \tag{4.316}$$

With $\mathbf{M} = \begin{bmatrix} m & 0 \\ 0 & I_d \end{bmatrix}$; $\mathbf{u} = \begin{Bmatrix} y \\ \varphi_x \end{Bmatrix}$; $\mathbf{f}_s = \begin{Bmatrix} f_y \\ M_{yz} \end{Bmatrix}$; $\mathbf{f}_{unb} = \begin{Bmatrix} me\omega^2 \\ 0 \end{Bmatrix} e^{j\omega t} = \mathbf{F}_{unb} e^{j\omega t}$;

Noting $\mathbf{f}_s = \mathbf{F}_s e^{j\omega t}$ and $\mathbf{u} = \mathbf{U}_s e^{j\omega t}$, the equations of motion take the following form

$$-\omega^2 \mathbf{MU}_s + \mathbf{F}_s = \mathbf{F}_{unb} \tag{4.317}$$

Noting Eq. (4.313), Eq. (4.317) becomes

$$-\omega^2 \mathbf{MU}_s + \mathbf{K}_{bs}\mathbf{U}_s = \mathbf{F}_{unb} \tag{4.318}$$

which gives

$$\mathbf{U}_s = \mathbf{HF}_{unb} \quad \text{with} \quad \mathbf{H} = \left(-\omega^2 \mathbf{M} + \mathbf{K}_{bs}\right)^{-1} \tag{4.319}$$

where $\mathbf{H}^{-1} = \left(-\omega^2 \mathbf{M} + \mathbf{K}_{bs}\right)$ is the equivalent dynamic stiffness matrix, as experienced by the disc, of the shaft and the bearing system. Once the response of the disc has been obtained, from the aforementioned equation for a given unbalance force, the loading applied to the shaft by the disc can be obtained by Eq. (4.310). Then from Eq. (4.300), we can get shaft end deflections $\mathbf{Y}_b$ at each bearing, which is substituted in Eq. (4.294) to get bearing forces $\mathbf{F}_b$. Alternately, bearing forces can be used directly from Eq. (4.297). Displacements and forces have the complex form; the amplitude and the phase information can be extracted from the real and imaginary parts. Amplitudes will be the modulus of complex numbers, and phase angles of all these displacements can be evaluated by calculating arctangent of the ratio of the imaginary to real components. For the free vibration analysis, Eq. (4.318) will take the following form

$$\left(-\omega_{nf}^2 \mathbf{MU} + \mathbf{K}_{bs}\right)\mathbf{U}_s = 0 \tag{4.320}$$

with $\mathbf{K}_{bs} = \mathbf{D}^{-1} = (\boldsymbol{\alpha} + \mathbf{C})^{-1} = \begin{bmatrix} \alpha_{11} + \beta_{11} & \alpha_{12} + \beta_{12} \\ \alpha_{21} + \beta_{21} & \alpha_{22} + \beta_{22} \end{bmatrix}^{-1} = \dfrac{1}{\Delta} \begin{bmatrix} \alpha_{22} + \beta_{22} & -(\alpha_{12} + \beta_{12}) \\ -(\alpha_{21} + \beta_{21}) & \alpha_{11} + \beta_{11} \end{bmatrix}$

$$\Delta = (\alpha_{11} + \beta_{11})(\alpha_{22} + \beta_{22}) - (\alpha_{12} + \beta_{12})(\alpha_{21} + \beta_{21})$$

$$\begin{bmatrix} \beta_{11} & \beta_{12} \\ \beta_{21} & \beta_{22} \end{bmatrix} = \begin{bmatrix} \frac{1}{k_1}\left(1-\frac{a}{l}\right)^2+\frac{1}{k_2}\left(\frac{a}{l}\right)^2 & \frac{1}{k_1}\left(1-\frac{a}{l}\right)\left(\frac{1}{l}\right)+\frac{1}{k_2}\frac{a}{l}\left(-\frac{1}{l}\right) \\ \frac{1}{k_1}\frac{1}{l}\left(1-\frac{a}{l}\right)+\frac{1}{k_2}\left(-\frac{1}{l}\right)\left(\frac{a}{l}\right) & \frac{1}{k_1}\left(\frac{1}{l}\right)^2+\frac{1}{k_2}\left(-\frac{1}{l}\right)^2 \end{bmatrix}$$

$$\alpha = \begin{bmatrix} \alpha_{11} & \alpha_{12} \\ \alpha_{21} & \alpha_{22} \end{bmatrix} = \frac{1}{3EIl}\begin{bmatrix} a^2b^2 & \left(3a^2l-2a^3-al^2\right) \\ ab(b-a) & -\left(3al-3a^2-l^2\right) \end{bmatrix}$$

It should be noted that the present analysis is valid for simply supported end conditions only. Since now all the matrices are expressed in expanded form, for chosen numerical values the unbalanced response can be simulated to get critical speeds. It is left to readers to do the same.

Exercise 4.19 Obtain the transverse natural frequencies of a rotor-bearing system as shown in Figure 4.62. Consider the shaft to be rigid. Both bearings have identical moment springs (of stiffness k_b) to resist tilting (or transverse rotational motion) and identical linear springs (of stiffness k) to resist translational motions. Take $a=0.7$ m, $b=0.3$ m, $k_b=100$ kN-m/rad and $k=1$ kN/m. The disc has $m=5$ kg and $I_d=0.02$ kg-m². Consider only the single-plane motion and ignore the gyroscopic couple effect.

***Solution*:** Let the generalised co-ordinate be $\left(x_G,\varphi_y\right)$ (refer Figure 4.63) at the disc location, where the whole mass is concentrated (assuming there is no mass of the shaft).
The kinetic energy (KE) of the rotor system for the present case is

$$T = \tfrac{1}{2}m\dot{x}_G^2 + \tfrac{1}{2}I_{dG}\dot{\varphi}_y^2 \tag{4.321}$$

where m is the mass of the rotor and I_{dG} is the diametral mass moment of inertia of the rotor about its centre of gravity, G. The potential energy (PE) of the rotor system is

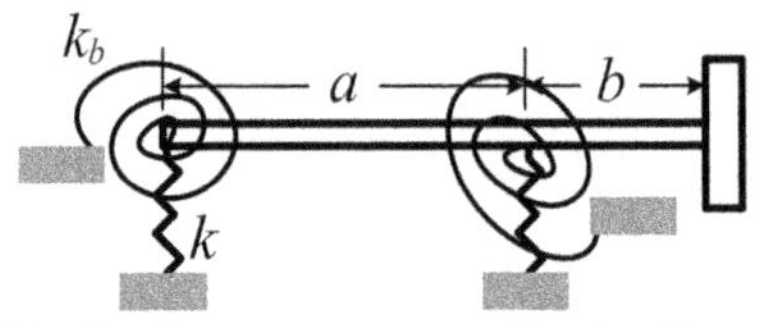

FIGURE 4.62 A rigid rotor mounted on flexible supports.

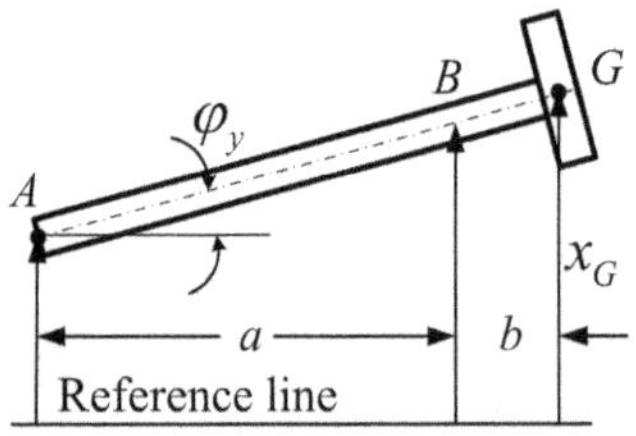

FIGURE 4.63 Deflected rigid shaft with generalized coordinates (x_G, φ_y).

$$U = \tfrac{1}{2}k_A\left(x_G - l_{AG}\varphi_y\right)^2 + \tfrac{1}{2}k_B\left(x_G - l_{BG}\varphi_y\right)^2 + \tfrac{1}{2}k_{bA}\varphi_y^2 + \tfrac{1}{2}k_{bB}\varphi_y^2 \tag{4.322}$$

$$\text{with} \quad l_{AG} = a + b \quad \text{and} \quad l_{BG} = b$$

where k_A and k_B are the stiffness of support at bearings A and B, respectively, and l_{AG} and l_{BG} are the distances AG and BG, respectively. For getting equations of motion from Lagrange's equation for the generalised coordinate, x_G, we have,

$$\frac{d}{dt}\left(\frac{\partial T}{\partial \dot{x}_G}\right) - \frac{\partial T}{\partial x_G} + \frac{\partial U}{\partial x_G} = 0 \tag{4.323}$$

On substituting Eqs. (4.321) and (4.322) into parts of Eq. (4.323), we get

$$\frac{d}{dt}\left(\frac{\partial T}{\partial \dot{x}_G}\right) = \frac{d}{dt}\left\{\frac{\partial}{\partial \dot{x}_G}\left(\tfrac{1}{2}m\dot{x}_G^2 + \tfrac{1}{2}I_{dG}\dot{\varphi}_y^2\right)\right\} = m\ddot{x}_G \tag{4.324}$$

and

$$\begin{aligned}\frac{\partial U}{\partial x_G} &= \frac{\partial}{\partial x_G}\left\{\tfrac{1}{2}k_A\left(x_G - l_{AG}\varphi_y\right)^2 + \tfrac{1}{2}k_B\left(x_G - l_{BG}\varphi_y\right)^2 + \tfrac{1}{2}k_{bA}\varphi_y^2 + \tfrac{1}{2}k_{bB}\varphi_y^2\right\} \\ &= k_A\left(x_G - l_{AG}\varphi_y\right) + k_B\left(x_G - l_{BG}\varphi_y\right) \\ &= \left(k_A + k_B\right)x_G - \left\{k_A l_{AG} + k_B l_{BG}\right\}\varphi_y\end{aligned} \tag{4.325}$$

On substituting Eqs. (4.324) and (4.325) into Eq. (4.323), we get

$$m\ddot{x}_G + \left(k_A + k_B\right)x_G - \left(k_A l_{AG} + k_B l_{BG}\right)\varphi_y = 0 \tag{4.326}$$

Similarly, for the generalised coordinate, φ_y, we have

$$\frac{d}{dt}\left(\frac{\partial T}{\partial \dot{\varphi}_y}\right) - \frac{\partial T}{\partial \varphi_y} + \frac{\partial U}{\partial \varphi_y} = 0 \tag{4.327}$$

On substituting Eqs. (4.321) and (4.322) into parts of Eq. (4.327), we get

$$\frac{d}{dt}\left(\frac{\partial T}{\partial \dot{\varphi}_y}\right) = \frac{d}{dt}\left\{\frac{\partial}{\partial \dot{\varphi}_y}\left(\tfrac{1}{2}m\dot{x}_G^2 + \tfrac{1}{2}I_{dG}\dot{\varphi}_y^2\right)\right\} = I_{dG}\ddot{\varphi}_y \tag{4.328}$$

and

$$\begin{aligned}\frac{\partial U}{\partial \varphi_y} &= \frac{\partial}{\partial \varphi_y}\left\{\tfrac{1}{2}k_A\left(x_G - l_{AG}\varphi_y\right)^2 + \tfrac{1}{2}k_B\left(x_G - l_{BG}\varphi_y\right)^2 + \tfrac{1}{2}k_{bA}\varphi_y^2 + \tfrac{1}{2}k_{bB}\varphi_y^2\right\} \\ &= \left\{k_A\left(x_G - l_{AG}\varphi_y\right)\left(-l_{AG}\right) + k_B\left(x_G - l_{BG}\varphi_y\right)\left(-l_{BG}\right) + k_{bA}\varphi_y + k_{bB}\varphi_y\right\} \\ &= -\left(k_A l_{AG} + k_B l_{BG}\right)x_G + \left(k_A l_{AG}^2 + k_B l_{BG}^2 + k_{bA} + k_{bB}\right)\varphi_y\end{aligned} \tag{4.329}$$

On substituting Eqs. (4.328) and (4.329) into Eq. (4.327), we get

$$I_{dG}\ddot{\varphi}_y - (k_A l_{AG} + k_B l_{BG})x_G + \left(k_A l_{AG}^2 + k_B l_{BG}^2 + k_{bA} + k_{bB}\right)\varphi_y = 0 \tag{4.330}$$

Equations (4.326) and (4.330), in the matrix form can be written as,

$$\begin{bmatrix} m & 0 \\ 0 & I_{dG} \end{bmatrix}\begin{Bmatrix} \ddot{x}_G \\ \ddot{\varphi}_y \end{Bmatrix} + \begin{bmatrix} (k_A + k_B) & -(k_A l_{AG} + k_B l_{BG}) \\ -(k_A l_{AG} + k_B l_{BG}) & \left(k_A l_{AG}^2 + k_B l_{BG}^2 + k_{bA} + k_{bB}\right) \end{bmatrix}\begin{Bmatrix} x_G \\ \varphi_y \end{Bmatrix} = \begin{Bmatrix} 0 \\ 0 \end{Bmatrix} \tag{4.331}$$

Given rotor data are: mass of disc, $m=5\,\text{kg}$; diametral moment of inertia, $I_d=0.02$ kg-m^2; stiffness of linear spring, $k_A=k_B=k=1$ kN/m; stiffness of moment spring, $k_{bA}=k_{bB}=k_b=100$ kN-m/rad; $a=0.7\,\text{m}$; $b=0.3\,\text{m}$, $l_{AG}=1.0\,\text{m}$ and $l_{BG}=0.3\,\text{m}$. On substituting the given values in equations on motion (4.331), we get

$$\begin{bmatrix} 5 & 0 \\ 0 & 0.02 \end{bmatrix}\begin{Bmatrix} \ddot{y} \\ \ddot{\varphi}_x \end{Bmatrix} + \begin{bmatrix} 2{,}000 & -2{,}000(0.7+0.3) \\ -2{,}000(0.7+0.3) & 2{,}000(1^2+0.3^2)+200\times 10^3 \end{bmatrix}\begin{Bmatrix} y \\ \varphi_x \end{Bmatrix} = \begin{Bmatrix} 0 \\ 0 \end{Bmatrix} \tag{4.332}$$

For obtaining the natural frequencies, we have

$$\left|\mathbf{K} - \omega_{nf}^2\mathbf{M}\right| = 0 \quad \text{or} \quad \left|\mathbf{M}^{-1}\mathbf{K} - \omega_{nf}^2\mathbf{I}\right| = 0 \tag{4.333}$$

On substituting values in Eq. (4.333), we get

$$\begin{vmatrix} 400 - \omega_{nf}^2 & -260 \\ -65{,}000 & 10{,}054{,}500 - \omega_{nf}^2 \end{vmatrix} = 0 \tag{4.334}$$

$$\text{or} \quad \omega_{nf}^4 - 10{,}054{,}900\omega_{nf}^2 + 4.0049\times 10^9 = 0 \tag{4.335}$$

By solving the aforementioned equation, we get

$$\omega_{nf1} = 19.96\ \text{rad/s} \quad \text{and} \quad \omega_{nf2} = 3{,}170.88\ \text{rad/s}$$

Exercise 4.20 Derive the equations of motion for transverse vibrations (in a single plane motion only) of a rotor-bearing-coupling system (Figure 4.64), which consists of two identical Jeffcott rotors connected by a coil spring (stiffness, k_φ in the plane shown). The coil spring resists the relative slopes of the shafts near bearings. Let the shaft has flexural rigidity of EI, length of each shaft is l and each disc have mass, m (= P/g). Ignore the effect of overhung part of the shaft at coupling and consider the masses to be symmetrically placed in each rotor. Figure 4.65 gives basic relations of the load and deflections (linear and slope). State assumptions in deriving equations of motion. Obtain the natural frequencies of the rotor system.

***Solution*:** First the assumptions made in the derivation of equations of motion are stated. The coupling is flexible and the disc is heavy. Under these conditions, the weight of the disc contributes more to the static slopes at the coupling location than does the reaction moment of the coupling due

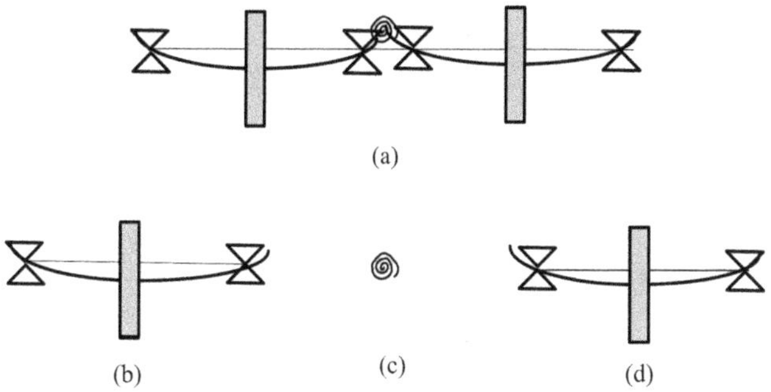

FIGURE 4.64 A rotor-coupling system (a) two Jeffcott rotor connected by a coupling (b) rotor 1 (c) the coupling (coil spring) (d) rotor 2.

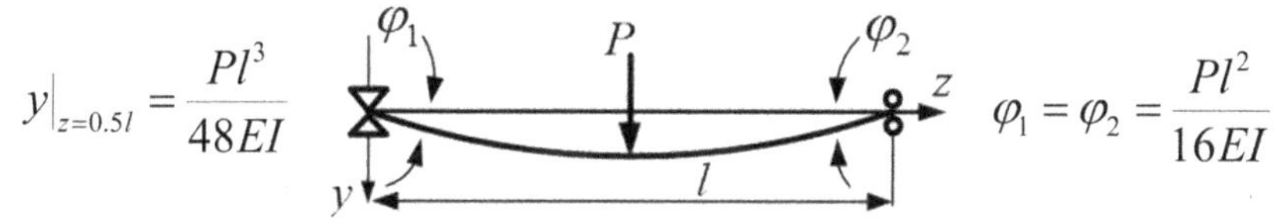

FIGURE 4.65 A simply supported shaft with a concentrated load P at the mid-span.

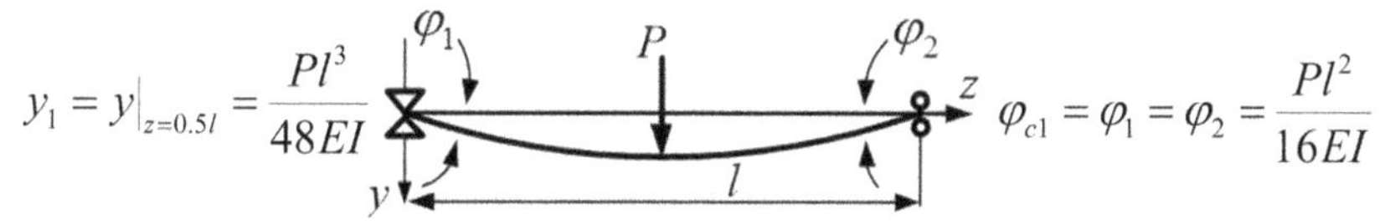

FIGURE 4.66 Translational displacement at mid span and angular displacement at coupling.

to dynamic forces. This makes it possible to establish linear relation between slopes at the coupling and translational displacements at the disc location.

For the simply supported shaft (Rotors 1 and 2) with a point load, P, at mid span in y–z plane (Figure 4.66), we have

$$y_1 = y_2 = \frac{Pl^3}{48EI}; \quad \varphi_{x1c} = \frac{Pl^2}{16EI} = \frac{3}{l}\frac{Pl^2}{48EI} = \frac{3}{l}y_1; \quad \varphi_{x2c} = \frac{3}{l}y_2 \tag{4.336}$$

where y_1 and y_2 are disc translatory displacements in y–z plane for rotor 1 and rotor 2, respectively; ϕ_{xc1} and ϕ_{xc2} are shaft rotational displacements at coupling in y–z plane for rotor 1 and rotor 2, respectively, l is the length of shaft in each rotor, and EI is the flexural rigidity of each shaft. Similarly, for z–x plane for rotor 1 and rotor 2 since the rotor is symmetric, we have

$$x_1 = x_2 = \frac{Pl^3}{48EI}; \quad \varphi_{y1c} = \frac{Pl^3}{16EI} = 3\frac{Pl^2}{48EI} = 3x_1; \quad \varphi_{y2c} = 3x_2 \tag{4.337}$$

Working under this assumption, the slope components in vector $\boldsymbol{\varphi}_{1c}$ at the coupling location for rotor-1 is given by $(\varphi_{yc1}, \varphi_{xc2})$ and the translational displacement components in vector $\boldsymbol{\eta}_1$ is given by (x_1, y_1) at disc-1 location are linearly related by a constant λ_1, as (refer Eqs. (4.336) and (4.337))

$$\varphi_{y1c} = 3x_1/l \quad \text{and} \quad \varphi_{x1c} = 3y_1/l \tag{4.338}$$

which can be combined as

$$\varphi_{1c} = \lambda_1 \alpha_1 \tag{4.339}$$

$$\text{with} \quad \varphi_{1c} = \begin{bmatrix} \varphi_{y1c} & \varphi_{x1c} \end{bmatrix}^T; \quad \alpha_1 = \begin{bmatrix} x_1 & y_1 \end{bmatrix}^T; \quad \lambda_1 = (3/l).$$

Likewise, for rotor-2, the corresponding relation can be written as

$$\varphi_{2c} = \lambda_2 \eta_2 \tag{4.340}$$

$$\text{with} \quad \varphi_{2c} = \begin{bmatrix} \varphi_{y2c} & \varphi_{x2c} \end{bmatrix}^T; \quad \eta_2 = \begin{bmatrix} x_2 & y_2 \end{bmatrix}^T; \quad \lambda_2 = (3/l).$$

Equations of motion for the coupled rotor-train system shall be derived from energy equations. The kinetic energy of rigid discs on rotor-1 and rotor-2 due to the translation and rotation motion is given by

$$T = \tfrac{1}{2} m_1 \left(\dot{x}_1^2 + \dot{y}_1^2\right) + \tfrac{1}{2} I_{p1}\omega^2 + \tfrac{1}{2} m_2 \left(\dot{x}_2^2 + \dot{y}_2^2\right) + \tfrac{1}{2} I_{p2}\omega^2 \tag{4.341}$$

The potential energy in the shafts and coupling is given by

$$V = \tfrac{1}{2} k_{1xx} x_1^2 + \tfrac{1}{2} k_{1yy} y_1^2 + \tfrac{1}{2} k_{2xx} x_2^2 + \tfrac{1}{2} k_{2yy} y_2^2 + \tfrac{1}{2} k_{\varphi_{xc}} \left(\varphi_{x1c} + \varphi_{x2c}\right)^2 + \tfrac{1}{2} k_{\varphi_{yc}} \left(\varphi_{y1c} + \varphi_{y2c}\right)^2 \tag{4.342}$$

On using Eqs. (4.339) and (4.340), we get

$$V = \tfrac{1}{2} k_{1xx} x_1^2 + \tfrac{1}{2} k_{1yy} y_1^2 + \tfrac{1}{2} k_{2xx} x_2^2 + \tfrac{1}{2} k_{2yy} y_2^2 + \tfrac{1}{2} k_{\varphi_{xc}} \left(\lambda_1 y_1 + \lambda_2 y_2\right)^2 + \tfrac{1}{2} k_{\varphi_{yc}} \left(\lambda_1 x_1 + \lambda_2 x_2\right)^2 \tag{4.341}$$

$$\text{with} \quad k_{1xx} = \frac{P}{x_1} = k_{1yy} = \frac{P}{y_1} = \frac{48EI}{l^3}; \quad k_{2xx} = \frac{P}{x_2} = k_{2yy} = \frac{P}{y_2} = \frac{48EI}{l^3}$$

where $k_{\varphi_{xc}}$ and $k_{\varphi_{yc}}$ are moment (torsional) stiffness of coupling in y–z and z–x planes, respectively. In the present case both are equal to $k_{\varphi c}$. Likewise, the Rayleigh's dissipative function due to viscous damping in shafts can be written as

$$\Im = \tfrac{1}{2} c_{1xx} \dot{x}_1^2 + \tfrac{1}{2} c_{1yy} \dot{y}_1^2 + \tfrac{1}{2} c_{2xx} \dot{x}_2^2 + \tfrac{1}{2} c_{2yy} \dot{y}_2^2 \tag{4.344}$$

where c_{1ij} and c_{2ij} ($i, j = x, y$ with $i = j$) are the viscous damping of rotor 1 and rotor 2, respectively. The external forces will give the non-conservative work done, as

$$\delta W_{nc} = \sum_{i=1}^{4} Q_i \delta \eta_i$$

or

$$\begin{aligned} \delta W_{nc} = &\left\{m_1 e_1 \omega^2 \cos(\omega t + \beta_1)\right\} \delta x_1 + \left\{m_1 e_1 \omega^2 \sin(\omega t + \beta_1) + k_{1yy} y_{10}\right\} \delta y_1 \\ &\left\{m_2 e_2 \omega^2 \cos(\omega t + \beta_2)\right\} \delta x_2 + \left\{m_2 e_2 \omega^2 \sin(\omega t + \beta_2) + k_{2yy} y_{20}\right\} \delta y_2 \end{aligned} \tag{4.345}$$

$$\text{with} \quad k_{1yy} y_{10} = m_1 g \quad \text{and} \quad k_{2yy} y_{20} = m_2 g$$

Herein, Q_i is the generalised force, and η_i is the generalised coordinates. Unbalances of disc 1 and disc 2 are given as $m_1 e_1$ and $m_2 e_2$, respectively, and β_1 and β_2 are respective phases. Static deflections at disc 1 and disc 2 due to disc weights are represented as y_{10} and y_{20}, respectively. To derive equations of motion, we use Lagrange's equation given by

$$\frac{d}{dt}\left(\frac{\partial T}{\partial \dot{\eta}_i}\right)-\frac{\partial T}{\partial \eta_i}+\frac{\partial V}{\partial \eta_i}+\frac{\partial \Im}{\partial \dot{\eta}_i}=Q_i \quad (i=1,2,\ldots,n) \tag{4.346}$$

The vector of generalised coordinates (so that $n = 4$) is given by

$$\eta_i=\begin{bmatrix} x_1 & y_1 & x_2 & y_2 \end{bmatrix}^T \tag{4.347}$$

Herein, it should be noted that since we have linear relations between translatory displacements and rotational displacements so only four translatory displacements have been taken as generalised coordinates. Applying Eq. (4.346) on energy expressions given by Eqs. (4.341), (4.343), (4.344) and (4.345) sequentially, the equations of motion corresponding to each DOF is obtained. Thus there are four equations for four DOFs, given by

$$m_1\ddot{x}_1+k_{1xx}x_1+\lambda_1^2k_{\varphi_{yc}}x_1+\lambda_1\lambda_2k_{\varphi_{yc}}x_2+c_{1xx}\dot{x}_1=m_1e_1\omega^2\cos(\omega t+\beta_1) \tag{4.348}$$

$$m_1\ddot{y}_1+k_{1yy}y_1+\lambda_1^2k_{\varphi_{xc}}y_1+\lambda_1\lambda_2k_{\varphi_{xc}}y_2+c_{1yy}\dot{y}_1=m_1e_1\omega^2\sin(\omega t+\beta_1)+k_{1yy}\delta_{y10} \tag{4.349}$$

$$m_2\ddot{x}_2+k_{2xx}x_2+\lambda_2^2k_{\varphi_{yc}}x_2+\lambda_1\lambda_2k_{\varphi_{yc}}x_1+c_{2xx}\dot{x}_2=m_2e_2\omega^2\cos(\omega t+\beta_2) \tag{4.350}$$

$$m_2\ddot{y}_2+k_{2yy}y_2+\lambda_2^2k_{\varphi_{xc}}y_2+\lambda_1\lambda_2k_{\varphi_{yc}}y_1+c_{2yy}\dot{y}_2=m_2e_2\omega^2\sin(\omega t+\beta_2)+k_{2yy}\delta_{y20} \tag{4.351}$$

As such there is no coupling between two transverse orthogonal planes. Parameters λ_1 and λ_2 are defined in eq. (4.354). The EOMs in a single plane (z–x plane) are

$$m_1\ddot{x}_1+k_{1xx}x_1+\lambda_1^2k_{\varphi_{yc}}x_1+\lambda_1\lambda_2k_{\varphi_{yc}}x_2+c_{1xx}\dot{x}_1=m_1e_1\omega^2\cos(\omega t+\beta_1) \tag{4.352}$$

$$m_2\ddot{x}_2+k_{2xx}x_2+\lambda_2^2k_{\varphi_{yc}}x_2+\lambda_1\lambda_2k_{\varphi_{yc}}x_1+c_{2xx}\dot{x}_2=m_2e_2\omega^2\cos(\omega t+\beta_2) \tag{4.353}$$

The governing differential equation for the undamped free vibration, in a matrix form, is

$$\begin{bmatrix} m_1 & 0 \\ 0 & m_2 \end{bmatrix}\begin{Bmatrix} \ddot{x}_1 \\ \ddot{x}_2 \end{Bmatrix}+\begin{bmatrix} k_{1xx}+\lambda_1^2k_{\varphi_{yc}} & \lambda_1\lambda_2k_{\varphi_{yc}} \\ \lambda_1\lambda_2k_{\varphi_{yc}} & k_{2xx}+\lambda_2^2k_{\varphi_{yc}} \end{bmatrix}\begin{Bmatrix} x_1 \\ x_2 \end{Bmatrix}=\begin{Bmatrix} 0 \\ 0 \end{Bmatrix} \quad \text{with} \quad \lambda_1=\lambda_2=\left(\frac{3}{l}\right) \tag{4.354}$$

Equation (4.353) can be written as

$$\begin{bmatrix} m_1 & 0 \\ 0 & m_2 \end{bmatrix}\begin{Bmatrix} \ddot{x}_1 \\ \ddot{x}_2 \end{Bmatrix}+\begin{bmatrix} k_{1xx}+(3/l)^2k_{\varphi_{yc}} & (3/l)^2k_{\varphi_{yc}} \\ (3/l)^2k_{\varphi_{yc}} & k_{2xx}+(3/l)^2k_{\varphi_{yc}} \end{bmatrix}\begin{Bmatrix} x_1 \\ x_2 \end{Bmatrix}=\begin{Bmatrix} 0 \\ 0 \end{Bmatrix} \tag{4.355}$$

The natural frequencies, ω_{nf}, can be obtained by solving the eigenvalue problem, which is of the following form

$$(\mathbf{A}-\lambda\mathbf{I})\mathbf{x}=0 \quad \text{with} \quad \mathbf{A}=\mathbf{M}^{-1}\mathbf{K} \quad \text{and} \quad \lambda=\omega_{nf}^2 \tag{4.356}$$

On taking the determinant of aforementioned equation, we get

$$|\mathbf{A}-\lambda\mathbf{I}|=0 \quad \text{or} \quad |\mathbf{K}-\lambda\mathbf{M}|=0 \tag{4.357}$$

$$\text{or} \quad \left| \begin{bmatrix} k_{1xx} + (3/l)^2 k_{\varphi_{yc}} & (3/l)^2 k_{\varphi_{yc}} \\ (3/l)^2 k_{\varphi_{yc}} & k_{2xx} + (3/l)^2 k_{\varphi_{yc}} \end{bmatrix} - \lambda \begin{bmatrix} m_1 & 0 \\ 0 & m_2 \end{bmatrix} \right| = 0$$

$$\text{or} \quad \begin{aligned} &\lambda^2 (m_1 m_2) - \lambda \left[m_1 \left\{ k_{2xx} + (3/l)^2 k_{\varphi_{yc}} \right\} + m_2 \left\{ k_{1xx} + (3/l)^2 k_{\varphi_{yc}} \right\} \right] \\ &+ \left\{ k_{1xx} + (3/l)^2 k_{\varphi_{yc}} \right\} \left\{ k_{2xx} + (3/l)^2 k_{\varphi_{yc}} \right\} - (3/l)^4 k_{\varphi_{yc}}^2 = 0 \end{aligned} \tag{4.358}$$

which is the required characteristic equation and can be solved to get two natural frequencies of the rotor system. Equations of motion in y-z plane will also give the same result.

Exercise 4.21 Obtain equations of motion of a two-spool rotor system (two co-centric shafts connected by an inner shaft bearing at annulus space with stiffness of k_i) and modelled as shown in Figure 4.67. Derive equations of motion and write in a matrix form in terms of chosen generalised coordinates. Shaft 1 is supported on two bearings at ends that are connected with the rigid foundation. Shaft 2 is supported by bearings, one connected with the rigid foundation and the other to Shaft 1 at its midspan by the intershaft bearing. Consider one-plane motion of the shafts. Treat the shafts as uniform rigid bars with the transverse translational and rotational motions only. The spring stiffness of bearing between the two shafts has stiffness k_i and other support (bearing) springs have a stiffness of k. Masses of the shafts are m_1 and m_2, and the diametral mass moments of inertia of I_{d1} and I_{d2}. Length of the shafts are l_1 and l_2 with $l_1 = 2l_2$, and radius of r_1 and r_2.

***Solution*:** Shafts 1 and 2 both are rigid members, so DOFs of each shaft will be two. Let the generalised coordinates at the centre of gravity of each shaft are (x_1, φ_1) and (x_2, φ_2). Let for shaft 1, the left and right ends are designated as A and B, respectively, the centre of shaft 1 as C, and for shaft 2, the left and right ends are designated as D and E, respectively as shown in Figure 4.68a. We have following relations

$$\begin{aligned} &x_A = x_1 - 0.5l_1\varphi_{y1}; \quad x_B = x_1 + 0.5l_1\varphi_{y1}; \quad x_C = x_1; \\ &x_D = x_2 - 0.5l_2\varphi_{y2}; \quad x_E = x_2 + 0.5l_2\varphi_{y2} \end{aligned} \tag{4.359}$$

where x_A, x_B and x_C are displacements of shaft 1 at points A, B and C, respectively, and x_D and x_E are displacements of shaft 2 at points D and E, respectively. Points C and F are the centre of gravity of shaft 1 and shaft 2, respectively. The lengths of shaft 1 and shaft 2 are l_1 and l_2, respectively.

The KE of the system is given as (Refer Figure 4.68b)

$$T = \frac{1}{2} m_1 \dot{x}_1^2 + \frac{1}{2} I_{d1} \dot{\varphi}_{y1}^2 + \frac{1}{2} m_2 \dot{x}_2^2 + \frac{1}{2} I_{d2} \dot{\varphi}_{y2}^2 \tag{4.360}$$

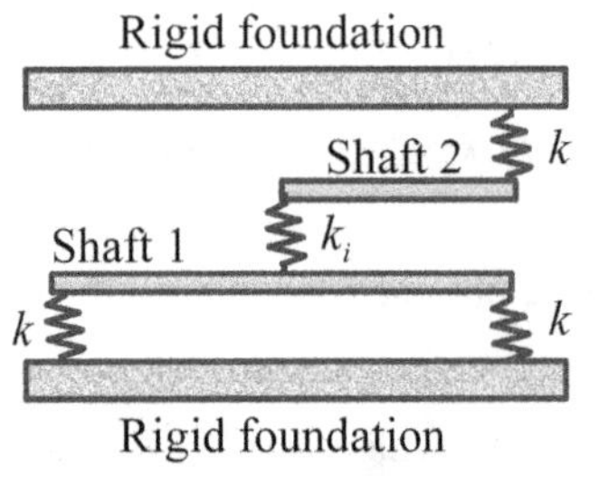

FIGURE 4.67 A two-spool rotor model with an intershaft bearing.

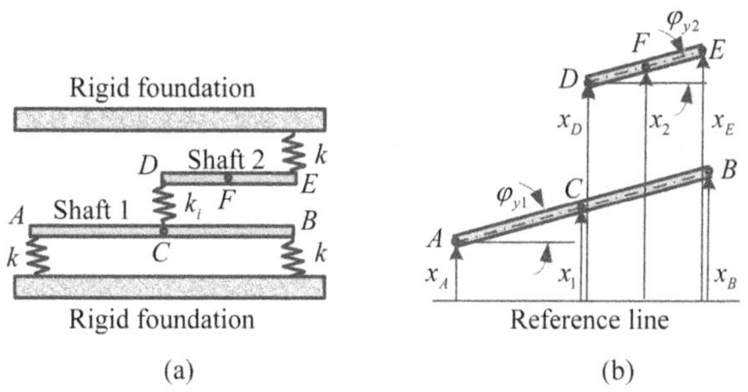

FIGURE 4.68 Rotor system with (a) with labelled location (b) generalised coordinates.

where m_1 is the mass of rotor 1, I_{d1} is the diametral mass moment of inertia about the centre of mass of the rotor 1, m_2 is the mass of rotor 2, and I_{d2} is the diametral mass moment of inertia about the centre of mass of rotor 2. The PE of the system is given as

$$U = \tfrac{1}{2}k_A x_A^2 + \tfrac{1}{2}k_B x_B^2 + \tfrac{1}{2}k_i(x_D - x_C)^2 + \tfrac{1}{2}k_E x_E^2 \tag{4.361}$$

On substituting Eq. (4.359) into Eq. (4.361), we get

$$\begin{aligned} U &= \tfrac{1}{2}k_A(x_1 - 0.5l_1\varphi_{y1})^2 + \tfrac{1}{2}k_B(x_1 + 0.5l_1\varphi_{y1})^2 \\ &\quad + \tfrac{1}{2}k_i\{(x_2 - 0.5l_2\varphi_{y2}) - x_1\}^2 + \tfrac{1}{2}k_E(x_2 + 0.5l_2\varphi_{y2})^2 \end{aligned} \tag{4.362}$$

For getting the equation of motion from the Lagrange's equation for the generalised coordinate, x_1, we have,

$$\frac{d}{dt}\left(\frac{\partial T}{\partial \dot{x}_1}\right) - \frac{\partial T}{\partial x_1} + \frac{\partial U}{\partial x_1} = 0 \tag{4.363 }$$

On substituting Eqs. (4.360) and (4.362) into parts of Eq. (4.363), we get

$$\frac{d}{dt}\left(\frac{\partial T}{\partial \dot{x}_1}\right) = \frac{d}{dt}\left\{\frac{\partial}{\partial \dot{x}_1}\left(\tfrac{1}{2}m_1\dot{x}_1^2 + \tfrac{1}{2}I_{d1}\dot{\varphi}_{y1}^2 + \tfrac{1}{2}m_2\dot{x}_2^2 + \tfrac{1}{2}I_{d2}\dot{\varphi}_{y2}^2\right)\right\} = m_1\ddot{x}_1 \tag{4.364}$$

and

$$\begin{aligned} \frac{\partial U}{\partial x_1} &= \frac{\partial}{\partial x_1}\left\{\begin{matrix} \tfrac{1}{2}k_A(x_1 - 0.5l_1\varphi_{y1})^2 + \tfrac{1}{2}k_B(x_1 + 0.5l_1\varphi_{y1})^2 \\ + \tfrac{1}{2}k_i\{(x_2 - 0.5l_2\varphi_{y2}) - x_1\}^2 + \tfrac{1}{2}k_E(x_2 + 0.5l_2\varphi_{y2})^2 \end{matrix}\right\} \\ &= k_A(x_1 - 0.5l_1\varphi_{y1}) + k_B(x_1 + 0.5l_1\varphi_{y1}) + k_i\{(x_2 - 0.5l_2\varphi_{y2}) - x_1\}(-1) \\ &= (k_A + k_B + k_i)x_1 - 0.5l_1\{k_A - k_B\}\varphi_{y1} - k_i x_2 + 0.5l_2 k_i\varphi_{y2} \end{aligned} \tag{4.365}$$

On substituting Eqs. (4.364) and (4.365) into Eq. (4.363), it simplifies to

$$m_1\ddot{x}_1 + (k_A + k_B + k_i)x_1 - 0.5l_1(k_A - k_B)\varphi_{y1} - k_i x_2 + 0.5l_2 k_i\varphi_{y2} = 0 \tag{4.366}$$

Similarly, for the generalised coordinate, φ_{y1}, we have

$$\frac{d}{dt}\left(\frac{\partial T}{\partial \dot{\varphi}_{y1}}\right) - \frac{\partial T}{\partial \varphi_{y1}} + \frac{\partial U}{\partial \varphi_{y1}} = 0 \tag{4.367}$$

On substituting Eqs. (4.360) and (4.362) into parts of Eq. (4.367), we get

$$\frac{d}{dt}\left(\frac{\partial T}{\partial \dot{\varphi}_{y1}}\right)=\frac{d}{dt}\left\{\frac{\partial}{\partial \dot{\varphi}_{y1}}\left(\tfrac{1}{2}m_1\dot{x}_1^2+\tfrac{1}{2}I_{d1}\dot{\varphi}_{y1}^2+\tfrac{1}{2}m_2\dot{x}_2^2+\tfrac{1}{2}I_{d2}\dot{\varphi}_{y2}^2\right)\right\}=I_{d1}\ddot{\varphi}_{y1} \tag{4.368}$$

and

$$\begin{aligned}\frac{\partial U}{\partial \varphi_{y1}}&=\frac{\partial}{\partial \varphi_{y1}}\left\{\begin{matrix}\tfrac{1}{2}k_A\left(x_1-0.5l_1\varphi_{y1}\right)^2+\tfrac{1}{2}k_B\left(x_1+0.5l_1\varphi_{y1}\right)^2\\+\tfrac{1}{2}k_i\left\{\left(x_2-0.5l_2\varphi_{y2}\right)-x_1\right\}^2+\tfrac{1}{2}k_E\left(x_2+0.5l_2\varphi_{y2}\right)^2\end{matrix}\right\}\\&=k_A\left(x_1-0.5l_1\varphi_{y1}\right)(-0.5l_1)+k_B\left(x_1+0.5l_1\varphi_{y1}\right)(0.5l_1)\\&=-0.5l_1\left(k_A-k_B\right)x_1+\left(0.5l_1\right)^2\left(k_A+k_B\right)\varphi_{y1}\end{aligned} \tag{4.369}$$

On substituting Eqs. (4.368) and (4.369) into Eq. (4.367), it simplifies to

$$I_{d1}\ddot{\varphi}_{y1}-0.5l_1\left(k_A-k_B\right)x_1+\left(0.5l_1\right)^2\left(k_A+k_B\right)\varphi_{y1}=0 \tag{4.370}$$

For getting the equation of motion from Lagrange's equation for the generalised coordinate, x_2, we have,

$$\frac{d}{dt}\left(\frac{\partial T}{\partial \dot{x}_2}\right)-\frac{\partial T}{\partial x_2}+\frac{\partial U}{\partial x_2}=0 \tag{4.371}$$

On substituting Eqs. (4.360) and (4.362) into parts of Eq. (4.371), we get

$$\frac{d}{dt}\left(\frac{\partial T}{\partial \dot{x}_2}\right)=\frac{d}{dt}\left\{\frac{\partial}{\partial \dot{x}_2}\left(\frac{1}{2}m_1\dot{x}_1^2+\frac{1}{2}I_{d1}\dot{\varphi}_{y1}^2+\frac{1}{2}m_2\dot{x}_2^2+\frac{1}{2}I_{d2}\dot{\varphi}_{y2}^2\right)\right\}=m_2\ddot{x}_2 \tag{4.372}$$

and

$$\begin{aligned}\frac{\partial U}{\partial x_2}&=\frac{\partial}{\partial x_2}\left\{\begin{matrix}\frac{1}{2}k_A\left(x_1-0.5l_1\varphi_{y1}\right)^2+\frac{1}{2}k_B\left(x_1+0.5l_1\varphi_{y1}\right)^2\\+\frac{1}{2}k_i\left\{\left(x_2-0.5l_2\varphi_{y2}\right)-x_1\right\}^2+\frac{1}{2}k_E\left(x_2+0.5l_2\varphi_{y2}\right)^2\end{matrix}\right\}\\&=k_i\left\{\left(x_2-0.5l_2\varphi_{y2}\right)-x_1\right\}+k_E\left(x_2+0.5l_2\varphi_{y2}\right)\\&=-k_ix_1+\left(k_i+k_E\right)x_2-0.5l_2\left(k_i-k_E\right)\varphi_{y2}\end{aligned} \tag{4.373}$$

On substituting Eqs. (4.372) and (4.373) into Eq. (4.371), it simplifies to

$$m_2\ddot{x}_2-k_ix_1-0.5l_2\left(k_i-k_E\right)\varphi_{y2}+\left(k_i+k_E\right)x_2=0 \tag{4.374}$$

Similarly, for the generalised coordinate, φ_{y2}, we have

$$\frac{d}{dt}\left(\frac{\partial T}{\partial \dot{\varphi}_{y2}}\right)-\frac{\partial T}{\partial \varphi_{y2}}+\frac{\partial U}{\partial \varphi_{y2}}=0 \tag{4.375}$$

On substituting Eqs. (4.360) and (4.362) into parts of Eq. (4.375), we get

$$\frac{d}{dt}\left(\frac{\partial T}{\partial \dot{\varphi}_{y2}}\right)=\frac{d}{dt}\left\{\frac{\partial}{\partial \dot{\varphi}_{y2}}\left(\frac{1}{2}m_1\dot{x}_1^2+\frac{1}{2}I_{d1}\dot{\varphi}_{y1}^2+\frac{1}{2}m_2\dot{x}_2^2+\frac{1}{2}I_{d2}\dot{\varphi}_{y2}^2\right)\right\}=I_{d2}\ddot{\varphi}_{y2} \tag{4.376}$$

and

$$\frac{\partial U}{\partial \varphi_{y2}} = \frac{\partial}{\partial \varphi_{y2}} \left\{ \begin{array}{l} \frac{1}{2}k_A\left(x_1 - 0.5l_1\varphi_{y1}\right)^2 + \frac{1}{2}k_B\left(x_1 + 0.5l_1\varphi_{y1}\right)^2 \\ +\frac{1}{2}k_i\left\{\left(x_2 - 0.5l_2\varphi_{y2}\right) - x_1\right\}^2 + \frac{1}{2}k_E\left(x_2 + 0.5l_2\varphi_{y2}\right)^2 \end{array} \right\}$$
$$= k_i\left\{\left(x_2 - 0.5l_2\varphi_{y2}\right) - x_1\right\}\left(-0.5l_2\right) + k_E\left(x_2 + 0.5l_2\varphi_{y2}\right)\left(0.5l_2\right) \tag{4.377}$$
$$= 0.5l_2k_ix_1 + 0.5l_2\left(-k_i + k_E\right)x_2 + \left(0.5l_2\right)^2\left(k_i + k_E\right)\varphi_{y2}$$

On substituting Eqs. (4.376) and (4.377) into Eq. (4.375), it simplifies to

$$I_{d2}\ddot{\varphi}_{y2} + 0.5l_2k_ix_1 + 0.5l_2\left(-k_i + k_E\right)x_2 + \left(0.5l_2\right)^2\left(k_i + k_E\right)\varphi_{y2} = 0 \tag{4.378}$$

On combining (4.366), (4.370), (4.374) and (4.378) in a matrix form, we get

$$\mathbf{M}\ddot{\boldsymbol{\eta}} + \mathbf{K}\boldsymbol{\eta} = \mathbf{0} \tag{4.379}$$

with

$$\boldsymbol{\eta} = \left\{ \begin{array}{c} x_1 \\ \varphi_{y1} \\ x_2 \\ \varphi_{y2} \end{array} \right\}; \quad \mathbf{M} = \begin{bmatrix} m_1 & 0 & 0 & 0 \\ 0 & I_{d1} & 0 & 0 \\ 0 & 0 & m_2 & 0 \\ 0 & 0 & 0 & I_{d2} \end{bmatrix};$$

$$\mathbf{K} = \begin{bmatrix} k_A + k_B + k_i & -0.5l_1\left\{k_A - k_B\right\} & -k_i & 0.5l_2k_i \\ -0.5l_1\left(k_A - k_B\right) & \left(0.5l_1\right)^2\left(k_A + k_B\right) & 0 & 0 \\ -k_i & 0 & k_i + k_E & -0.5l_2\left(k_i - k_E\right) \\ 0.5l_2k_i & 0 & -0.5l_2\left(k_i - k_E\right) & \left(0.5l_2\right)^2\left(-k_i + k_E\right) \end{bmatrix}$$

For the present system $k_A = k_B = k_E = k$, it takes the following form

$$\mathbf{K} = \begin{bmatrix} 2k + k_i & 0 & -k_i & 0.5l_2k_i \\ 0 & 0.5l_1^2k & 0 & 0 \\ -k_i & 0 & k_i + k & -0.5l_2\left(k_i - k\right) \\ 0.5l_2k_i & 0 & -0.5l_2\left(k_i - k\right) & \left(0.5l_2\right)^2\left(-k_i + k\right) \end{bmatrix} \tag{4.380}$$

It should be noted that the mass matrix is a diagonal matrix, the stiffness matrix is a symmetric. Any other convenient generalised coordinates can be choosen and corresponding linear relations with the present chosen generalised coordinates at centers of gravity of shafts can be used to get EOMs in new generalised coordinates by substituting these relations to expressions of the KE and the PE in terms of new chosen generaliased coordinates or directly transformation matrix can be used in the mass and stiffness matrices to get respective matrices in new generaliased coordinates. It is left to the reader to do such exercise and for a chosen numberical values it can be checked whether all forms of EOMs are giving same natural frequencies as it was illustrated in Exercises 4.8 and 4.9.

Exercise 4.22 Choose a single correct answer from the multiple-choice questions:

i. A rigid long rotor supported on flexible anisotropic bearings can have how many transverse natural frequencies?
   A. 1   B. 2   C. 3   D. 4   E. more than 4

Solution: A rigid body (i.e. a long rotor) can have six DOFs. But in the present case, torsional and axial vibrations have been ignored. So, for transverse vibration, the rotor will have four DOFs. Two transverse translational displacements and two transverse rotational displacements. So the rotor-bearing system will have four natural frequencies.

ii. A rigid long rotor supported on flexible anisotropic bearings can have a reversal of the orbit direction as the spin speed of the rotor is increased.
 A. true B. false

Solution: In a rigid long rotor supported on flexible anisotropic bearings there will be four critical speeds. Usually, two transverse translatory modes will have lower critical speed values as compared to two transverse rotational modes. In between two transverse translatory mode critical speeds there will be backward whirl and similarly between two transverse rotational mode critical speeds there will be the backward whirl. In the remaining regions, it will have the forward whirl.

iii. A symmetrical rigid rotor mounted on fluid-film bearings would have coupling of motions in
 A. translational displacements (x, y) only B. rotational displacements (φ_x, φ_y) only
 C. between the translational and rotational displacements (x and φ_y) and (y and φ_x)
 D. both (A) and (B)

Solution: Fluid-film bearings do not couple motion between the translational and rotational displacements, if the rotor is symmetric. However, it couples the two translational displacements (x, y) as well as the two rotational displacements (φ_x, φ_y).

iv. A flexible rotor (e.g., a Jeffcott rotor with an offset disc) mounted on rigid bearings would have coupling of motions in
 A. translational displacements (x, y) only B. rotational displacements (φ_x, φ_y) only
 C. between the translational and rotational displacements (x and φ_y) and (y and φ_x)
 D. both (A) and (B)

Solution: In a flexible shaft the coupling between the translational and rotational displacements (x and φ_y) and (y and φ_x) take place. Cross-coupled influence coefficients couple these two motions in the same plane.

v. A flexible rotor (e.g., a Jeffcott rotor with a disc at the midspan) mounted on rigid bearings would have coupling of motions in
 A. translational displacements (x, y) only B. rotational displacements (φ_x, φ_y) only
 C. between the translational and rotational displacements (x and φ_y) and (y and φ_x)
 D. none of the displacement would be coupled

Solution: When the disc is at the midspan then cross-coupled influence coefficients are zero. So, none of the displacement would be coupled. In fact, tilting of disc will not take place for the first mode, when translatory displacements take place. Similarly, pure titling is possible in the second mode.

vi. For a flexible rotor (e.g., a Jeffcott rotor with offset disc) mounted on fluid-film bearings would have coupling of motions in
 A. translational displacements (x, y) only B. rotational displacements (φ_x, φ_y) only
 C. between the translational and rotational displacements (x and φ_y) and (y and φ_x)
 D. all translational and rotational displacements would be coupled

Solution: The flexible shaft with offset disc couples the translational and rotational DOFs in the same plane, where the fluid-film bearings couples translational DOFs in two orthogonal planes. So in totality, all four DOFs will be coupled.

vii. Obtain the transverse natural frequency of the rotor system shown in Figure 4.69. Consider the shaft to be rigid. It is assumed that it oscillates (precesses) about its centre of gravity while whirling (i.e. pure tilting without translational motion). The stiffness of each bearing is k_b, and the distance between the bearings is l. Let us assume the centre of gravity lies from the right bearing towards the left at one-quarter of l. The diametral mass moment of inertia of the rotor is I_d. Ignore the gyroscopic effects.

A. $\sqrt{\dfrac{5l^2 k_b}{8I_d}}$ B. $\sqrt{\dfrac{8l^2 k_b}{5I_d}}$ C. $\sqrt{\dfrac{l^2 k_b}{2I_d}}$ D. $\sqrt{\dfrac{2l^2 k_b}{I_d}}$

Solution: The moment due to bearings at its centre of gravity will be

$$M = \left(k_b \frac{3}{4} l\varphi_x\right)\frac{3}{4} l + \left(k_b \frac{1}{4} l\varphi_x\right)\frac{l}{4} \quad \text{or} \quad M = \left(\frac{5}{8} k_b l^2\right)\varphi_x \tag{4.381}$$

Hence, the tilting stiffness imparted by the bearings is

$$k_t = \frac{M}{\varphi_x} = \left(\frac{5}{8} k_b l^2\right) \tag{4.382}$$

Hence, the natural frequency of the system is

$$\omega_{nf} = \sqrt{\frac{k_t}{I_d}} = \sqrt{\frac{5k_b l^2}{8I_d}} \tag{4.383}$$

viii. A long symmetrical rigid rotor with mass, m, is mounted on anisotropic flexible bearings with k_x and k_y as effective stiffness in two orthogonal directions (with $k_y > k_x$). For a rotor speed of $\sqrt{k_x/m} < \omega < \sqrt{k_y/m}$, the direction of pure translational whirl due to the unbalance would be

A. synchronous
B. anti-synchronous
C. asynchronous (forward)
D. asynchronous (backward)

Solution: In a long symmetrical rigid rotor supported on flexible anisotropic bearings there will be four critical speeds. Usually, two transverse translatory modes will have lower critical speed values as compared to two transverse rotational modes. In between two transverse translatory mode critical speeds $\sqrt{k_x/m} < \omega < \sqrt{k_y/m}$, there will be backward whirl and similarly between two transverse rotational mode critical speeds $\sqrt{0.5k_x l^2/I_d} < \omega < \sqrt{0.5k_y l^2/I_d}$ there will be the backward whirl. In the remaining regions ($\omega < \sqrt{k_x/m}$, $\omega > \sqrt{k_y/m}$, $\omega < \sqrt{0.5k_x l^2/I_d}$, and $\omega > \sqrt{0.5k_y l^2/I_d}$), it will have the forward whirl.

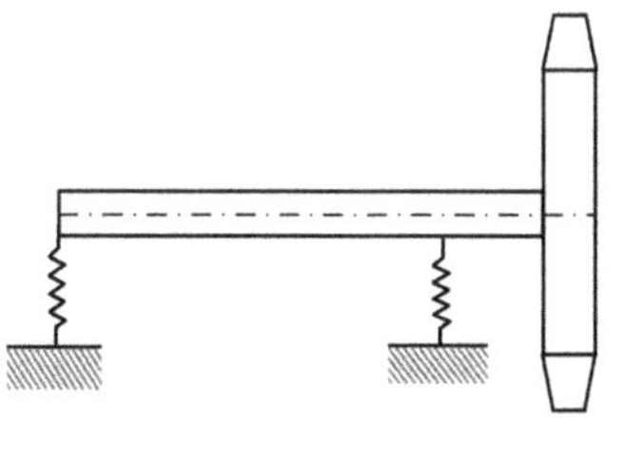

FIGURE 4.69 Rigid rotor mounted on flexible bearings.

ix. A rotor mounted on anisotropic bearings has an elliptical orbit, with the major axis tilted with respect to the horizontal or vertical axis. This tilt of the orbit is mainly due to
   A. the difference in the direct stiffness coefficients in the two orthogonal directions
   B. the cross-coupled damping coefficients
   C. the cross-coupled stiffness coefficients
   D. the difference in the direct damping coefficients in the two orthogonal directions

   Solution: The shape of the orbit is governed by stiffness terms. The direct stiffness terms change the relative ratio of major and minor axis of an elliptical orbit, whereas the cross-coupled stiffness changes the orientation of the elliptical orbit. A circular or elliptical orbit is possible if the rotor has only unbalance (i.e. a single excitation frequency). In the presence of other kinds of faults (like misalignment and crack), multiple excitation frequencies may be present and the orbit will have several loops. The damping has a role in reducing the amplitude with time, especially in transient response. In steady-state response, it may change the shape of orbit if damping coefficients are anisotropic in nature.

x. A rigid symmetrical rotor mounted on isotropic bearings, but having cross-ccoupled stiffness terms, can have how many transverse critical speeds?
   A. 1   B. 2   C. 3   D. 4

   Solution: Since the bearing is isotropic but coupling terms are not zero so the rotor will have four DOFs. In the absence of cross-coupled term, it will have two DOFs. Since the rotor is symmetric so two modes will be uncoupled and orbit will be circular.

xi. For a rigid rotor supported on flexible bearings, when it is undergoing conical whirl, then two ends of the shaft would have
   A. the same direction of whirl   B. opposite direction of whirl
   C. a direction dependent upon whether it is synchronous or antisynchronous whirl
   D. a direction dependent upon the asynchronous whirl condition only

   Solution: For a rigid rotor supported on flexible bearings, when it is undergoing a conical whirl, then two ends of the shaft would have a direction dependent upon whether it is synchronous or antisynchronous whirl. In fact, as the spin speed of the rotor is increased up to first conical whirl critical speed it will have forward whirl and in between two conical whirl critical speeds it will have backward whirl, and after second critical speed, it will have again forward whirl with some phase difference.

xii. The whirling of the shaft shown in Figure 4.70 (the black dot represents the centre of gravity of the disc) represents the case of
   A. forward asynchronous whirl   B. synchronous whirl
   C. backward asynchronous whirl   D. anti-synchronous whirl

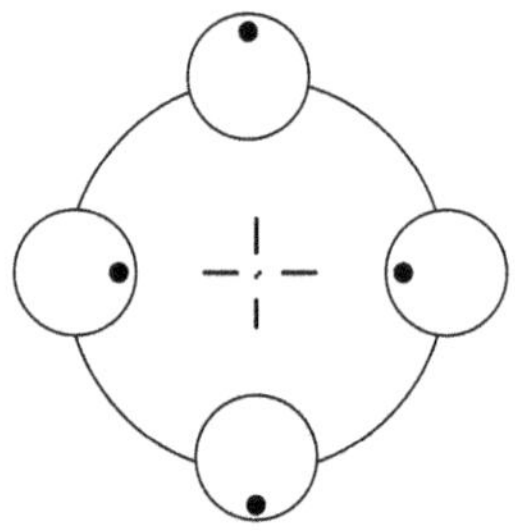

FIGURE 4.70 A shaft in the orbital motion.

Solution: If we assume the whirl direction (orbital motion direction) is clockwise, then from the top, the shaft position at 3 o'clock can be achieved if the shaft is rotating in a counter-clock direction (i.e., the anti-synchronous whirl). Herein, we need to observe the position of the black spot on the shaft. Similarly, it can be checked if the whirl direction is counter-clockwise, and then from top, the shaft position at 9 o'clock can be achieved if shaft is rotating in the clock direction.

xiii. A rigid long rotor mounted on flexible anisotropic springs (with no cross-coupled terms and no damping) would have how many frequency bands in which it has synchronous backward whirls?

A. 0 B. 1 C. 2 D. 3

Solution: When a long rotor is mounted on anisotropic bearings, it will have four critical speeds (four DOFs). Since cross-coupled terms are zero, and if the rotor is symmetric, then all four modes will be independent. In such a case, there will be two critical speeds corresponding to translational modes and two conical modes. In between these critical speeds of the translational modes, there will be one backward mode and, similarly, another for the conical whirl. So, a total of two backward whirl mode regions will be present. The remaining four regions will have forward whirl motion.

xiv. A rigid tapered shaft rotating about its geometrical axis (consider the centre of the rigid shaft and its centre of gravity at the same location, i.e. the centre of gravity axis and geometrical axis to be collinear and ignore the gravity effect) will have

A. unbalance force B. unbalance moment
C. both unbalance force and moment D. no unbalance force or moment

Solution: Since the centre of gravity and centre of shaft (or centre of rotation of the shaft) are the same, there will not be any unbalanced force. Also, since the shaft axis line and its rotational line are colinear, there will not be any moment. The shaft asymmetry due to taper will have an effect in coupling the translational and rotational motions in a plane.

xv. In the conical whirl of a rigid long rotor mounted on anisotropic bearings**,** if the rotor is rotating below the first conical whirl critical speed, then the two ends of the rotor will have

A. synchronous whirl with respect to the spin speed
B. asynchronous whirl with respect to the spin speed
C. anti-synchronous whirl with respect to the spin speed
D. none of the above with respect to the spin speed

Solution: Below the first conical whirl critical speed, the motion of shaft ends (i.e. the orbit) will be synchronous whirl with respect to the spin speed. After crossing the first conical whirl critical speed it will be anti-synchronous whirl with respect to the spin speed and subsequently, after crossing second conical whirl critical speed it will be again synchronous whirl with respect to the spin speed with some phase.

xvi. For a rotor system shown in Figure 4.71, if each sphere has mass of m and polar mass moment of inertia as I_p and connected by a massless rigid shaft of length $2l$. The bearing supports are identical and isotropic, and having stiffness of k. The natural frequency of the system in pure conical whirl would be

A. $\sqrt{\dfrac{kl^2}{I_p + ml^2}}$ B. $\sqrt{\dfrac{2kl^2}{I_p + 2ml^2}}$ C. $\sqrt{\dfrac{kl^2}{I_p + 2ml^2}}$ D. $\sqrt{\dfrac{kl^2}{2I_p + 2ml^2}}$

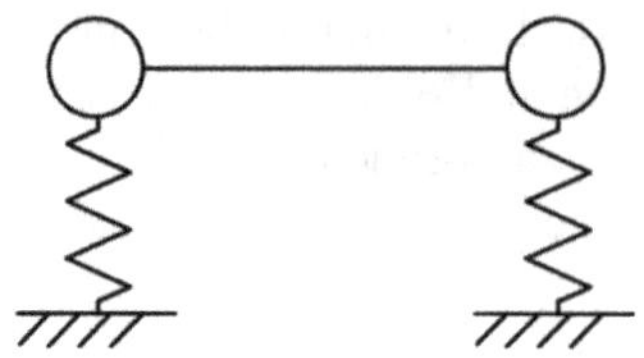

FIGURE 4.71 A rigid rotor supported on flexible bearings.

Solution: The effective diametral mass moment of the rotor system for pure conical whirl will be

$$I_{deff} = 2\{I_p + ml^2\} = 2I_p + 2ml^2 \quad (4.384)$$

The effective bending stiffness of bearings, in one of planes, will be

$$k_{beff} = 2(kl^2) = 2kl^2 \quad (4.385)$$

So, the natural frequency is given as

$$\omega_{nf} = \sqrt{\frac{k_{beff}}{I_{deff}}} = \sqrt{\frac{2kl^2}{2I_p + 2ml^2}} = \sqrt{\frac{kl^2}{I_p + ml^2}} \quad (4.386)$$

xvii. For a simple rotor-bearing system with a disc at the centre of the shaft, the left portion of the shaft stiffness (i.e. half of the shaft transverse stiffness) is assumed to be connected to the left bearing stiffness in series, and similarly in the right half. Then these stiffnesses (left and right sides) so obtained are connected in parallel to the disc. The equivalent stiffness the disc experiences from shaft as well as bearing will be

A. $k_{eq} = \dfrac{2}{2/k_{\text{shaft}} + 1/k_{\text{bearing}}}$ B. $k_{eq} = \dfrac{1}{2/k_{\text{shaft}} + 1/k_{\text{bearing}}}$

C. $k_{eq} = \dfrac{2}{1/k_{\text{shaft}} + 1/k_{\text{bearing}}}$ D. $k_{eq} = \dfrac{1}{1/k_{\text{shaft}} + 1/k_{\text{bearing}}}$

Solution: The equivent stiffness of the shaft and the bearing (connected in series) in one of the end is

$$k_{eq} = \frac{1}{2/k_{\text{shaft}} + 1/k_{\text{bearing}}} \quad (4.387)$$

The total effective stiffness from both ends of the shaft (connected in parallel) will be

$$k_{eq} = \frac{2}{2/k_{\text{shaft}} + 1/k_{\text{bearing}}} \quad (4.388)$$

xviii. A rigid rotor is mounted at the ends by springs of stiffness $k = 1$ kN/m each; the rotor has a length of 1 m and a diameter of 3 cm. The shaft material density is 7,800 kg/m^3. The pure rotational transverse mode natural frequency (in rad/s) of the rotor-bearing system would be

A. 24 B. 27 C. 31 D. 33

Solution: The mass of the rigid rotor is given as

$$m = \rho \frac{\pi d^2}{4} l = 7{,}800 \frac{\pi \times 0.03^2}{4} \times 1 = 5.5135 \text{ kg} \tag{4.389}$$

The diametral mass moment of inertia of the rotor is given as

$$I_d = \tfrac{1}{16} mD^2 + \tfrac{1}{12} ml^2 = 5.5135 \times \left(\tfrac{1}{16} \times 0.03^2 + \tfrac{1}{12} \times 1^2 \right) = 0.4598 \text{ kg-m}^2 \tag{4.390}$$

The natural frequency of conical whirl (eq. (4.6) of Tiwari (2013)) is given as

$$\omega_{cr3,4} = \sqrt{\frac{0.5kl^2}{I_d}} = \sqrt{\frac{0.5 \times 10^3 \times 1^2}{0.4598}} = 32.98 \approx 33 \text{ rad/s} \tag{4.391}$$

Similarly, the natural frequency of the cylindrical whirl (eq. (4.6) of Tiwari (2017)) is given as

$$\omega_{cr1,2} = \sqrt{\frac{2k}{m}} = \sqrt{\frac{2 \times 10^3}{5.5135}} = 19.0459 \approx 19.1 \text{ rad/s} \tag{4.392}$$

Since the spring stiffness is same in two orthogonal directions hence we will have only two whirl natural frequencies. For anisotropic bearing it becomes four.

xix For a long unbalanced symmetrical rigid rotor mounted on anisotropic undamped bearings, the path of a point on the shaft during the whirling motion would be
A. elliptical B. circular C. spiral D. a straight line

Solution: For a long unbalanced symmetrical rigid rotor mounted on anisotropic bearings the path of any point on the shaft will have an elliptical in shape. However, for isotropic bearing, it will be circular.

xx A flexible rotor-bearing mounted on a flexible foundation would have critical speeds as compared to the same rotor-bearing with a rigid foundation.
A. less B. more C. same
D. more or less depending upon the flexibility of the rotor and foundation

Solution: Due to the flexibility of the foundation, which is usually in series with the bearing flexibility, it will reduce the overall stiffness of the shaft support so the critical speed will also be reduced.

xxi The pure conical whirl is built in a long symmetrical rigid rotor supported by flexible bearings due to
A. pure radial unbalance B. pure axial unbalance
C. both radial and axial unbalances D. neither radial nor axial unbalances

Solution: The pure conical whirl builds in a symmetrical rigid rotor due to a unbalanced moment. The pure axial unbalance gives a unbalanced moment to excite the conical whirl. The pure unbalance moment will develop due to two factors (i) two similar radial unbalance axially offset in opposite directions (ii) when the principal polar axis of the rotor is inclined with the axis of spinning of the shaft. The radial unbalance (without axial offset) only excites the cylindrical whirl in a symmetrical rigid rotor supported on flexible bearings.

xxii Because of the bearings in a rotor system, its critical speed would
A. increase B. decrease C. remain same D. either increase or decrease

Solution: The bearing increases overall flexibility as experienced by a rotor so the system critical speed will reduce due to bearings. The foundation flexibility further reduces the system critical speed.

xxiii. Indicate the anti-synchronous conical whirl in Figure 4.72:

A. Option A B. Option B C. Option C D. Option D

Solution: The conical whirl is due to pure tilting of an unbalanced rigid long rotor and when the spin speed and the whirl direction is opposite then we have anti-synchronous whirl. For the cylindrical whirl, the rotor has pure translatory motion. It should be noted that in all kinds of modes, every particle of the shaft has an eliipical orbit (or circular for isotropic bearings).

xxiv. A flexible long rotor supported on flexible anisotropic bearings can have how many transverse natural frequencies?

A. 1 B. 2 C. 3 D. 4 E. Infinite

Solution: A flexible continuous shaft has infinite number of transverse natural frequencies. So even if it is supported on bearings it will have infinite transverse natural frequencies.

xxv. A rigid long rotor supported on flexible anisotropic bearings can have a reversal of the orbit direction as the spin speed of the rotor is increased.

A. True B. False

Solution: It can have a reversal of orbit direction in two regions, one between two translational critical speeds and another between two rotational critical speeds.

xxvi. In an unbalanced rigid long rotor supported on anisotropic flexible bearings with k_x and k_y stiffness of each bearing. No other bearing coefficients are present and the gyroscopic effect is ignored. For a pure translatory motion, the whirl direction between two critical speeds of the system will be

A. backward but non-synchronous B. forward but non-synchronous
C. synchronous D. anti-synchronous

Solution: Between two critical speeds (either in the cylindrical mode or the conical whirl) the rotor will have anti-synchronous whirl.

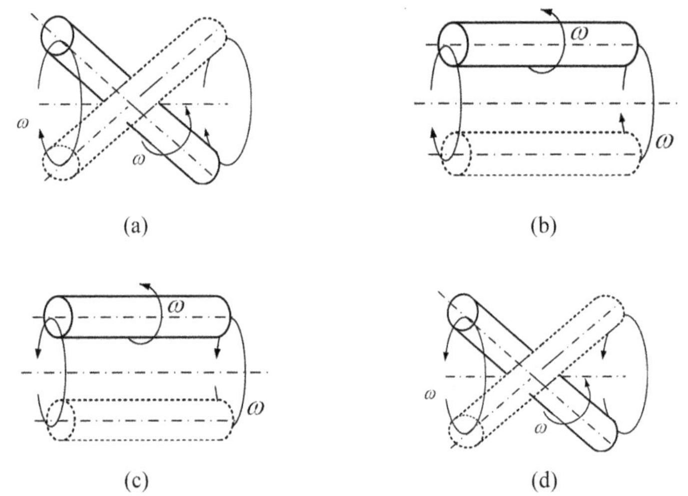

FIGURE 4.72 Whirling of a rigid rotor mounted on flexible bearings (a) option A (b) option B (c) option C (d) option D.

xxvii. For a rigid long rotor supported on anisotropic flexible bearings, how many critical speeds related to transverse vibration will be present?
A. 1 B. 2 C. 3 D. 4

Solution: For a rigid rotor on flexible supports for the transverse vibration, there will be four critical speeds. A rigid rotor has four-DOF in transverse plane.

xxviii. For a long uniform rigid rotor supported on bearings, each having eight linearised stiffness and damping coefficients. The coupling will be present in
A. all four DOFs B. only in translatory motion C. only in rotational motion
D. between translatory motions and between rotational motions

Solution: For a uniform rotor, there will not be coupling between the translatory and rotational motions. For the anisotropic bearings any way the two plane motions will be coupled.

xxix. For a long non-uniform rigid rotor supported on bearings, each having eight linearised stiffness and damping coefficients. The coupling will be present in
A. all four DOFs B. only in translatory motion C. only in rotational motion
D. between translatory motions and between rotational motions

Solution: For a non-uniform rotor, there will be coupling in the translatory and rotational motions. For the anisotropic bearings any way the two plane motions will be coupled. So, all four DOFs will be coupled.

xxx. For an unbalanced non-uniform rotor with anisotropic bearings (having both direct and cross-coupled stiffness and damping terms), the shape of shaft centre orbit will be (with y-axis in vertical direction and x-axis in horizontal direction)
A. elliptical with major and minor axes along x and y axes B. circle
C. elliptical with major and minor axes inclined with x and y axes
D. non-elliptical

Solution: For an unbalanced non-uniform rotor with anisotropic bearings, the shape of shaft centre orbit will be elliptical with major and minor axes inclined with x and y axes. The unbalance gives single frequency so no loops of the orbit will form and since bearings are anisotropic so orbit would not be circular. When both direct and cross-coupled stiffness and damping terms are present in the anisotropic bearing then the elliptical shape will not have major and minor axis along the vertical and horizontal directions.

xxxi. For a symmetrical long rigid rotor supported by the linear and angular springs at the ends, the natural frequency in a pure tilting will be (with k and k_a are the linear and angular (tilt) direct stiffness, respectively. L is the length of rigid rotor, m is the mass and I_d is the diametral mass moment of inertia of the rotor about centre of gravity)

A. $\sqrt{\frac{(0.5kL^2 + 2k_a)}{I_d}}$ B. $\sqrt{\frac{(0.5kL^2 + k_a)}{I_d}}$ C. $\sqrt{\frac{(kL^2 + 2k_a)}{I_d}}$ D. $\sqrt{\frac{(0.5kL^2 + k_a)}{I_d + mL^2}}$

Solution: The moment on the rotor due to bearings at its centre of gravity will be

$$M = 2\left\{\left(k\frac{L}{2}\varphi_x\right)\frac{L}{2} + k_a\right\} = \left(0.5kL^2 + 2k_a\right)\varphi_x \tag{4.393}$$

Hence, the tilting stiffness imparted by the bearings is

$$k_t = \frac{M}{\varphi_x} = 0.5kL^2 + 2k_a \tag{4.394}$$

Hence, the corresponding natural frequency of the rotor system is

$$\omega_{nf} = \sqrt{\frac{k_t}{I_d}} = \sqrt{\frac{0.5kL^2 + 2k_a}{I_d}} \tag{4.395}$$

xxxii. For a long symmetrical rotor of the mass, m, diametral mass moment of inertia, I_d, and length L, is supported on anisotropic bearings at its ends with each has stiffness of k_h and k_v in the horizontal and vertical directions, respectively; with $k_v > k_h$. The system's second natural frequency is same as third natural frequency, the condition for this to happen will be (it is to be noted that always natural frequencies are arranged in ascending order)

A. $\frac{k_v}{k_h} = \frac{mL^2}{4I_d}$ B. $\frac{k_v}{k_h} = \frac{mL^2}{2I_d}$ C. $\frac{k_v}{k_h} = \frac{mL^2}{I_d}$ D. $\frac{k_v}{k_h} = \frac{mL^2}{8I_d}$

Solution: Since the long rotor is symmetric so its translational and rotational motions are uncoupled. Also, since the bearing stiffness has no cross-coupled terms, so the vertical and horizontal motion are uncoupled. Since $k_v > k_h$, so from Example 4.2 (Tiwari, 2017) squares of the second and third natural frequencies, respectively, are

$$\omega_{nf2}^2 = \frac{2k_v}{m} \quad \text{and} \quad \omega_{nf3}^2 = \frac{k_h l^2}{2I_d} \tag{4.396}$$

On equating them, we get

$$\omega_{nf2}^2 = \omega_{nf3}^2 = \frac{2k_v}{m} = \frac{k_h l^2}{2I_d} \Rightarrow \frac{k_v}{k_h} = \frac{ml^2}{4I_d} \tag{4.397}$$

It should be noted that squares of the first and fourth natural frequencies, respectively, are

$$\omega_{nf1}^2 = \frac{2k_h}{m} \quad \text{and} \quad \omega_{nf4}^2 = \frac{k_v l^2}{2I_d} \tag{4.398}$$

FINAL REMARKS

The main focus of this chapter is to analyse simple rotor-bearing-foundation system using lumped mass approach for transverse vibration. Mostly, analytical expressions are derived, which are easily to interpret without much numerical analysis. In this chapter, mainly rigid rotor mounted on flexible bearings are considered for transverse vibration. Few cases are considered with shaft also flexible along with bearings. The foundation flexibility is also considered in one of the examples and the anlysis for such cases are quite involved and one need to resort to computation procedure. The bearing model considered is linearised model with four stiffness and four damping coefficients, which has both direct and cross-coupled terms. Few simple cases of rotor train (turbine-generator-coupling) model are considered. A very representative model of two-spool rotor system is also considered. In analysis both free and forced vibrations are considered for transvesre vibration without gyroscopic effect, which will be covered in Chapter 5. A large number of MCQs are certainly helps in clearing very fundamental concepts and steps involved in the rotor-bearing transverse vibration analysis.

REFERENCE

Tiwari, R., 2017, *Rotor Systems: Analysis and Identification*. Boca Raton, FL: CRC Press.

ANSWERS TO MCQs

Exercise 4.22

i. D	ii. A	iii. D	iv. C	v. D	vi. D
vii. A	viii. B	ix. C	x. D	xi. C	xii. D
xiii. C	xiv. D	xv. A	xvi. A	xvii. A	xviii. D
xix. A	xx. A	xxi. B	xxii. B	xxiii. A	xxiv. E
xxv. A	xxvi. D	xxvii. D	xxviii. D	xxix. A	xxx. C
xxxi. A	xxxii. A.				

5 Transverse Vibrations of Simple Rotor Systems with Gyroscopic Effects

Exercise 5.1 Obtain the forward and backward synchronous transverse critical speeds for a general motion of a rotor as shown in Figure 5.1. The rotor is assumed to be fixed and supported at one end. Take mass of the thin disc m is 5 kg and its radius is 15 cm. The shaft is assumed to be massless, and its length and diameter are 0.2 m and 0.02 m, respectively. Take shaft Young's modulus $E = 2.1 \times 10^{11}$ N/m^2. Obtain first two forward and backward synchronous transverse critical speeds by drawing the Campbell diagram. Consider the gyroscopic effects.

***Solution*:** Critical speeds for the synchronous forward and backward whirls need to be obtained for a rotor system having cantilever boundary conditions. For the present rotor system, the following data are given

$$l = 0.2\text{ m};\quad d = d_s = 0.02\text{ m};\quad E = 2.1\times 10^{11}\text{ N/m}^2;\quad m = 2\text{ kg};\quad r = r_d = 0.15\text{ m}.$$

So that the disc (polar mass moment of inertia and diametral mass moment of inertia) and shaft (second moment of area) properties are

$$I_p = \frac{1}{2}mr^2 = 0.0225\text{ kg-m}^2;\quad I_d = \frac{1}{4}mr^2 = 0.01125\text{ kg-m}^2;\quad I = \frac{\pi d^4}{64} = 7.854\times 10^{-9}\text{ m}^4 \tag{5.1}$$

Influence coefficients are obtained as

$$\alpha_{11} = \frac{l^3}{3EI} = 1.6168\times 10^{-6}\text{ m/N};\quad \alpha_{12} = \alpha_{21} = \frac{l^2}{2EI} = 1.2126\times 10^{-5}\text{ 1/N};$$
$$\alpha_{22} = \frac{l}{EI} = 1.2126\times 10^{-4}\text{ 1/m-N} \tag{5.2}$$

Now, the disc effect and elastic coupling parameters are obtained as

$$\mu = \frac{I_d\alpha_{22}}{m\alpha_{11}} = 0.4219 \quad\text{and}\quad \bar{\alpha} = \frac{\alpha_{12}^2}{\alpha_{11}\alpha_{22}} = 0.75 \tag{5.3}$$

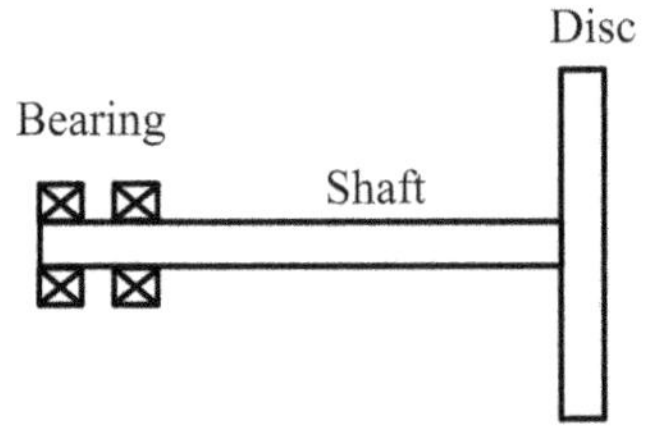

FIGURE 5.1 A cantilever rotor.

DOI: 10.1201/9781032638218-5

The frequency equation (refer eq. (5.78) of Tiwari (2017)), for a general motion, is given as

$$\bar{\nu}^4 - 2\bar{\omega}\bar{\nu}^3 + \frac{\mu+1}{\mu(\bar{\alpha}-1)}\bar{\nu}^2 - \frac{2\bar{\omega}}{\bar{\alpha}-1}\bar{\nu} - \frac{1}{\mu(\bar{\alpha}-1)} = 0 \tag{5.4}$$

where $\bar{\omega}$ and $\bar{\nu}$ are speed ratio and frequency ratio, respectively. On substituting the values in Eq. (5.4), we get

$$\bar{\nu}^4 - 2\bar{\omega}\bar{\nu}^3 - 13.48151\bar{\nu}^2 + 8\bar{\omega}\bar{\nu} + 9.4815 = 0 \tag{5.5}$$

For the forward whirl, we have $\bar{\nu} = +\bar{\omega}$ (the frequency ratio is equal to the speed ratio), on substituting in Eq. (5.5), we get

$$-\bar{\nu}^4 - 5.48151\bar{\nu}^2 + 9.4815 = 0 \tag{5.6}$$

which gives

$$\bar{\nu}^2 = \frac{5.4815 \pm \sqrt{(-5.4815)^2 - 4\times(-1)\times 9.4815}}{2\times(-1)} = 1.382, -6.863 \tag{5.7}$$

On neglecting the negative value (since it will give complex critical speed) and considering the positive value, we get

$$\bar{\nu}^2 = 1.382 \quad \Rightarrow \bar{\nu} = 1.174 = \left(\omega_{cr}^F\sqrt{\alpha_{11}m}\right) \quad \Rightarrow \omega_{cr}^F = 653.6 \text{ rad/s} \tag{5.8}$$

For the backward whirl, we have $\bar{\nu} = -\bar{\omega}$, so that on substituting in Eq. (5.5), we get

$$3\bar{\nu}^4 - 21.4815\bar{\nu}^2 + 9.4815 = 0 \tag{5.9}$$

$$\text{which gives} \quad \bar{\nu} = 2.586 \quad \text{and} \quad 0.6874 \tag{5.10}$$

Both values are positive; therefore, two backward whirls are possible, as

For $\bar{\nu} = 2.586$, we have

$$\omega_{cr2}^B = \frac{2.586}{\sqrt{1.6168\times 10^{-6}\times 2}} = 1428.0 \text{ rad/s} \tag{5.11}$$

and for $\bar{\nu} = 0.6874$, we have

$$\omega_{cr1}^B = \frac{0.6874}{\sqrt{1.6168\times 10^{-6}\times 2}} = 382.3 \text{ rad/s} \tag{5.12}$$

Corresponding Campbell diagram is shown in Figure 5.2. Herein, two solid curves represent the forward whirl natural frequency (non-dimensional) corresponding to two modes (i.e., the first and second modes) and two dashed curves represent backward whirl natural frequency (non-dimensional) corresponding to respectively two modes (i.e., the first and second modes). It should be noted that both the forward whirl natural frequencies and the backward whirl natural frequencies are identical at zero non-dimensional speed. This is due to the fact that the gyroscopic effect is absent when shaft is not rotating and two orthogonal plane motions are uncoupled and natural frequency in these planes are identical. The forward whirl natural frequency has increasing trend with the speed ratio, whereas for the backward whirl natural frequency it has decreasing trend. It can also be observed

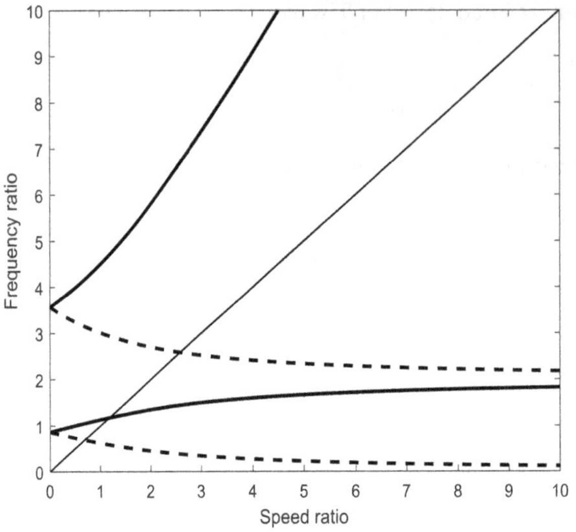

FIGURE 5.2 Campbell diagram showing the non-dimensional forward and backward critical speeds.

that the separation between forward and backward for a particular mode increases with the speed ratio due to the fact that the gyroscopic effect is function of speed. Moreover, the separation in higher modes is more as compared to lower (first) mode for the cantilever rotor configuration, and this trend may be different for other boundary conditions of the rotor. A line at 45° is drawn to get intercepts with frequency curves to get critical speeds of the system. It can be seen that with the forward natural frequency curve, which are two in numbers, only one intercept is feasible corresponding to one forward critical speed as observed in Eq (5.8). However, for the backward natural frequency curves, it has two intercepts corresponding to two backward critical speeds as given in Eqs. (5.11) and (5.12).

Exercise 5.2 Consider a rotor system as shown in Figure 5.3. The mass of the thin disc, m, is 5 kg and the diametral mass moment of inertia, I_d, is 0.02 kg-m^2. The shaft lengths are a = 0.3 m and b = 0.7 m. The diameter of the shaft is 10 mm. Obtain first two forward and backward synchronous transverse critical speeds by drawing the Campbell diagram. Consider the gyroscopic effects. Take $E = 2.1 \times 10^{11}$ N/m^2 for the shaft material.

Solution: Refer chapter 8 of Tiwari (2017) for the calculation procedure of influence coefficients (readers can refer to basic strength of material book also for the same). At present, closed-form analytical expressions available (refer Appendix 2.1) will be used for obtaining critical speeds of the system. For the present rotor, the following data are given

$$a = 0.3\,\text{m};\quad b = 0.7\,\text{m};\quad d = 0.01\,\text{m};\quad E = 2.1\times10^{11}\ \text{N/m}^2;\quad m = 5\,\text{kg}.$$

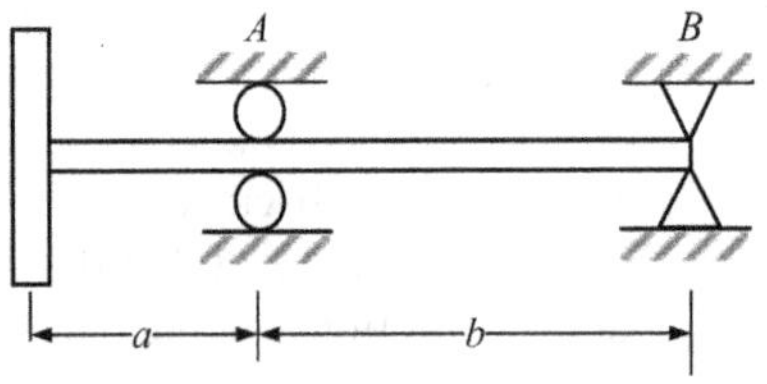

FIGURE 5.3 An overhung rotor system.

so that, we can obtain disc and shaft properties, as

$$I_d = 0.02 \text{ kg-m}^2; \quad I_p = 2I_d = 0.04 \text{ kg-m}^2; \quad I = \frac{\pi d^4}{64} = 4.9087 \times 10^{-10} \text{ m}^4$$

Influence coefficients (from Exercise 2.4 and also from Appendix 2.1) are given by

$$\alpha_{11} = \frac{a^2(a+b)}{3EI} = 2.9103 \times 10^{-4} \text{ m/N}; \quad \alpha_{12} = \alpha_{21} = \frac{a(3a+2b)}{6EI} = 1.1 \times 10^{-3} \text{ N}^{-1}$$

$$\alpha_{22} = \frac{(3a+b)}{3EI} = 5.2 \times 10^{-3} \text{ m}^{-1}\text{N}^{-1}.$$

Now, the disc effect and elastic coupling parameters can be obtained as

$$\mu = \frac{I_d \alpha_{22}}{m\alpha_{11}} = 0.0711; \quad \bar{\alpha} = \frac{\alpha_{12}^2}{\alpha_{11}\alpha_{22}} = 0.8266. \tag{5.13}$$

The frequency equation (refer eq. (5.78) of Tiwari (2017)) for a general motion is given as

$$\bar{\nu}^4 - 2\bar{\omega}\bar{\nu}^3 + \frac{\mu+1}{\mu(\bar{\alpha}-1)}\bar{\nu}^2 - \frac{2\bar{\omega}}{\bar{\alpha}-1}\bar{\nu} - \frac{1}{\mu(\bar{\alpha}-1)} = 0 \tag{5.14}$$

where $\bar{\omega}$ and $\bar{\nu}$ are speed ratio and frequency ratio. On substituting the values, we get

$$\bar{\nu}^4 - 2\bar{\omega}\bar{\nu}^3 - 86.8468\bar{\nu}^2 + 11.5315\bar{\omega}\bar{\nu} + 81.0811 = 0 \tag{5.15}$$

For the forward whirl, we have $\bar{\nu} = +\bar{\omega}$, so that Eq. (5.15) gives

$$\bar{\nu}^4 - 2\bar{\nu}^4 - 86.8468\bar{\nu}^2 + 11.5315\bar{\nu}^2 + 81.0811 = 0 \quad \Rightarrow \bar{\nu}^4 + 75.3153\bar{\nu}^2 - 81.0811 = 0$$

which gives $\bar{\nu}^2 = 1.062, -76.38$.

On neglecting the negative value since otherwise frequency will be complex, which is not feasible, so we get the forward whirl critical speed, as

$$\bar{\nu}^2 = 1.062 = \left(\omega_{cr}^F \sqrt{\alpha_{11} m}\right)^2 \quad \Rightarrow \omega_{cr}^F = 27.01 \text{ rad/s}.$$

For the backward whirl, we have $\bar{\nu} = -\bar{\omega}$, so that Eq. (5.15) gives

$$\bar{\nu}^4 + 2\bar{\nu}^4 - 86.8468\bar{\nu}^2 - 11.5315\bar{\nu}^2 + 81.0811 = 0 \quad \Rightarrow \bar{\nu}^4 - 32.7927\bar{\nu}^2 + 27.027 = 0$$

which gives

$$\bar{\nu} = 5.652 \text{ and } 0.9198$$

Herein, both values are positive; therefore, two backward whirls are possible, as

$$\bar{\nu} = \omega_{cr}^B \sqrt{\alpha_{11} m} \quad \Rightarrow \omega_{cr}^B = \frac{\bar{\nu}}{\sqrt{\alpha_{11} m}}$$

$$\text{For} \quad \bar{\nu} = 5.652 \quad \omega_{cr2}^B = \frac{5.652}{\sqrt{2.9103 \times 10^{-4} \times 5}} = 148.2 \text{ rad/s}$$

$$\text{For} \quad \bar{\nu} = 0.9198 \quad \omega_{cr1}^{B} = \frac{0.9198}{\sqrt{2.9103 \times 10^{-4} \times 5}} = 24.11 \text{ rad/s}.$$

Corresponding Campbell diagram is shown in Figure 5.4. Herein, solid lines are the forward whirl and dashed lines are the backward whirl. Intercepts of these curve with line $\nu = \omega$ (synchronous condition) give corresponding critical speeds.

Exercise 5.3 Consider a rotor system as shown in Figure 5.5 (In the figure, all dimensions are in cm). Consider the shaft as massless and is made of steel with the Young's modulus, $E = 2.1 \times 10^{11}$ N/m^2. The thin disc has the mass of 10 kg and a radius of 15 cm. The shaft is simply supported at ends. Obtain the first two forward and backward synchronous transverse critical speeds by drawing the Campbell diagram. Consider the gyroscopic effects.

***Solution*:** Refer example 8.2 of Tiwari (2017) for basic steps to obtain influence coefficients. However, in that example, only translational displacement has been considered so only single influence coefficient is defined. But in the present case, both translational and rotational displacements have been considered and we will have four influence coefficients.

For the present rotor the following data are given:

$$l_1 = 0.6 \text{ m}; \quad l_2 = 0.4 \text{ m}; \quad d_1 = 0.01 \text{ m}; \quad d_2 = 0.03 \text{ m}; \quad E = 2.1 \times 10^{11} \text{ N/m}^2; \quad m = 10 \text{ kg}; \quad r = 0.15 \text{ m}$$

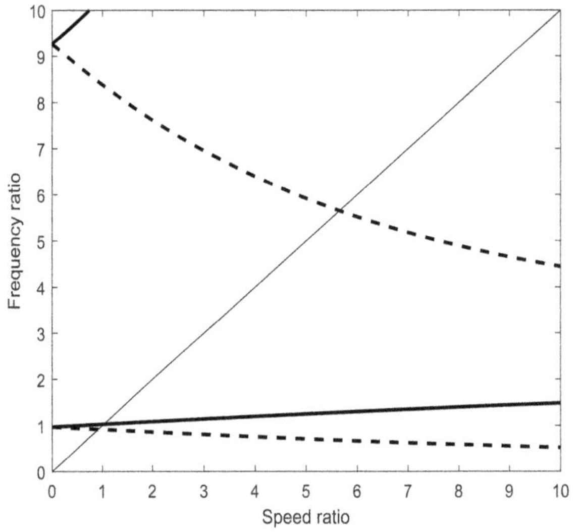

FIGURE 5.4 Campbell diagram showing critical speeds at intercepts with synchronous (or anti-synchronous for backward whirl) whirl line.

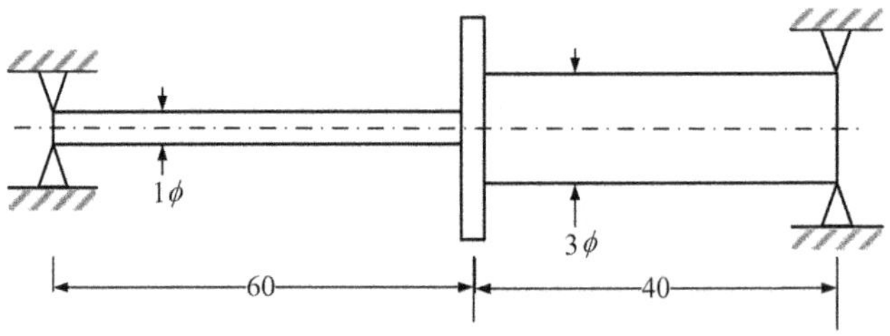

FIGURE 5.5 A stepped shaft with a thin disc and simply supported at ends.

so that we have disc and shaft properties as

$$I_d = \frac{1}{4}mr^2 = 0.05625 \text{ kg-m}^2; \quad I_p = 2I_d = 0.1125 \text{ kg-m}^2; \tag{5.16}$$

$$I_1 = \frac{\pi d_1^4}{64} = 4.9087 \times 10^{-10} \text{ m}^4; \quad I_2 = \frac{\pi d_2^4}{64} = 3.9761 \times 10^{-8} \text{ m}^4 \tag{5.17}$$

Due to a downward force, *F*, and a CW moment, *M*, at disc location (refer Figure 5.6) will give reactions at left (end *A*) and right (end *B*) supports as

$$F_A = 0.4000F - 1.6667M \tag{5.18}$$

$$\text{and} \quad F_B = 0.6000F + 2.5000M \tag{5.19}$$

Bending moments in $0 < z < a$ and $a < z < L$ regions are given as (refer Figure 5.6a and b)

$$M_{x1} = z(0.4000F - 1.6667M) \tag{5.20}$$

$$\text{and} \quad M_{x2} = z(0.4000F - 1.6667M) - F(z - 0.6000) \tag{5.21}$$

The strain energy stored in the shaft from bending moments can be obtained as

$$U = \int_0^a \frac{M_{x1}^2 dz}{2EI_1} + \int_a^b \frac{M_{x2}^2 dz}{2EI_2} \tag{5.22}$$

From the *Castigliano's theorem*, the translational displacement can be obtained as

$$\delta = \frac{\partial U}{\partial F} = \int_0^a \frac{M_{x1} \frac{\partial M_{x1}}{\partial F} dz}{EI_1} + \int_a^b \frac{M_{x2} \frac{\partial M_{x2}}{\partial F} dz}{2EI_2} \tag{5.23}$$

On substituting Eqs. (5.20) and (5.21) into Eq. (5.23), we get

$$\delta = \delta_1 + \delta_2 = \left(1.1175 \times 10^{-4} F - 4.6564 \times 10^{-4} M\right) + \left(9.1979 \times 10^{-7} F - 7.0261 \times 10^{-6} M\right)$$
$$\text{or} \quad \delta = 1.1267 \times 10^{-4} F - 4.7267 \times 10^{-4} M \equiv \alpha_{11} F + \alpha_{12} M \tag{5.24}$$

and rotational displacement can be obtained as

$$\varphi = \frac{\partial U}{\partial M} = \int_0^a \frac{M_{x1} \frac{\partial M_{x1}}{\partial M} dz}{EI_1} + \int_a^b \frac{M_{x2} \frac{\partial M_{x2}}{\partial M} dz}{2EI_2} \tag{5.25}$$

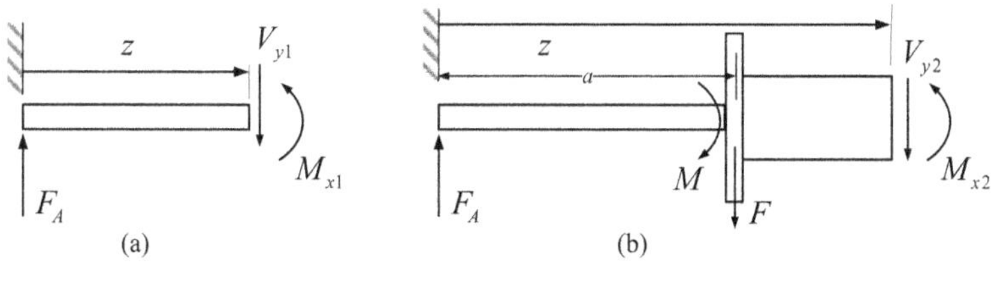

FIGURE 5.6 Free-body diagram of shaft segments (a) $0 < z < a$ (b) $a < z < L$.

On substituting Eqs. (5.20) and (5.21) into Eq. (5.25), we get

$$\varphi = \varphi_1 + \varphi_2 = \left(0.0019M - 4.6564\times10^{-4}F\right) + \left(8.6940\times10^{-5}M - 7.0261\times10^{-6}F\right)$$
$$\text{or} \quad \varphi = -4.7267\times10^{-4}F + 0.0020M \equiv \alpha_{21}F + \alpha_{22}M \tag{5.26}$$

So from Eqs. (5.24) and (5.26), influence coefficients are given by

$$\alpha_{11} = 1.1267\times10^{-4}\ \text{m/N}; \quad \alpha_{12} = \alpha_{21} = -4.7267\times10^{-4}\ \text{N}^{-1}; \quad \alpha_{22} = 0.0020\ \text{m}^{-1}\text{N}^{-1}.$$

Also, we have

$$\mu = \frac{I_d\alpha_{22}}{m\alpha_{11}} = 0.1012 \quad \text{and} \quad \bar{\alpha} = \frac{\alpha_{12}^2}{\alpha_{11}\alpha_{22}} = 0.9782.$$

The frequency equation (eq. (5.78) of Tiwari (2017)) for a general motion is given as

$$\bar{\nu}^4 - 2\bar{\omega}\bar{\nu}^3 + \frac{\mu+1}{\mu(\bar{\alpha}-1)}\bar{\nu}^2 - \frac{2\bar{\omega}}{\bar{\alpha}-1}\bar{\nu} - \frac{1}{\mu(\bar{\alpha}-1)} = 0 \tag{5.27}$$

On substituting the values in Eq. (5.27), we get

$$\bar{\nu}^4 - 2\bar{\omega}\bar{\nu}^3 + \frac{0.1012+1}{0.1012(0.9782-1)}\bar{\nu}^2 - \frac{2\bar{\omega}}{(0.9782-1)}\bar{\nu} - \frac{1}{0.1012(0.9782-1)} = 0$$
$$\text{or} \quad \bar{\nu}^4 - 2\,\bar{\omega}\bar{\nu}^3 - 498.30091\,\bar{\nu}^2 + 91.5869\,\bar{\omega}\bar{\nu} + 452.5075 = 0 \tag{5.28}$$

For the forward whirl, we have $\bar{\nu} = +\bar{\omega}$, so that Eq. (5.28) gives

$$\bar{\nu}^4 - 2\,\bar{\nu}^4 - 498.30091\,\bar{\nu}^2 + 91.5869\,\bar{\nu}^2 + 452.5075 = 0 \tag{5.29}$$

which gives $\bar{\nu}^2 = 1.1096, -407.8236$.

On neglecting the negative value since it will give a complex frequency ratio, which is not feasible, we get

$$\bar{\nu}_1^2 = 1.1096 \quad \Rightarrow \bar{\nu}_1 = 1.0534 \tag{5.30}$$

On using the non-dimensional frequency expression, we get the forward critical speeds, as

$$\bar{\nu}_1 = 1.0534 = \left(\omega_{cr1}^F\sqrt{\alpha_{11}m}\right) \quad \Rightarrow \omega_{cr1}^F = \frac{1.0534}{\alpha_{11}m} \quad \Rightarrow \omega_{cr1}^F = 31.38\ \text{rad/s} \tag{5.31}$$

For the backward whirl, we have $\bar{\nu} = -\bar{\omega}$, so that Eq. (5.28) gives

$$\bar{\nu}^4 + 2\,\bar{\nu}^4 - 498.30091\,\bar{\nu}^2 - 91.5869\,\bar{\nu}^2 + 452.5075 = 0 \tag{5.32}$$

$$\text{or} \quad \bar{\nu}_{1,2}^2 = 0.7701 \quad \text{and} \quad 195.8592 \quad \Rightarrow \bar{\nu}_{1,2} = 0.8776 \quad \text{and} \quad 13.9950$$

Both values are positive; therefore, two backward whirls are possible, as

$$\bar{\nu} = \omega_{cr}^{B}\sqrt{\alpha_{11}m} \quad \Rightarrow \omega_{cr}^{B} = \frac{\bar{\nu}}{\sqrt{\alpha_{11}m}}$$

$$\text{For} \quad \bar{\nu} = 0.8776 \quad \Rightarrow \omega_{cr1}^{B} = 26.1438 \text{ rad/s} \tag{5.33}$$

$$\text{and} \quad \text{For} \quad \bar{\nu} = 13.9950 \quad \Rightarrow \omega_{cr2}^{B} = 416.93 \text{ rad/s} \tag{5.34}$$

Corresponding Campbell diagram is shown in Figure 5.7. The solid lines are forward whirl frequency and dashed lines are backward whirl frequency. Since there is no intercept of the second mode forward natural frequency curve with $\bar{\nu} = \bar{\omega}$ line so it is not shown in the plot so that lower (fundamental) mode critical speeds and second backward critical speed are distinctly seen. Three intercepts show these critical speeds.

Exercise 5.4 Formulate the standard eigen value problem for the following equations of motion

$$\mathbf{M}\ddot{\mathbf{x}} + (\mathbf{C} - \omega\mathbf{G})\dot{\mathbf{x}} + \mathbf{K}\mathbf{x} = \mathbf{f}$$

where **M**, **C**, **G** and **K** are the mass, damping, gyroscopic and stiffness matrices, respectively; **x** and **f** are the response and force vectors, respectively; and ω is the spin speed of the rotor. Discuss characteristics of eigen values and interpret them physically for the present case.

***Solution*:** Equations of motion are given as

$$\mathbf{M}\ddot{\mathbf{x}} + (\mathbf{C} - \omega\mathbf{G})\dot{\mathbf{x}} + \mathbf{K}\mathbf{x} = \mathbf{f} \tag{5.35}$$

To convert the EOM into the state space form, we define

$$\mathbf{v} = \dot{\mathbf{x}} \tag{5.36}$$

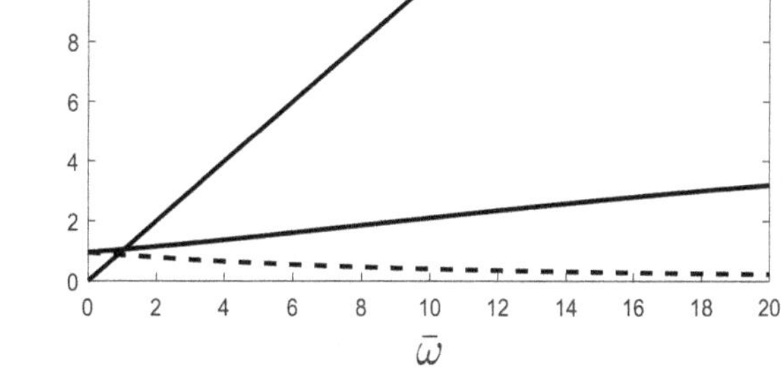

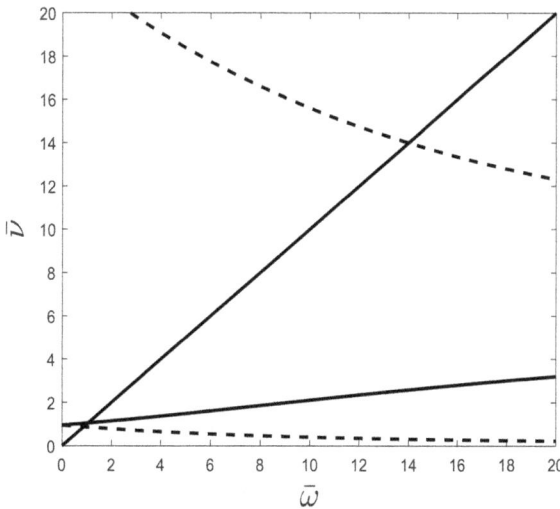

FIGURE 5.7 Campbell diagram showing critical speeds of the rotor system.

So that Eq. (5.35) can be written as

$$\mathbf{M}\dot{\mathbf{v}}+(\mathbf{C}-\omega\mathbf{G})\mathbf{v}+\mathbf{K}\mathbf{x}=\mathbf{f} \quad \Rightarrow \mathbf{M}\dot{\mathbf{v}}=\mathbf{f}-(\mathbf{C}-\omega\mathbf{G})\mathbf{v}-\mathbf{K}\mathbf{x}$$
$$\Rightarrow \dot{\mathbf{v}}=-\mathbf{M}^{-1}\mathbf{K}\mathbf{x}-\mathbf{M}^{-1}(\mathbf{C}-\omega\mathbf{G})\mathbf{v}+\mathbf{M}^{-1}\mathbf{f} \tag{5.37}$$

On combining Eqs. (5.36) and (5.37), we get

$$\left\{\begin{array}{c}\dot{\mathbf{x}}\\ \dot{\mathbf{v}}\end{array}\right\}=\left[\begin{array}{cc}\mathbf{0} & \mathbf{1}\\ -\mathbf{M}^{-1}\mathbf{K} & -\mathbf{M}^{-1}(\mathbf{C}-\omega\mathbf{G})\end{array}\right]\left\{\begin{array}{c}\mathbf{x}\\ \mathbf{v}\end{array}\right\}+\left\{\begin{array}{c}\mathbf{0}\\ \mathbf{M}^{-1}\mathbf{f}\end{array}\right\} \tag{5.38}$$

Let us assume, for free vibration, the solution as

$$\left\{\begin{array}{c}\mathbf{v}\\ \mathbf{x}\end{array}\right\}=\left\{\begin{array}{c}\bar{\mathbf{V}}\\ \bar{\mathbf{X}}\end{array}\right\}e^{\lambda t} \quad \text{so that} \quad \left\{\begin{array}{c}\dot{\mathbf{v}}\\ \dot{\mathbf{x}}\end{array}\right\}=\lambda\left\{\begin{array}{c}\bar{\mathbf{V}}\\ \bar{\mathbf{X}}\end{array}\right\}e^{\lambda t} \tag{5.39}$$

where λ is the eigenvalue. Eigenvalue is expected to be complex due to damping present in the system. On substituting Eq. (5.39) into (5.38), for free vibration, we get

$$\lambda\left\{\begin{array}{c}\mathbf{x}\\ \mathbf{v}\end{array}\right\}=\left[\begin{array}{cc}\mathbf{0} & \mathbf{1}\\ -\mathbf{M}^{-1}\mathbf{K} & -\mathbf{M}^{-1}(\mathbf{C}-\omega\mathbf{G})\end{array}\right]\left\{\begin{array}{c}\mathbf{x}\\ \mathbf{v}\end{array}\right\} \tag{5.40}$$

where $\lambda=\alpha\pm j\beta$ is the complex eigenvalue and its form is $\lambda=-\zeta\omega_{nf}\pm j\omega_{nf}\sqrt{1-\zeta^2}$, where ζ is the damping ratio, ω_{nf} is the undamped natural frequency and $\omega_d=\omega_{nf}\sqrt{1-\zeta^2}$ is the damped natural frequency. Hence, we have

$$\omega_{nf}=\sqrt{\alpha^2+\beta^2} \quad \text{and} \quad \zeta=-\alpha/\omega_{nf} \tag{5.41}$$

The parameter of logarithmic decrement, δ, is defined as

$$\delta=-\frac{2\pi\alpha}{\beta}=\frac{2\pi\zeta}{\sqrt{1-\zeta^2}} \tag{5.42}$$

where δ represents the instability threshold when $\delta<0$. The response of a dynamic system is a decay function, which involves the damping term. To get a stable response, the amplitude of vibration should decay as time increases. This will happen if the damping index ($\alpha<0$) is negative. Hence, for $\delta>0$, the rotor is stable and for $\delta<0$ it is unstable. The stability can be checked by observing the sign of the logarithmic decrement. It should be noted that eigenvalue, in presence of gyroscopic effect, depends upon spin speed of the rotor (refer eq. 5.40). So, eigenvalue problem needs to be solved for different speeds within the operating speed range. Imaginary part of eigenvalue (i.e., variation of damped whirl natural frequency with speed) can be used to draw Campbell diagram. Eigenvectors for the present case will also be complex in nature, which will contain relative phase also due to presence of damping in the system (refer Chapter 12 of Tiwari (2017) for more detail).

Exercise 5.5 Find the transverse critical speeds of a rotor system shown in Figure 5.8. The shaft is massless, and Young's modulus $E=2.1\times10^{11}$ N/m^2. The thin disc properties about its principal axes are $I_p=0.02$ kg-m^2 and $I_d=0.01$ kg-m^2, and the tilt of disc from vertical axis is $\varphi_0=2°$. Discuss whether gyroscopic effects will be present or not. Obtain critical speeds accordingly. Obtain plots of transverse rotational displacements with spin speed due to the angular (couple) unbalance.

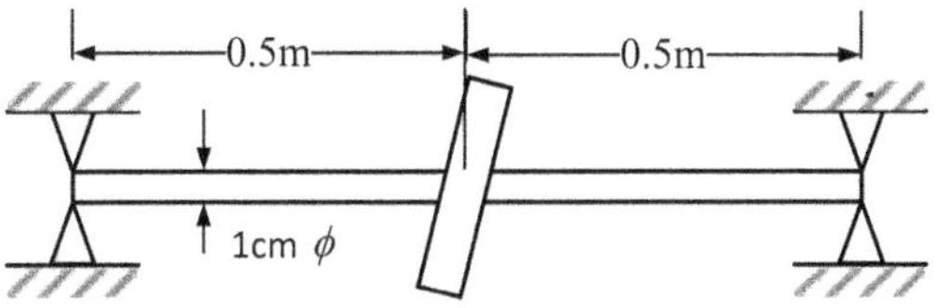

FIGURE 5.8 A simply supported rotor with a tilted thin disc.

***Solution*:** The disc is in the middle of the shaft. It has no radial eccentricity but has a tilt. The radial eccentricity gives unbalance force, which makes the rotor whirl about its bearing axis. Due to disc tilt, there will be a couple unbalance of magnitude $(I_p - I_d)\omega^2\varphi_0$. For the present case, the disc is at midspan of the shaft and this external unbalance couple will give pure tilting motion of the disc about its centre of gravity, which will lie on the bearing axis. So, a pure wobbling motion will take place. Since both spinning and wobbling (precision) motion is taking place so there will be a gyroscopic couple acting on the disc.

Equations of motion for a pure transverse rotational motion are given as (refer eqs. (5.141) and (5.142) of Tiwari (2017))

$$I_d\ddot{\varphi}_y - I_p\omega\dot{\varphi}_x + \alpha_{\varphi M}\varphi_y = -\left(I_p - I_d\right)\varphi_0\omega^2\cos\left(\omega t - \phi\right) \tag{5.43}$$

$$I_d\ddot{\varphi}_x + I_p\omega\dot{\varphi}_y + \alpha_{\varphi M}\varphi_x = -\left(I_p - I_d\right)\varphi_0\omega^2\sin\left(\omega t - \phi\right) \tag{5.44}$$

where I_p and I_d are the polar and diametral mass moments of inertia of the disc, φ_x and φ_y are disc tilt about the x and y directions, respectively, during the wobbling motion, ω is the spin speed of the shaft, $\alpha_{\varphi M}$ is the influence coefficient corresponding to the tilt motion, ϕ is the phase of unbalance moment due to initial tilt, φ_0, with respect to a fixed reference point on the shaft. For a simply supported shaft (refer Appendix 2.1), we have

$$\alpha_{22} = -\frac{\left(3al - 3a^2 - l^2\right)}{3EIl} \tag{5.45}$$

for $a = l/2$, we have

$$\alpha_{\varphi M} = \alpha_{22} = \frac{l}{12EI} \tag{5.46}$$

For the analysis, Eqs. (5.43) and (5.44) are combined by defining a complex angular displacement, as

$$\chi = \varphi_y + j\varphi_x \tag{5.47}$$

So, we get

$$\begin{aligned} &I_d\left(\ddot{\varphi}_y + j\ddot{\varphi}_x\right) + I_p\omega\left(-\dot{\varphi}_x + j\dot{\varphi}_y\right) + \left(1/\alpha_{\varphi M}\right)\left(\varphi_y + j\varphi_x\right) \\ &= -\left(I_p - I_d\right)\varphi_0\omega^2\left\{\cos\left(\omega t - \phi\right) + j\sin\left(\omega t - \phi\right)\right\} \end{aligned} \tag{5.48}$$

$$\text{or}\quad I_d\ddot{\chi} + jI_p\omega\dot{\chi} + \left(1/\alpha_{\varphi M}\right)\chi = -\left(I_p - I_d\right)\varphi_0\omega^2 e^{j(\omega t - \phi)} \tag{5.49}$$

Let us assume the response solution as

$$\chi(t) = \bar{\chi}e^{j\omega t} \tag{5.50}$$

where $\bar{\chi}$ is the complex response. Therefore, we have

$$\dot{\chi}(t) = j\omega\bar{\chi}e^{\omega t} \quad \text{and} \quad \ddot{\chi}(t) = -\omega^2\bar{\chi}e^{\omega t} \tag{5.51}$$

On substituting Eqs. (5.50) and (5.51) in Eq. (5.49), we get

$$\bar{\chi} = \frac{-(I_p - I_d)\varphi_0\omega^2}{\{(1/\alpha_{\varphi M}) - \omega^2(I_p + I_d)\}} e^{j\phi} \tag{5.52}$$

On equating the denominator to zero (for this condition the spin speed will be the critical speed), we get

$$\alpha_{\varphi M} - \omega_{cr}^2(I_p + I_d) = 0 \quad \Rightarrow \omega_{cr} = \sqrt{\frac{1/\alpha_{\varphi M}}{(I_p + I_d)}} = \sqrt{\frac{12EI}{(I_p + I_d)l}} = \sqrt{\frac{4EI}{I_d l}} \tag{5.53}$$

$$\text{with} \quad I_p = 2I_d$$

On plotting Eq. (5.52), with respect to rotor spin speed, we can expect resonance at the critical speed mentioned in Eq. (5.53). Similar exercise can be performed for other boundary conditions (with change in corresponding influence coefficients and other geometrical properties) also with single disc rotor system.

Exercise 5.6 A shaft of modulus of rigidity, EI, of total length, l, is simply supported at its ends. At a quarter length between the end bearings, the shaft carries a thin disc of mass, m, and of diameter moment of inertia, I_d. Find (a) the non-rotational natural frequency, and (b) the transverse whirl natural frequency.

***Solution*:** For simply supported end conditions with an axially off-set disc, (refer Figure 5.9) we have the following influence coefficients (refer Appendix 2.1)

$$\begin{Bmatrix} y \\ \varphi_x \end{Bmatrix} = \begin{bmatrix} \alpha_{11} & \alpha_{12} \\ \alpha_{21} & \alpha_{22} \end{bmatrix} \begin{Bmatrix} f \\ M \end{Bmatrix} \tag{5.54}$$

$$\text{with} \quad \alpha_{11} = \frac{a^2b^2}{3EIl}; \quad \alpha_{12} = -\frac{(3a^2l - 2a^3 - al^2)}{3EIl}; \quad \alpha_{21} = \frac{ab(b-a)}{3EIl}; \quad \alpha_{22} = -\frac{(3al - 3a^2 - l^2)}{3EIl}.$$

For the present case, we have $b = l/4$ and $a = 3l/4$, so that the influence coefficients become

$$\alpha_{11} = \frac{3l^3}{256EI}; \quad \alpha_{12} = \alpha_{21} = -\frac{l^2}{32EI}; \quad \alpha_{22} = \frac{7l}{48EI} \tag{5.55}$$

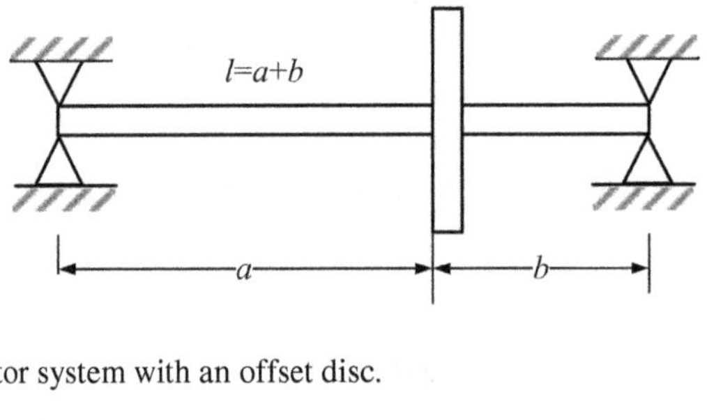

FIGURE 5.9 A simple rotor system with an offset disc.

So that the elastic coupling can be obtained as

$$\bar{\alpha} = \frac{\alpha_{12}^2}{\alpha_{11}\alpha_{22}} = \frac{4}{7} \tag{5.56}$$

and the disc effect takes the following form

$$\mu = \frac{I_d \alpha_{22}}{m\alpha_{11}} = \frac{112 I_d}{9ml^2} \tag{5.57}$$

Let us assume $m = 1$ kg, $l = 1$ m, $d = 0.01$ m and $I_d = 9/112$ kg-m^2. So that coupling effect becomes $\mu = 1$. The frequency equation (refer eq. (5.78) of Tiwari (2017)), for a general motion, is given as

$$\bar{v}^4 - 2\bar{\omega}\bar{v}^3 + \frac{\mu+1}{\mu(\bar{\alpha}-1)}\bar{v}^2 - \frac{2\bar{\omega}}{\bar{\alpha}-1}\bar{v} - \frac{1}{\mu(\bar{\alpha}-1)} = 0 \tag{5.58}$$

On substituting various assumed values, we get

$$\bar{v}^4 - 2\bar{\omega}\bar{v}^3 + \frac{1+1}{(4/7-1)}\bar{v}^2 - \frac{2\bar{\omega}}{(4/7-1)}\bar{v} - \frac{1}{1\times(4/7-1)} = 0$$

$$\text{or} \quad \bar{v}^4 - 2\bar{\omega}\bar{v}^3 - (14/3)\bar{v}^2 + (14/3)\bar{\omega}\bar{v} + (7/3) = 0 \tag{5.59}$$

Above frequency equation may be plotted to show the variation of the whirl natural frequency (i.e. the frequency ratio) with spin speed (i.e. the speed ratio) of the shaft.

Part (a): For non-rotating case, we have $\bar{\omega} = 0$, so that from Eqs. (5.58) and (5.59), we have

$$\bar{v}^4 + \frac{\mu+1}{\mu(\bar{\alpha}-1)}\bar{v}^2 - \frac{1}{\mu(\bar{\alpha}-1)} = 0 \tag{5.60}$$

$$\text{or} \quad \bar{v}^4 - (14/3)\bar{v}^2 + (7/3) = 0 \tag{5.61}$$

which gives $\bar{v}_{1,2} = 0.5695 \quad \text{and} \quad 4.0972$

So that we have

$$\bar{v} = \omega_{nf}\sqrt{\alpha_{11}m} \quad \Rightarrow \omega_{nf1} = \frac{\bar{v}_1}{\sqrt{\alpha_{11}m}} = 70.79 \text{ rad/s}; \quad \omega_{nf2} = \frac{\bar{v}_2}{\sqrt{\alpha_{11}m}} = 189.84 \text{ rad/s.}$$

which are non-rotating natural frequencies. These are two in numbers due to two-DOF of the system, and in the absence of the gyroscopic couple, the orthogonal transverse planes are uncoupled. In Part (b), when will consider the rotation of the shaft then due to the coupling of two orthogonal planes we will have 4-DOF and maximum of four natural frequencies are expected at a given speed. Of course, in that case, natural frequency will change with speed also.

Part (b): For obtaining critical speeds for the rotating case, the following steps can be performed. For the forward whirl critical speed, we have $\bar{v} = +\bar{\omega}$, so that from Eq. (5.59), we get

$$\bar{v}^4 - 2\bar{\omega}\bar{v}^3 - (14/3)\bar{v}^2 + (14/3)\bar{\omega}\bar{v} + (7/3) = 0 \quad \Rightarrow \bar{v}^4 - 2\bar{v}^4 - (14/3)\bar{v}^2 + (14/3)\bar{v}^2 + (7/3) = 0$$

$$\text{or} \quad -\bar{v}^4 + (7/3) = 0$$

which gives $\bar{\nu}^2 = \pm 2.3333$.

On neglecting the negative value, we get

$$\bar{\nu}^2 = 2.3333 = \left(\omega_{cr1}^F \sqrt{\alpha_{11} m}\right)^2 \quad \Rightarrow \omega_{cr1}^F = \sqrt{\frac{2.3333}{\alpha_{11} m}} = 115.92 \text{ rad/s}$$

For the backward whirl critical speed, we have $\bar{\nu} = -\bar{\omega}$, so that from Eq. (5.59), we get

$$\bar{\nu}^4 - 2\bar{\omega}\bar{\nu}^3 - (14/3)\bar{\nu}^2 + (14/3)\bar{\omega}\bar{\nu} + (7/3) = 0 \quad \Rightarrow \bar{\nu}^4 + 2\bar{\nu}^4 - (14/3)\bar{\nu}^2 - (14/3)\bar{\nu}^2 + (7/3) = 0$$

$$\text{or} \quad 3\bar{\nu}^4 - 28/3\bar{\nu}^2 + 7/3 = 0$$

which gives $\bar{\nu} = 0.5236$ and 1.6843

Both values are positive, therefore two backward whirls are possible, as

$$\bar{\nu} = \omega_{cr}^B \sqrt{\alpha_{11} m} \quad \Rightarrow \omega_{cr}^B = \frac{\bar{\nu}}{\sqrt{\alpha_{11} m}}$$

$$\text{For} \quad \bar{\nu} = 0.5236 \quad \Rightarrow \omega_{cr1}^B = 49.11 \text{ rad/s}; \quad \text{and} \quad \text{for} \quad \bar{\nu}_2 = 1.6843 \quad \Rightarrow \omega_{cr2}^B = 157.97 \text{ rad/s}.$$

Corresponding Campbell diagram is shown in Figure 5.10. The solid lines are forward whirl natural frequency and dashed lines are backward whirl frequencies. The synchronous whirl condition is represented by a line ($\nu = \omega$) and wherever it cuts frequency lines we have corresponding critical speeds (two backward critical speeds and one forward whirl critical speed).

Exercise 5.7 A shaft of total length l on the end bearings carries two discs at the quarter-length points. The discs have mass m and diametral mass moment of inertia I_d and the shaft modulus of rigidity is EI. (i) Derive equations for the whirling shaft, where the whirling frequency is equal to spin speed, (ii) make the frequency equation dimensionless in terms of the critical speed function

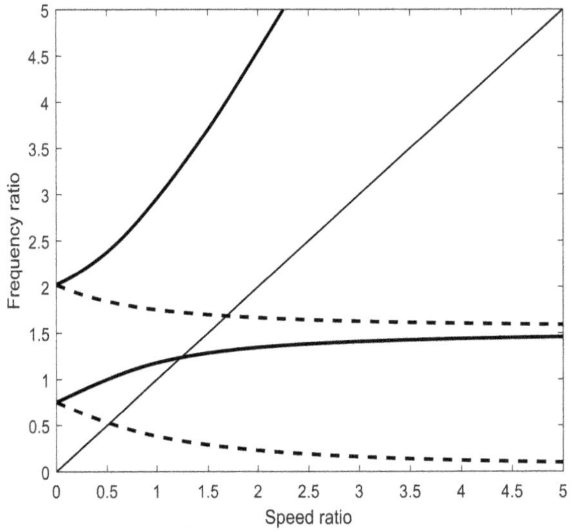

FIGURE 5.10 Campbell diagram for the rotor system.

$\bar{\omega}_{cr}^2 = \dfrac{ml^3\omega^2}{EI}$ and the disc effect $\mu = \dfrac{I_d}{ml^2}$, and (iii) find the whirling speed for the following three cases: (i) $\mu = 0$, (ii) $\mu \to \infty$, and (iii) $\mu = 1/12$.

***Solution*:** For the synchronous whirl case for each disc we have δ_1 and δ_2 translational displacements and φ_1 and φ_2 rotation displacements. We have the following forces and moments on these discs

$$F_{y1} = m\omega^2\delta_1; \quad M_{yz1} = I_d\omega^2\varphi_{x1} \quad F_{y2} = m\omega^2\delta_2; \quad M_{yz2} = I_d\omega^2\varphi_{x2} \tag{5.62}$$

These are related as

$$\begin{aligned} &\delta_1 = \alpha_{11}F_{y1} + \alpha_{12}M_{yz1} + \alpha_{13}F_{y2} + \alpha_{14}M_{yz2}; \quad \varphi_1 = \alpha_{21}F_{y1} + \alpha_{22}M_{yz1} + \alpha_{23}F_{y2} + \alpha_{24}M_{yz2} \\ &\delta_2 = \alpha_{31}F_{y1} + \alpha_{32}M_{yz1} + \alpha_{33}F_{y2} + \alpha_{34}M_{yz2}; \quad \varphi_2 = \alpha_{41}F_{y1} + \alpha_{42}M_{yz1} + \alpha_{43}F_{y2} + \alpha_{44}M_{yz2} \end{aligned} \tag{5.63}$$

But because of the reciprocal theorem, we have

$$\alpha_{12} = \alpha_{21}; \quad \alpha_{13} = \alpha_{31}; \quad \alpha_{14} = \alpha_{41}; \quad \alpha_{23} = \alpha_{32}; \quad \alpha_{24} = \alpha_{42}; \quad \alpha_{34} = \alpha_{43} \tag{5.64}$$

and because of symmetry (equal disc at identical location), we have

$$\alpha_{11} = \alpha_{33}; \quad \alpha_{22} = \alpha_{44}; \quad \alpha_{12} = \alpha_{34}; \quad \alpha_{23} = \alpha_{41}; \tag{5.65}$$

On combining expressions (5.64) and (5.65), we get

$$\begin{aligned} &\alpha_{11} = \alpha_{33} = \alpha_{y1y1}; \quad \alpha_{22} = \alpha_{44} = \alpha_{\varphi1\varphi1}; \quad \alpha_{12} = \alpha_{21} = \alpha_{34} = \alpha_{43} = \alpha_{y1\varphi1}; \\ &\alpha_{13} = \alpha_{31} = \alpha_{y1y2}; \quad \alpha_{24} = \alpha_{42} = \alpha_{\varphi1\varphi2}; \quad \alpha_{14} = \alpha_{41} = \alpha_{23} = \alpha_{32} = \alpha_{y1\varphi2} \end{aligned} \tag{5.66}$$

Therefore, from Eq. (55.63), we have

$$\begin{aligned} \delta_1 &= \alpha_{y1y1}m\omega^2\delta_1 + \alpha_{y1\varphi1}I_d\omega^2\varphi_{x1} + \alpha_{y1y2}m\omega^2\delta_2 + \alpha_{y1\varphi2}I_d\omega^2\varphi_{x2} \\ \varphi_1 &= \alpha_{y1\varphi1}m\omega^2\delta_1 + \alpha_{\varphi1\varphi1}I_d\omega^2\varphi_{x1} + \alpha_{y1\varphi2}m\omega^2\delta_2 + \alpha_{\varphi1\varphi2}I_d\omega^2\varphi_{x2} \\ \delta_2 &= \alpha_{y1y2}m\omega^2\delta_1 + \alpha_{y1\varphi2}I_d\omega^2\varphi_{x1} + \alpha_{y1y1}m\omega^2\delta_2 + \alpha_{y1\varphi1}I_d\omega^2\varphi_{x2} \\ \varphi_2 &= \alpha_{y1\varphi2}m\omega^2\delta_1 + \alpha_{\varphi1\varphi2}I_d\omega^2\varphi_{x1} + \alpha_{y1\varphi1}m\omega^2\delta_2 + \alpha_{\varphi1\varphi1}I_d\omega^2\varphi_{x2} \end{aligned} \tag{5.67}$$

which can be written as

$$\begin{bmatrix} \left(\alpha_{y1y1}m\omega^2 - 1\right) & \alpha_{y1\varphi1}I_d\omega^2 & \alpha_{y1y2}m\omega^2 & \alpha_{y1\varphi2}I_d\omega^2 \\ \alpha_{y1\varphi1}m\omega^2 & \left(\alpha_{\varphi1\varphi1}I_d\omega^2 - 1\right) & \alpha_{y1\varphi2}m\omega^2 & \alpha_{\varphi1\varphi2}I_d\omega^2 \\ \alpha_{y1y2}m\omega^2 & \alpha_{y1\varphi2}I_d\omega^2 & \left(\alpha_{y1y1}m\omega^2 - 1\right) & \alpha_{y1\varphi1}I_d\omega^2 \\ \alpha_{y1\varphi2}m\omega^2 & \alpha_{\varphi1\varphi2}I_d\omega^2 & \alpha_{y1\varphi1}m\omega^2 & \left(\alpha_{\varphi1\varphi1}I_d\omega^2 - 1\right) \end{bmatrix} \begin{Bmatrix} y_1 \\ \varphi_{x1} \\ y_2 \\ \varphi_{x2} \end{Bmatrix} = \begin{Bmatrix} 0 \\ 0 \\ 0 \\ 0 \end{Bmatrix} \tag{5.68}$$

Since both mass and diametral mass moment of inertia are given so for two-disc system total degrees of freedom will be 4. Hence, we need to relate four displacements with corresponding forces and moments using an influence coefficient matrix of 4 × 4 size. The frequency equation will be then of fourth-degree polynomial. It will be too cumbersome to solve in closed form even for obtaining critical speeds. So the idea of this exercise is to have feel of complexity in analysis of such a simple two-disc system with quasi-static analysis, and for such cases, the dynamic analysis is preferred. For forming eigenvalue problem (refer chapter 8 of Tiwari (2017)) from the aforementioned equation, we have

$$\left(\omega^2\begin{bmatrix}\alpha_{y1y1}m & \alpha_{y1\varphi1}I_d & \alpha_{y1y2}m & \alpha_{y1\varphi2}I_d\\ \alpha_{y1\varphi1}m & \alpha_{\varphi1\varphi1}I_d & \alpha_{y1\varphi2}m & \alpha_{\varphi1\varphi2}I_d\\ \alpha_{y1y2}m & \alpha_{y1\varphi2}I_d & \alpha_{y1y1}m & \alpha_{y1\varphi1}I_d\\ \alpha_{y1\varphi2}m & \alpha_{\varphi1\varphi2}I_d & \alpha_{y1\varphi1}m & \alpha_{\varphi1\varphi1}I_d\end{bmatrix}-\begin{bmatrix}1&0&0&0\\0&1&0&0\\0&0&1&0\\0&0&0&1\end{bmatrix}\right)\begin{Bmatrix}y_1\\ \varphi_{x1}\\ y_2\\ \varphi_{x2}\end{Bmatrix}=\begin{Bmatrix}0\\0\\0\\0\end{Bmatrix} \tag{5.69}$$

which can be written as

$$(\mathbf{A}-\lambda\mathbf{I})=\mathbf{0} \tag{5.70}$$

$$\text{with}\quad \lambda=1/\omega^2;\quad \mathbf{A}=\begin{bmatrix}\alpha_{y1y1}m & \alpha_{y1\varphi1}I_d & \alpha_{y1y2}m & \alpha_{y1\varphi2}I_d\\ \alpha_{y1\varphi1}m & \alpha_{\varphi1\varphi1}I_d & \alpha_{y1\varphi2}m & \alpha_{\varphi1\varphi2}I_d\\ \alpha_{y1y2}m & \alpha_{y1\varphi2}I_d & \alpha_{y1y1}m & \alpha_{y1\varphi1}I_d\\ \alpha_{y1\varphi2}m & \alpha_{\varphi1\varphi2}I_d & \alpha_{y1\varphi1}m & \alpha_{\varphi1\varphi1}I_d\end{bmatrix};\quad \boldsymbol{\eta}=\begin{Bmatrix}y_1\\ \varphi_{x1}\\ y_2\\ \varphi_{x2}\end{Bmatrix} \tag{5.71}$$

Transfer matrix method (TMM) (refer chapter 8 of Tiwari (2017)) and finite element method (FEM) (refer chapter 9 of Tiwari (2017)) can also be used to numerically solve this problem. A closed-form analytical expression will be too cumbersome to present here.

Exercise 5.8 Obtain the forward and backward transverse critical speeds of the rotor system shown in Figure 5.11. Take the shaft as rigid. It is assumed that it oscillates (precesses) about its centre of gravity while whirling (i.e., pure tilting without translational motion). The effective moment (tilt) stiffness of the bearings on the rotor is k_t = 2,000 N-m/rad, the polar and diametral mass moments of inertia of the rotor are 0.03 kg-m^2 and 0.2 kg-m^2, respectively. Consider the gyroscopic effects.

Solution: In the present case, since the whole rotor is having pure rotational motion, φ_x, with frequency, ν, in the transverse plane along with its spinning, ω, so the change in the angular momentum of the rotor, having polar mass moment of inertia I_p, is given as (refer eq. (5.59) of Tiwari (2017))

$$\frac{\Delta(I_p\omega)}{I_p\omega}=\varphi_x\nu dt \tag{5.72}$$

where dt is a small time duration. The rate of change of angular momentum is then given as

$$\frac{d(I_p\omega)}{dt}=I_p\omega\varphi_x\nu \tag{5.73}$$

which is nothing but the gyroscopic effect. Overall tilting moment, due to support stiffness, k_t, and the gyroscopic moment, is given as

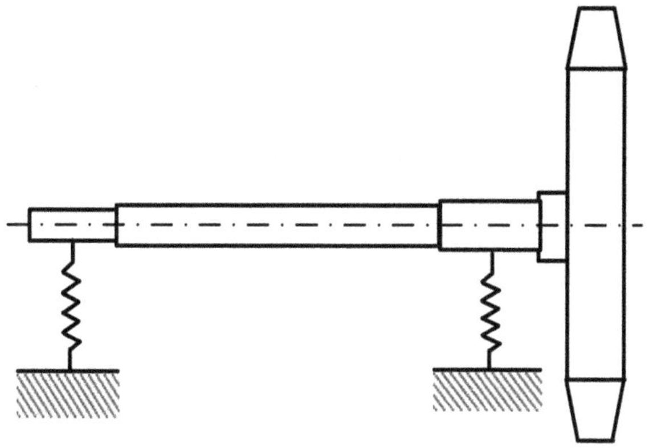

FIGURE 5.11 An overhung rotor supported on flexible bearings.

$$M_x = k_{eff}\varphi_x = k_t\varphi_x \pm I_P\omega\varphi_x\nu = (k_t \pm I_P\omega\nu)\varphi_x \tag{5.74}$$

where the positive sign for the forward whirl and the negative sign for the backward whirl. Overall tilting stiffness is given as

$$k_{eff} = M_x / \varphi_x = (k_t \pm I_P\omega\nu) \tag{5.75}$$

Whirl natural frequencies are given as

$$\nu^2 = \frac{(k_t \pm I_p\omega\nu)}{I_d} \tag{5.76}$$

For the critical speed, we have $\omega = \nu$, so that

$$I_d\nu^2 = (k_t \pm I_p\nu^2);\Rightarrow \quad (I_d \mp I_p)\nu^2 = k_t;\Rightarrow \quad \nu^2 = \left(\frac{k_t}{I_d \mp I_p}\right) \tag{5.77}$$

It should be noted that for the present case $I_d > I_p$ to have feasible forward whirl natural frequency. The following rotor data are given

$$k_{eff} = 2{,}000 \text{ Nm/rad}; \quad I_p = 0.03 \text{ kg-m}^2; \quad I_d = 0.2 \text{ kg-m}^2$$

On substituting values, we get

$$\nu_{cr}^F = \sqrt{\frac{2{,}000}{0.2 - 0.03}} = 108.47 \text{ rad/s} \quad \text{and} \quad \nu_{cr}^B = \sqrt{\frac{2{,}000}{0.23}} = 93.25 \text{ rad/s}$$

For the case when the gyroscopic effect is not considered then the whirl natural frequency will be constant and equal to the critical speed, we have

$$\nu = \sqrt{\frac{k_t}{I_d}} = \sqrt{\frac{2{,}000}{0.2}} = 100.00 \text{ rad/s} \tag{5.78}$$

It can be observed that the critical speed without the gyroscopic effect is bounded by the forward and backward critical speeds obtained with the gyroscopic effect. In fact, due to the gyroscopic effect splitting of natural frequencies (as well as critical speeds) can be observed.

Exercise 5.9 A massless shaft of total length, l between bearings (simply supports) carries at its centre a disc of diameteral mass moment of inertia, I_d. The disc is keyed on at a small angle φ_0. When rotating at constant angular speed, ω, the centrifugal forces tend to diminish the angle, φ_0, to a new value, $(\varphi_0 - \varphi)$. Find the ratio, φ/φ_0, as a function of the spin speed, ω.

***Solution*:** The disc is in the middle of the shaft. It has no radial eccentricity but has a tilt. The radial eccentricity gives an unbalance force, which makes the rotor to whirl about its bearing axis. Due to an initial tilt of the disc, there will be a couple unbalance of magnitude $(I_p - I_d)\omega^2\varphi_0$ (refer eq. (5.52) of Tiwari (2017) and eq. (5.32)). This external unbalance couple will give a pure tilting motion of the disc about its centre of gravity, which will lie on the bearing axis. So, a pure wobbling motion will take place. Since both spinning and wobbling (precisional) motion is taking place so there will be a gyroscopic couple acting on the disc. For a simply supported shaft (refer Appendix 2.1), we have

$$\alpha_{22} = \alpha_{\varphi M} = \frac{\varphi}{M} = -\frac{\left(3al - 3a^2 - l^2\right)}{3EIl} \tag{5.79}$$

for $a = l/2$, we have

$$\alpha_{22} = \alpha_{\varphi M} = -\frac{\left(3l^2/2 - 3l^2/4 - l^2\right)}{3EIl} = -\frac{\left(3l^2/2 - 3l^2/4 - l^2\right)}{3EIl} = -\frac{-l^2/4}{3EIl} = \frac{l}{12EI} \tag{5.80}$$

Equations of motion for a pure transverse rotational motion are given as (refer eqs. (5.141) and (5.142) of Tiwari (2017))

$$I_d\ddot{\varphi}_y - I_p\omega\dot{\varphi}_x + \left(1/\alpha_{\varphi M}\right)\varphi_y = -\left(I_p - I_d\right)\varphi_0\omega^2\cos\left(\omega t - \phi\right) \tag{5.81}$$

$$I_d\ddot{\varphi}_x + I_p\omega\dot{\varphi}_y + \left(1/\alpha_{\varphi M}\right)\varphi_x = -\left(I_p - I_d\right)\varphi_0\omega^2\sin\left(\omega t - \phi\right) \tag{5.82}$$

where I_p and I_d are the polar and diametral mass moments of inertia of the disc, φ_x and φ_y are disc tilts about x and y directions, respectively, during wobbling motion, ω is the spin speed of the shaft, $\alpha_{\varphi M}$ is the influence coefficient corresponding to tilt motion, ϕ is the phase of unbalance moment due to initial tilt of the disc, φ_0, with respect to a fixed reference point on the shaft. For analysis Eqs. (5.81) and (5.82) are combined by defining complex angular displacement $\chi = \varphi_y + j\varphi_x$, we get

$$I_d\left(\ddot{\varphi}_y + j\ddot{\varphi}_x\right) + I_p\omega\left(-\dot{\varphi}_x + j\dot{\varphi}_y\right) + \left(1/\alpha_{\varphi M}\right)\left(\varphi_y + j\varphi_x\right) = -\left(I_p - I_d\right)\varphi_0\omega^2\left\{\cos\left(\omega t - \phi\right) + j\sin\left(\omega t - \phi\right)\right\} \tag{5.83}$$

$$\text{with } \alpha_{\varphi M} = \frac{l}{12EI} \text{ and } I_p = 2I_d$$

$$\text{or} \quad I_d\ddot{\chi} + jI_p\omega\dot{\chi} + \left(1/\alpha_{\varphi M}\right)\chi = -\left(I_p - I_d\right)\varphi_0\omega^2 e^{j(\omega t - \phi)} \tag{5.84}$$

Let us assume the solution as

$$\chi(t) = \bar{\chi} e^{j\omega t} \tag{5.85}$$

where $\bar{\chi}$ is the complex response amplitude. So that, we have

$$\dot{\chi}(t) = j\omega\bar{\chi}e^{\omega t} \quad \text{and} \quad \ddot{\chi}(t) = -\omega^2\bar{\chi}e^{\omega t} \tag{5.86}$$

On substituting in Eq. (5.83), we get

$$\chi(\omega) = \frac{-\left(I_p - I_d\right)\varphi_0\omega^2}{\left\{\left(1/\alpha_{\varphi M}\right) - \omega^2\left(I_p + I_d\right)\right\}} e^{-j\phi} \tag{5.87}$$

which can be split into the real and imaginary parts, as

$$\frac{\varphi_x(\omega)}{\varphi_0} = \bar{\varphi}_x = \frac{-\left(I_p - I_d\right)\omega^2}{\left\{\left(1/\alpha_\varphi\right) - \omega^2\left(I_p + I_d\right)\right\}} \cos\phi \tag{5.88}$$

$$\text{and} \quad \frac{\varphi_y(\omega)}{\varphi_0} = \bar{\varphi}_y = \frac{\left(I_p - I_d\right)\omega^2}{\left\{\left(1/\alpha_\varphi\right) - \omega^2\left(I_p + I_d\right)\right\}} \sin\phi \tag{5.89}$$

From the denominator, the critical speed is given as

$$\omega_{cr} = \sqrt{\frac{1}{\alpha_\varphi\left(I_p + I_d\right)}} = \sqrt{\frac{EI}{12l\left(I_p + I_d\right)}} \tag{5.90}$$

For plotting purposes chosen data are

$I_p = 0.02$ kg-m^2; $I_d = I_p/2$ kg-m^2; $\varphi_0 = 30°$; $a = 0.5$ m; $b = 0.5$ m; $l = a + b$;

$d = 0.01$ m; $E = 2.1 \times 10^{11}$ N/m^2.

On plotting Eq. (5.89), with respect to rotor spin speed for its magnitude ratio of the disc tilt can be observed (see Figure 5.12). At critical speeds, we have the resonance condition.

On subtracting both sides by φ_0, we get

$$\varphi_x(\omega) - \varphi_0 = \frac{-\left(I_p - I_d\right)\varphi_0\omega^2}{\left\{\left(1/\alpha_\varphi\right) - \omega^2\left(I_p + I_d\right)\right\}} \cos\phi - \varphi_0 = \left[\frac{-\left(I_p - I_d\right)\omega^2 \cos\phi}{\left\{\left(1/\alpha_\varphi\right) - \omega^2\left(I_p + I_d\right)\right\}} - 1\right]\varphi_0$$

$$\text{or} \quad \varphi_x(\omega) - \varphi_0 = \left[\frac{-\left(I_p - I_d\right)\omega^2 \cos\phi - \left\{\left(1/\alpha_\varphi\right) - \omega^2\left(I_p + I_d\right)\right\}}{\left\{\left(1/\alpha_\varphi\right) - \omega^2\left(I_p + I_d\right)\right\}}\right]\varphi_0$$

$$\text{or} \quad \frac{\varphi_0 - \varphi_x(\omega)}{\varphi_0} = \left[\frac{\left(I_p - I_d\right)\omega^2 \cos\phi + \left\{\left(1/\alpha_\varphi\right) - \omega^2\left(I_p + I_d\right)\right\}}{\left\{\left(1/\alpha_\varphi\right) - \omega^2\left(I_p + I_d\right)\right\}}\right] \tag{5.91}$$

Exercise 5.10 Consider a Jeffcott rotor, a simply supported shaft with a central disc. Consider only the titling mode of vibration with no translational displacement. (i) Find the natural frequency of this motion for a non-rotating disc. (ii) Find the natural frequency (or frequencies), ω_{nf}, for the case

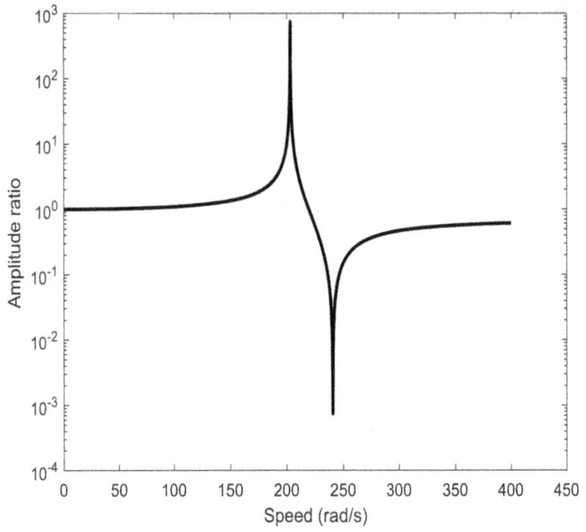

FIGURE 5.12 Variation of amplitude ratio with spin speed.

of a disc rotating with angular speed, ω. (iii) Plot the variation of the obtained natural frequency with respect to the angular speed of the rotor.

Solution*: *Case I: For non-rotating simply supported shaft with disc at mid-span and for a pure translational motion (refer Figure 5.13a) of disc (refer Case II in page 218 of Tiwari (2017)), the natural frequency is given as (refer Appendix 2.1 for influence coefficient)

$$\omega_{nf1} = \frac{1}{\sqrt{m\alpha_{11}}} \quad \text{with} \quad \alpha_{11} = \frac{l^3}{12EI} \tag{5.92}$$

For non-rotating simply supported shaft with disc at mid-span and for a pure tilting motion of disc (refer Figure 5.13b), the natural frequency is given as (refer Appendix 2.1 for influence coefficient)

$$\omega_{nf2} = \frac{1}{\sqrt{I_d\alpha_{22}}} \quad \text{with} \quad \alpha_{22} = \frac{l}{12EI} \tag{5.93}$$

***Case II*:** $\bar{\alpha} = 0$, i.e., no elastic coupling between the translational and rotational displacements. For this case, a force causes the translational deflection only without any rotational deflection φ_x, while a moment causes the rotational deflection φ_x only without any translational deflection y. For a shaft with simple supports and a disc at the mid-span, such a case is possible as shown in Figure 5.13.

For $\bar{\alpha} = 0$, eq. (5.78) of Tiwari (2017) reduces to

$$\bar{v}^4 - 2\bar{\omega}\bar{v}^3 - \frac{\mu+1}{\mu}\bar{v}^2 + 2\bar{\omega}\bar{v} + \frac{1}{\mu} = 0 \tag{5.94}$$

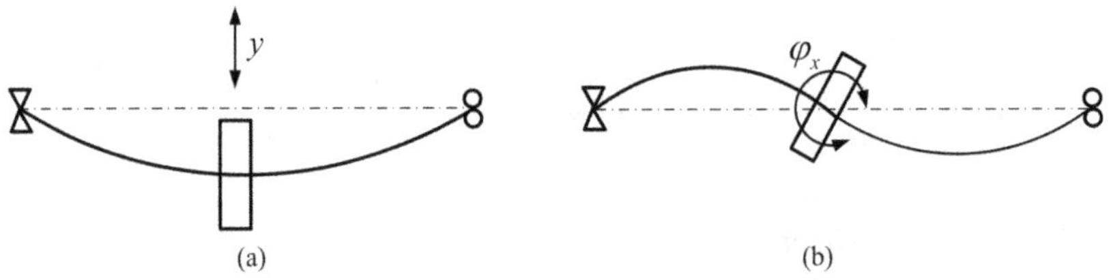

FIGURE 5.13 A Jeffcott rotor with the disc motion as (a) the pure translation and (b) the pure rotational.

Equation (5.94) can be simplified in the following steps

$$\bar{\nu}^4 - 2\bar{\omega}\bar{\nu}^3 + 2\bar{\omega}\bar{\nu} - \frac{\mu+1}{\mu}\bar{\nu}^2 + \frac{1}{\mu} = 0 \quad \text{or} \quad (\bar{\nu}+1)(\bar{\nu}-1)\left(\bar{\nu}^2 - 2\bar{\omega}\bar{\nu} - \frac{1}{\mu}\right) = 0 \tag{5.95}$$

Equation (5.95) gives whirl frequencies, as

$$\bar{\nu}_1 = +1; \quad \bar{\nu}_2 = -1 \tag{5.96}$$

$$\text{and} \quad \bar{\nu}_{3,4} = \bar{\omega} \pm \sqrt{\bar{\omega}^2 + \frac{1}{\mu}} \tag{5.97}$$

The non-dimensional frequency $\bar{\nu}$ is plotted against the speed $\bar{\omega}$ for a numerical value of $\mu = 1$, as shown in Figure 5.14. For $\bar{\omega} = 0$ (in fact for any other values) from the first and second solutions, i.e. Eq. (5.96), we have $\bar{\nu} = \pm 1$, which is the natural frequency related to the up and down motions of the disc without any tilting (refer Figure 5.13a). For $\bar{\omega} = 0$ (but not for $\bar{\omega} \neq 0$) from the third and fourth solutions, i.e. Eq. (5.97), we have $\bar{\nu} = \pm\sqrt{1/\mu}$ and since $\mu = 1$, hence $\bar{\nu} = \pm 1$. This is the natural frequency related to the precessional motion of the disc, i.e. wobbling without up and down motions (refer Figure 5.13b). From Eq. (5.97), it can be seen that for the present case, the second forward critical speed (for $\bar{\nu} = \bar{\omega} = \bar{\omega}^F_{cr_2}$) would not be feasible since it gives $\bar{\omega}^F_{cr_2} = \sqrt{-1/\mu}$, and only the second backward critical speed (for $\bar{\nu} = -\bar{\omega} = \bar{\omega}^B_{cr_2}$) would be $\bar{\omega}^B_{cr_2} = \sqrt{1/(3\mu)}$. That means the gyroscopic couple reduces the number of forward critical speeds (Tondl, 1965; Shravankumar and Tiwari, 2011; and refer chapter 11 of Tiwari (2017)).

The disc effect μ is so chosen or dimensioned such that the natural frequency of the translational up and down motions is the same as the natural frequency of the wobbling (tilting) without up and down motions at zero speed, i.e. $\bar{\omega} = 0$ (see in Figure 5.14). It should be noted that these kinds of uncoupling of the translational and rotational motions is a special case with $\bar{\alpha} = 0$ and, in general, such uncoupling of motions is not feasible and for such systems all natural frequencies both the translational and rotational motions would be present with different relative magnitudes and directions. From eq. (5.77) of Tiwari (2017), we have

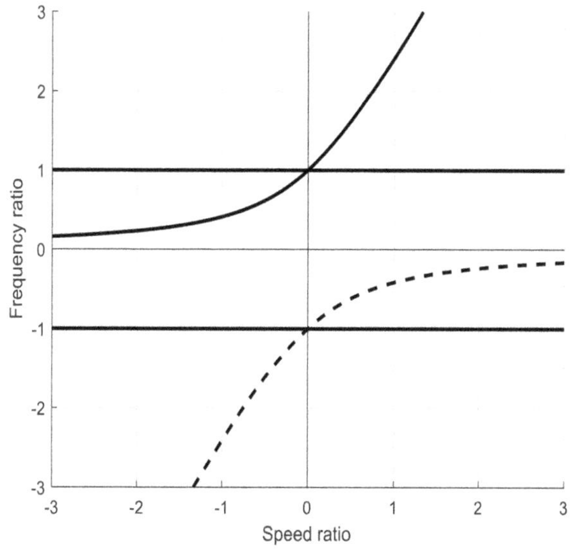

FIGURE 5.14 The variation of the whirl frequency versus the spin speed (for $\bar{\alpha} = 0, \mu = 1$).

$$\mu = \frac{I_d\alpha_{22}}{m\alpha_{11}} = 1 \quad \Rightarrow \frac{1}{\sqrt{m\alpha_{11}}} = \frac{1}{\sqrt{I_d\alpha_{22}}} \quad \Rightarrow \omega_{nf\text{linear}} = \omega_{nf\text{wobbling}} \tag{5.98}$$

Figure 5.14 shows that there are four natural frequencies at any given speed. For the positive value of $\bar{\omega}$, it could be seen that corresponding to the translational up and down motion, the natural whirl frequency remains constant and the same. Whereas, for the tilting motion corresponding to the forward whirl it increases, and for the backward whirl it decreases. This is because the gyroscopic couple is affected only by the tilting (transverse rotational) motion and there is no coupling of the transverse translational and rotational motions for the present case (i.e., $\bar{\alpha} = 0$). For $-\bar{\omega}$, the frequency curve is symmetric with respect to $\bar{\omega} = 0$ line, which means the natural frequency does not depend upon the direction of rotation of the symmetric rotor. For the zero speed, the four natural frequencies is reduced to two $\bar{\nu} = \pm 1$, which is actually one frequency only because $\bar{\nu} = +1$ corresponding to the forward whirl and $\bar{\nu} = -1$ corresponding to the backward whirl and the non-rotating shaft cannot distinguish the forward and backward whirls.

Exercise 5.11 Obtain the transverse critical speed for the synchronous motion of a rotor as shown in Figure 5.15 by considering the gyroscopic moment effects. The shaft is assumed to be simply supported at both ends, and the disc has a diameter of D and a thickness of t. It is assumed that even when the thickness of the disc is large, the disc is attached to the shaft at the location shown without interfering with the motion of the shaft. Consider the dimensions of the disc for the following four cases (i) $D = 0.2$ m, $t = 0.0082$ m; (ii) $D = 0.0721$ m, $t = 0.0628$ m; (iii) $D = 0.0689$ m, $t = 0.0689$ m; and (iv) $D = 0.0547$ m, $t = 0.1093$ m. The shaft is assumed to be massless, and its length and diameter are 1 m and 0.01 m, respectively, with $a = 0.75$ m. Let Young's modulus of the shaft material be 2.1×10^{11} N/m^2 and the density of the material be 7,800 kg/m^3.

For the simply supported shaft, the influence coefficients are defined as (refer Appendix 2.1)

$$\begin{Bmatrix} y \\ \varphi_x \end{Bmatrix} = \begin{bmatrix} \alpha_{11} & \alpha_{12} \\ \alpha_{21} & \alpha_{22} \end{bmatrix} \begin{Bmatrix} f \\ M \end{Bmatrix}$$

with $\alpha_{11} = \dfrac{a^2b^2}{3EIl}$; $\alpha_{12} = -\dfrac{\left(3a^2l - 2a^3 - al^2\right)}{3EIl}$; $\alpha_{21} = \dfrac{ab(b-a)}{3EIl}$; $\alpha_{22} = -\dfrac{\left(3al - 3a^2 - l^2\right)}{3EIl}$.

Solution: The translational and rotational displacements at the disc location of the simply supported rotor system as shown in Figure 5.16 will be (refer eqs. (5.33) and (5.34) of Tiwari (2017))

$$\delta = \alpha_{11}F_y + \alpha_{12}M_{yz} = \alpha_{11}\left(m\omega^2\delta\right) + \alpha_{12}\left(-I_d\omega^2\varphi_x\right) \tag{5.99}$$

and $$\varphi_x = \alpha_{21}F_y + \alpha_{22}M_{yz} = \alpha_{21}\left(m\omega^2\delta\right) + \alpha_{22}\left(-I_d\omega^2\varphi_x\right) \tag{5.100}$$

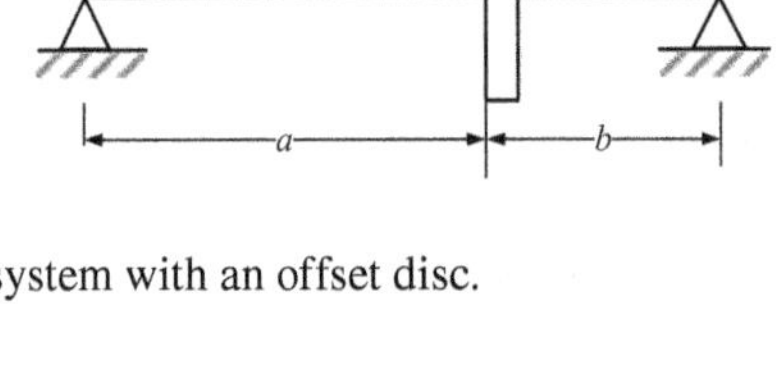

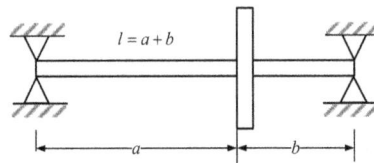

FIGURE 5.15 A Jeffcott rotor system with an offset disc.

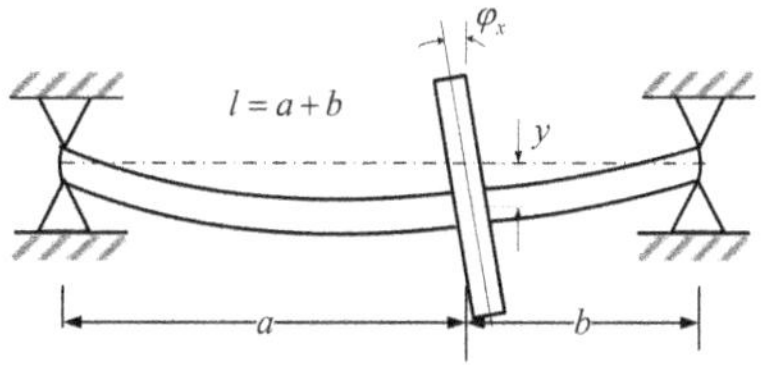

FIGURE 5.16 Simply supported rotor with an offset-disc.

$$\text{with} \quad \alpha_{11} = \frac{a^2 b^2}{3EIl}; \quad \alpha_{12} = -\frac{\left(3a^2 l - 2a^3 - al^2\right)}{3EIl}; \quad \alpha_{21} = \frac{ab(b-a)}{3EIl}; \quad \alpha_{22} = -\frac{\left(3al - 3a^2 - l^2\right)}{3EIl}.$$

Equations (5.99) and (5.100) are based on synchronous conditions. So directly critical speeds can be obtained using the aforementioned equations.

Let us retain the influence coefficients in a general form and derive the frequency equation. Equations (5.99) and (5.100) can be written as

$$\left(m\omega^2\alpha_{11} - 1\right)\delta + \left(-I_d\omega^2\alpha_{12}\right)\varphi_x = 0 \tag{5.101}$$

$$\text{and} \quad \left(m\omega^2\alpha_{21}\right)\delta + \left(-I_d\omega^2\alpha_{22} - 1\right)\varphi_x = 0 \tag{5.102}$$

which can be arranged in a matrix form, as

$$\begin{bmatrix} \left(m\omega^2\alpha_{11} - 1\right) & \left(-I_d\omega^2\alpha_{12}\right) \\ \left(m\omega^2\alpha_{21}\right) & \left(-I_d\omega^2\alpha_{22} - 1\right) \end{bmatrix} \begin{Bmatrix} \delta \\ \varphi_x \end{Bmatrix} = 0 \tag{5.103}$$

This homogeneous set of equations can have a non-trivial solution for δ and φ_x, only when the following determinant vanishes, so we get

$$\begin{vmatrix} \left(m\omega^2\alpha_{11} - 1\right) & \left(-I_d\omega^2\alpha_{12}\right) \\ \left(m\omega^2\alpha_{21}\right) & \left(-I_d\omega^2\alpha_{22} - 1\right) \end{vmatrix} = 0 \tag{5.104}$$

which gives the frequency equation (now replacing ω with ω_{cr}) as

$$\left(m\omega_{cr}^2\alpha_{11} - 1\right)\left(-I_d\omega_{cr}^2\alpha_{22} - 1\right) - \left(m\omega_{cr}^2\alpha_{21}\right)\left(-I_d\omega_{cr}^2\alpha_{12}\right) = 0$$

$$\text{or} \quad \omega_{cr}^4\left\{mI_d\left(\alpha_{11}\alpha_{22} - \alpha_{21}\alpha_{12}\right)\right\} + \omega_{cr}^2\left(m\alpha_{11} - I_d\alpha_{22}\right) - 1 = 0 \tag{5.105}$$

For a general case of disc thickness in the aforementioned equation I_d should be $I_{deff} = I_1 - I_2$, where I_1 and I_2 are the polar and diametral mass moments of inertia, respectively. For thin disc $I_{deff} = I_1 - I_2 = 2I_d - I_d = I_d$.

For a general motion (asynchronous) case from (5.76) of Tiwari (2017) (please note that it is valid for thin disc only with the use of $I_p = 2I_d$ it has been derived), we have

$$\nu^4\left(-m\alpha_{11}\alpha_{22}I_d+m\alpha_{12}^2I_d\right)+\nu^3\left(2m\alpha_{11}\alpha_{22}I_d\omega-2m\alpha_{12}^2I_d\omega\right)+\nu^2\left(\alpha_{22}I_d+m\alpha_{11}\right)+\nu\left(-\alpha_{22}I_d2\omega\right)-1=0 \tag{5.106}$$

But here also, for a general case of disc thickness, in equation I_d should be $I_{deff}=I_1-I_2$. This is valid only to get the critical speeds ($\nu=\omega$) and not for the whirl frequencies at any general speed (refer eqs. (5.67) through (5.70) of Tiwari (2017)). Hence, for the synchronous case $\nu=\omega$, so the aforementioned equation becomes

$$\nu^4mI_d\left(-\alpha_{11}\alpha_{22}+\alpha_{12}^2\right)+\nu^4mI_d\left(2\,\alpha_{11}\alpha_{22}-2\alpha_{12}^2\right)+\nu^2\left(\alpha_{22}I_d+m\alpha_{11}\right)+\nu^2\left(-2\alpha_{22}I_d\right)-1=0$$

$$\text{or}\quad \nu^4mI_d\left(\alpha_{11}\alpha_{22}-\alpha_{12}^2\right)+\nu^2\left(m\alpha_{11}-\alpha_{22}I_d\right)-1=0 \tag{5.107}$$

which is same as Eq. (5.105) with ($\nu=\omega_{cr}$).

For a pure tilting motion, taking $\alpha_{11}=\alpha_{12}=0$, so that

$$\omega_{crF}^2\left(\alpha_{22}I_d\right)-1=0\quad\Rightarrow\omega_{crF}=\sqrt{\frac{1}{I_d\alpha_{22}}} \tag{5.108}$$

Now introducing four non-dimensional variables, as

$$\begin{aligned}&\text{The dimensionless frequency:}\bar{\nu}=\nu\sqrt{\alpha_{11}m};\qquad \text{The disc effect:}\mu=\frac{I_d\alpha_{22}}{m\alpha_{11}};\\&\text{The elastic coupling:}\bar{\alpha}=\frac{\alpha_{12}^2}{\alpha_{11}\alpha_{22}};\qquad \text{The dimensionless speed:}\bar{\omega}=\omega\sqrt{m\alpha_{11}}\end{aligned} \tag{5.109}$$

Equation (5.107) can be written as

$$\omega_{cr}^4\left\{mI_d\left(\alpha_{11}\alpha_{22}-\alpha_{21}\alpha_{12}\right)\right\}+\omega_{cr}^2\left(m\alpha_{11}-I_d\alpha_{22}\right)-1=0$$

$$\text{or}\quad \bar{\omega}_{cr}^4+\bar{\omega}_{cr}^2\frac{(1-\mu)}{\mu(1-\bar{\alpha})}-\frac{1}{\mu(1-\bar{\alpha})}=0 \tag{5.110}$$

For a general motion, the frequency equation in the non-dimensional form from eq. (5.79) of Tiwari (2017) is given as

$$\bar{\nu}^4-2\bar{\omega}\bar{\nu}^3+\frac{\mu+1}{\mu(\bar{\alpha}-1)}\bar{\nu}^2-\frac{2\bar{\omega}}{\bar{\alpha}-1}\bar{\nu}-\frac{1}{\mu(\bar{\alpha}-1)}=0 \tag{5.111}$$

For synchronous case $\bar{\nu}=\bar{\omega}$, so the aforementioned equation becomes

$$\bar{\nu}^4-2\bar{\nu}^4+\frac{\mu+1}{\mu(\bar{\alpha}-1)}\bar{\nu}^2-\frac{2}{\bar{\alpha}-1}\bar{\nu}^2-\frac{1}{\mu(\bar{\alpha}-1)}=0$$

$$\text{or}\quad \bar{\nu}^4+\left(\frac{1-\mu}{\mu(1-\bar{\alpha})}\right)\bar{\nu}^2-\frac{1}{\mu(1-\bar{\alpha})}=0 \tag{5.112}$$

which is the same as Eq. (5.110). Equation (5.112) can be solved as

$$\bar{\omega}_{cr}^2 = -\frac{(1-\mu)}{2\mu(1-\bar{\alpha})} \pm \frac{1}{2}\sqrt{\left\{\frac{(1-\mu)}{\mu(1-\bar{\alpha})}\right\}^2 + \frac{4}{\mu(1-\bar{\alpha})}}$$

$$\text{or} \quad \bar{\omega}_{cr}^2 = -\frac{(1-\mu)}{2\mu(1-\bar{\alpha})} \pm \frac{1}{2\mu(1-\bar{\alpha})}\sqrt{1+2\mu(1-2\bar{\alpha})+\mu^2} \tag{5.113}$$

We have the following rotor data

$l = 1$ m, $d = 0.01$ m, $a = 0.75$ m, $b = 0.25$ m, $E = 2.1 \times 10^{11}$ N/m^2, and $\rho = 7{,}800$ kg/m^3.

$$I = \frac{\pi d^4}{64} = \frac{\pi \times 0.01^4}{64} = 4.9087 \times 10^{-10} \text{ m}^4.$$

Influence coefficients (refer Appendix 2.1) can be obtained as

$$\alpha_{11} = \frac{a^2 b^2}{3EIl} = 1.1368 \times 10^{-4} \text{ m/N}; \qquad \alpha_{12} = -\frac{\left(3a^2 l - 2a^3 - al^2\right)}{3EIl} = -3.0315 \times 10^{-4} \text{ 1/N};$$

$$\alpha_{21} = \frac{ab(b-a)}{3EIl} = -3.0315 \times 10^{-4} \text{ 1/N}; \qquad \alpha_{22} = -\frac{\left(3al - 3a^2 - l^2\right)}{3EIl} = 1.4147 \times 10^{-3} \text{ 1/N-m}.$$

The following cases are considered for obtaining critical speeds:

(i) $D = 0.2$ m, $t = 0.0082$ m

$$m = \frac{\pi \rho D^2 t}{4} = 2.0094 \text{ kg}; \quad I_d = \frac{mD^2}{16} + \frac{mt^2}{12} = 0.0050 \text{ kg-m}^2; \quad I_p = \frac{mD^2}{8} = 0.0100 \text{ kg-m}^2;$$

$$I_{deff} = I_p - I_d = \frac{mD^2}{16} - \left(\frac{mD^2}{16} + \frac{mt^2}{12}\right) = \frac{mD^2}{16} - \frac{mt^2}{12} = 0.0050 \text{ kg-m}^2;$$

$$\mu = \frac{I_{deff}\alpha_{22}}{m\alpha_{11}} = 0.0310; \quad \bar{\alpha} = \frac{\alpha_{12}^2}{\alpha_{11}\alpha_{22}} = 0.5714;$$

Ignoring the negative sign in Eq. (5.113), we get

$$\bar{\omega}_{cr} = 1.0089; \quad \Rightarrow \omega_{cr} = \frac{\bar{\omega}_{cr}}{\sqrt{m\alpha_{11}}} = 66.7538 \text{ rad/s}.$$

(ii) $D = 0.0721$ m, $t = 0.0628$ m

$$m = \frac{\pi \rho D^2 t}{4} = 1.9999 \text{ kg}; \quad I_{deff} = \frac{mD^2}{16} - \frac{mt^2}{12} = 0.0000 \text{ kg-m}^2; \quad \mu = \frac{I_{deff}\alpha_{22}}{m\alpha_{11}} = 0.0000;$$

$$\bar{\alpha} = \frac{\alpha_{12}^2}{\alpha_{11}\alpha_{22}} = 0.5714;$$

Ignoring the negative sign in Eq. (5.113), we get

$$\bar{\omega}_{cr} = 1.0000 \quad \Rightarrow \omega_{cr} = \frac{\bar{\omega}_{cr}}{\sqrt{m\alpha_{11}}} = 66.3194 \text{ rad/s.}$$

(iii) $D = 0.0689$ m, $t = 0.0689$ m

$$m = \frac{\pi \rho D^2 t}{4} = 2.0037 \text{ kg}; I_{deff} = \frac{mD^2}{16} - \frac{mt^2}{12} = -0.0002 \text{ kg-m}^2;$$

$$\mu = \frac{I_d \alpha_{22}}{m\alpha_{11}} = -0.0012; \quad \bar{\alpha} = \frac{\alpha_{12}^2}{\alpha_{11}\alpha_{22}} = 0.5714.$$

Ignoring the negative sign in Eq. (5.113), we get

$$\bar{\omega}_{cr} = 0.9996 \quad \omega_{cr} = \frac{\bar{\omega}_{cr}}{\sqrt{m\alpha_{11}}} = 66.2339 \text{ rad/s.}$$

(iv) $D = 0.0547$ m, $t = 0.1093$ m

$$m = \frac{\pi \rho D^2 t}{4} = 2.0035 \text{ kg}; \quad I_{deff} = \frac{mD^2}{16} - \frac{mt^2}{12} = -0.0016 \text{ kg-m}^2;$$

$$\mu = \frac{I_d \alpha_{22}}{m\alpha_{11}} = -0.0101; \quad \bar{\alpha} = \frac{\alpha_{12}^2}{\alpha_{11}\alpha_{22}} = 0.5714.$$

Ignoring the negative sign in Eq. (5.113), we get

$$\bar{\omega}_{cr} = 0.9971 \quad \Rightarrow \omega_{cr} = \frac{\bar{\omega}_{cr}}{\sqrt{m\alpha_{11}}} = 66.0715 \text{ rad/s.}$$

Since the given parameters are such that change in μ is negligible and $\bar{\alpha}$ remain the same so hardly any appreciable change in critical speeds is observed.

Exercise 5.12 A shaft of modulus of rigidity *EI*, length *l*, and negligible mass is simply supported at both ends. At one end it carries a thin disc of mass *m* and diametral mass moment of inertia I_d. This disc is very close to the bearing: assume it to be at the bearing so that it can tilt, but not displace its centre of gravity. (i) For the case of no rotation, find the natural frequency, (ii) Derive the general frequency equation in terms of a rotational speed, ω, and the whirl frequency, ν; make this equation dimensionless in terms of $\bar{\omega}$ and $\bar{\nu}$-functions, $\bar{\omega} = \dfrac{\omega}{\sqrt{EI/(I_d l)}}$ and $\bar{\nu} = \dfrac{\nu}{\sqrt{EI/(I_d l)}}$. Solve for the non-dimensional whirl frequency $\bar{\nu}$ in terms of the non-dimensional spin frequency ω; (iii) Is the critical speed exist for the aforementioned system?

***Solution*:** Figure 5.17 shows the rotor configuration for the present case. Towards the right end bearing a thin rigid disc mounted such that it can have only tilting motion without interfering with the right support.

For the disc placed at the right bearing, the influence coefficient is defined as (refer table 8.15 and figure 8.72 in Tiwari (2017) and Appendix 2.1 of the present book)

$$\alpha = \frac{\varphi_{x2}}{M} = \frac{l}{3EI} \tag{5.114}$$

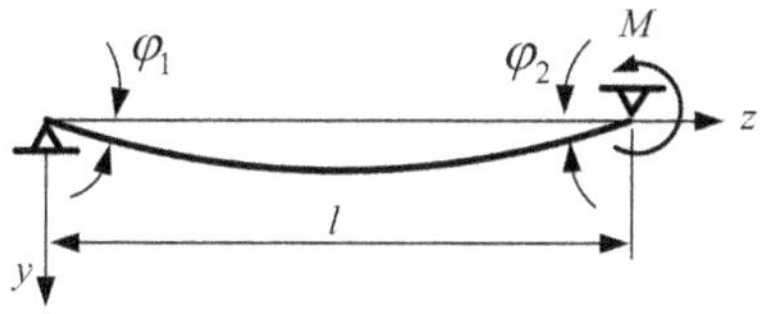

FIGURE 5.17 A simply supported shaft with a concentrated moment M at one end point.

i. For the non-rotating case, the natural frequency of the rotor system will be given as

$$\omega_{nf} = \sqrt{\frac{1/\alpha}{I_d}} = \sqrt{\frac{3EI}{I_d l}} \tag{5.115}$$

ii. For a general (synchronous) motion, since the present system has a pure rotational motion so the whirl frequency is expressed as (refer eq. (5.62) of Tiwari (2017))

$$\nu = \sqrt{\frac{k_t \pm I_p\omega\nu}{I_d}} \quad \text{with} \quad k_t = \frac{1}{\alpha} = \frac{3EI}{l} \tag{5.116}$$

$$\text{or} \quad \nu^2 I_d = \frac{3EI}{l} \pm I_p\omega\nu \quad \Rightarrow \nu^2 I_d \mp I_p\omega\nu - \frac{3EI}{l} = 0$$

$$\text{or} \quad \left(\bar{\nu}\sqrt{\frac{EI}{I_d l}}\right)^2 I_d \mp I_p\left(\bar{\omega}\sqrt{\frac{EI}{I_d l}}\right)\left(\bar{\nu}\sqrt{\frac{EI}{I_d l}}\right) - \frac{3EI}{l} = 0 \Rightarrow \bar{\nu}^2\left(\frac{EI}{I_d l}\right) I_d \mp I_p\bar{\omega}\bar{\nu}\left(\frac{EI}{I_d l}\right) - \frac{3EI}{l} = 0 \tag{5.117}$$

We have the following two non-dimensional parameters:

$$\bar{\omega} = \frac{\omega}{\sqrt{EI/(I_d l)}} \quad \text{and} \quad \bar{\nu} = \frac{\nu}{\sqrt{EI/(I_d l)}} \tag{5.118}$$

On taking common terms out and noting Eq. (5.118), Eq. (5.117) simplifies to

$$\bar{\nu}^2 \mp 2\bar{\nu}\bar{\omega} - 3 = 0 \quad \text{with} \quad \text{for thin disc} \quad I_p = 2I_d \tag{5.119}$$

which can be solved to give

$$\bar{\nu} = \pm\bar{\omega} \pm \sqrt{\bar{\omega}^2 + 3} \tag{5.120}$$

On taking only the positive sign (the negative sign will give the same values but with the negative sign) from outside the square root, we get

$$\bar{\nu} = \pm\bar{\omega} + \sqrt{\bar{\omega}^2 + 3} \quad \Rightarrow \bar{\nu}_1 = \bar{\omega} + \sqrt{\bar{\omega}^2 + 3} \quad \text{and} \quad \bar{\nu}_2 = -\bar{\omega} + \sqrt{\bar{\omega}^2 + 3} \tag{5.121}$$

iii. For the critical speed, we have $\bar{\nu} = \bar{\omega}$ so that from Eq. (5.120), we have

$$\bar{\omega}_{cr}^2 \mp 2\bar{\omega}_{cr}^2 - 3 = 0 \quad \Rightarrow \bar{\omega}_{cr1}^2 + 2\bar{\omega}_{cr1}^2 - 3 = 0 \quad \text{and} \quad \bar{\omega}_{cr2}^2 - 2\bar{\omega}_{cr2}^2 - 3 = 0$$

$$\text{or} \quad 3\bar{\omega}_{cr1}^2 - 3 = 0 \quad \text{and} \quad -\bar{\omega}_{cr2}^2 - 3 = 0$$

$$\text{or} \quad \bar{\omega}_{cr1}^2 = 1 \quad \text{and} \quad \bar{\omega}_{cr2}^2 = -3 \tag{5.120}$$

So, the feasible critical speed is

$$\bar{\omega}_{cr1}^2 = 1 \quad \Rightarrow \frac{\omega_{cr1}^2}{EI/(I_d l)} = 1 \quad \Rightarrow \omega_{cr1} = \sqrt{\frac{EI}{I_d l}} \tag{5.121}$$

The critical speed is lower than the non-rotating case, which means it is a backward whirl critical speed. The forward whirl critical speed does not exist.

Exercise 5.13 For the synchronous backward whirl condition of a cantilever rotor system, derive the frequency equation in terms of the critical speed function $\bar{\omega}_{cr}^2 = \dfrac{ml^3\omega^2}{EI}$, and the disc effect $\mu = \dfrac{I_d}{ml^2}$. Show a plot of $\bar{\omega}_{cr}^2$ versus μ and show limiting cases (i) $\mu = 0$ (ii) $\mu \to \infty$.

***Solution*:** Since for the present case, the synchronous backward (or anti-synchronous) whirl condition is prevailing and as we know the gyroscopic effect only affects the moment. The visualisation of the gyroscopic moment is difficult for anti-synchronous case. For this case, we will take help from the momentum concept of Section 5.5 (Tiwari, 2017). From eq. (5.70) (refer Tiwari, 2017), we have the gyroscopic moment for asynchronous case as follows

$$M_{yz} = I_d\varphi_x(2\omega - \nu)\nu \tag{5.122}$$

Now for the synchronous case we have $\nu = \omega \equiv \omega_{crF}$, which gives the gyroscopic couple as

$$M_{yz} = I_d\varphi_x(2\omega - \omega)\omega = I_d\varphi_x\omega^2 \equiv I_d\varphi_x\omega_{crF}^2 \tag{5.123}$$

which is the same as eq. (5.32) of Tiwari (2017). However, for the anti-synchronous case, we have $\nu = -\omega \equiv -\omega_{crB}$, which gives the gyroscopic couple as

$$M_{yz} = I_d\varphi_x(2\omega + \omega)(-\omega) = -3I_d\varphi_x\omega^2 \equiv -3I_d\varphi_x\omega_{crB}^2 \tag{5.124}$$

So not only it is giving opposite direction gyroscopic moment but its magnitude is three times the forward whirl case. And this may be the reason that the backward whirl natural frequency decreases much faster than the forward whirl frequency (but these occur at different speeds). So, in the present case, there will not only be a change in magnitude of the moment but also the direction will be reversed ($M_{yz} = -3I_d\omega^2\varphi_x$).

The translational and rotational displacements of the free end of the cantilever (i.e., the fixed-free end conditions) beam will be (Timoshenko and Young, 1968; also refer section 5.3.1 of Tiwari (2017))

$$\delta = \frac{F_y l^3}{3EI} - \frac{M_{yz} l^2}{2EI} = \frac{\left(m\omega^2\delta\right)l^3}{3EI} - \frac{\left(-3I_d\omega^2\varphi_x\right)l^2}{2EI} \tag{5.125}$$

$$\text{and} \quad \varphi_x = \frac{F_y l^2}{2EI} - \frac{M_{yz} l}{EI} = \frac{\left(m\omega^2\delta\right)l^2}{3EI} - \frac{\left(-3I_d\omega^2\varphi_x\right)l}{EI} \tag{5.126}$$

It should be noted here that these relations can also be developed for single disc rotors with other boundary conditions, such as the simply supported, fixed-fixed and fixed-hinged (refer Appendix 2.1). It requires the calculation of influence coefficients by using the deflection theory of strength of materials (refer Appendix 2.1). Now the aforementioned relations can be used to find critical speeds for the fixed-free boundary condition. Equations (5.125) and (5.126) can be rearranged as

$$\left(m\omega^2\frac{l^3}{3EI}-1\right)\delta-\left(3I_d\omega^2\frac{l^2}{2EI}\right)\varphi_x=0 \tag{5.127}$$

and
$$\left(-m\omega^2\frac{l^2}{2EI}\right)\delta+\left(3I_d\omega^2\frac{l}{EI}+1\right)\varphi_x=0 \tag{5.128}$$

This homogeneous set of equations can have a non-trivial solution for δ and φ_x, only when the following determinant vanishes

$$\begin{vmatrix}\left(m\omega^2\frac{l^3}{3EI}-1\right) & \left(-3I_d\omega^2\frac{l^2}{2EI}\right)\\ \left(-m\omega^2\frac{l^2}{2EI}\right) & \left(3I_d\omega^2\frac{l}{EI}+1\right)\end{vmatrix}=0 \tag{5.129}$$

which gives the frequency equation (now replacing ω with ω_{crB}) as

$$\left(m\omega_{crB}^2\frac{l^3}{3EI}-1\right)\left(3I_d\omega_{crB}^2\frac{l}{EI}+1\right)-\left(-m\omega_{crB}^2\frac{l^2}{2EI}\right)\left(-3I_d\omega_{crB}^2\frac{l^2}{2EI}\right)=0$$

or
$$\omega_{crB}^4+\frac{4EI}{mI_dl^3}\left(\frac{ml^2}{3}-3I_d\right)\omega_{crB}^2-\frac{4E^2I^2}{mI_dl^4}=0 \tag{5.130}$$

This is the condition at which the rotor would have anti-synchronous whirling and it will take place at critical speeds obtained from the aforementioned condition. To reduce the number of parameters involved in the aforementioned equation, defining the non-dimensional *backward critical speed function*, $\bar{\omega}_{crB}$ and the *disc mass effect*, μ, as

$$\bar{\omega}_{crB}=\omega_{crB}\sqrt{\frac{ml^3}{EI}} \quad \text{and} \quad \mu=\frac{I_d}{ml^2} \tag{5.131}$$

where ω_{crB} is the *backward synchronous critical speed.* Equation (5.130) can be written as

$$\omega_{crB}^4+\frac{4EI}{mI_dl^3}\left(\frac{ml^2}{3}-3I_d\right)\omega_{crB}^2-\frac{4E^2I^2}{mI_dl^4}=0$$

or
$$\left(\bar{\omega}_{crB}\sqrt{\frac{EI}{ml^3}}\right)^4+\frac{4EI}{mI_dl^3}\left(\frac{ml^2}{3}-3I_d\right)\left(\bar{\omega}_{crB}\sqrt{\frac{EI}{ml^3}}\right)^2-\frac{4E^2I^2}{mI_dl^4}=0$$

or
$$\bar{\omega}_{crB}^4+\left[\frac{4}{I_d}\left(\frac{ml^2}{3}+I_d\right)\right]\bar{\omega}_{crB}^2-\frac{4ml^2}{I_d}=0 \tag{5.132}$$

On using the disc parameter expression, we get

$$\bar{\omega}_{crB}^4 - \left(\frac{4}{3\mu} + 12\right)\bar{\omega}_{crB}^2 + \frac{4}{\mu} = 0 \tag{5.133}$$

The solution of which is

$$\bar{\omega}_{crB}^2 = \left(\frac{4}{3\mu} + 12\right) \pm \sqrt{\left(\frac{4}{3\mu} + 12\right)^2 - \frac{8}{\mu}} \tag{5.134}$$

Both the positive and negative signs will give a positive value for $\bar{\omega}_{crB}^2$ or a real value for $\bar{\omega}_{crB}$ (as against the negative sign gives a complex value of the critical speed for the forward whirl, refer eq. (5.43) of Tiwari (2017)). This characteristic, we have observed in the general asynchronous whirl formulation also where only a single forward critical speed and two backward critical speeds are observed. Hence, we will have two backward critical speeds for a particular value of μ, so both positive and negative sign need to be considered. Plot of $\bar{\omega}_{crB}^2$ versus μ is given in Figure 5.18. For the horizonal axis μ is varied from 0 to 4. It can be seen that limiting cases of $\mu \rightarrow 0$ and $\mu \rightarrow \infty$, the $\bar{\omega}_{cr}^2$ approaches 3 and 0, respectively, as discussed now analytically also.

It should be noted that the critical speed of the rotor decreases with the disc mass effect, μ. That means the effective stiffness of the rotor system decreases due to the thin disc as compared to a point-mass disc. This can be seen from Figure 5.18 that the effect of centrifugal forces is to increase the tilting of the disc thereby decreasing the effective stiffness of the rotor system. Overall, for the anti-synchronous whirl condition ($\nu = -\omega$) due to the gyroscopic effect the (backward) critical speed of the system decreases. It has been seen that for the synchronous whirl condition ($\nu = +\omega$) due to the gyroscopic effect the (forward) critical speed of the system increases. Two limiting cases of Figure 5.17 are discussed as follows:

Case 1: A disc having point-mass

For the disc effect $\mu = 0$ (i.e., the concentrated mass of the disc) from Eq. (5.132), we have

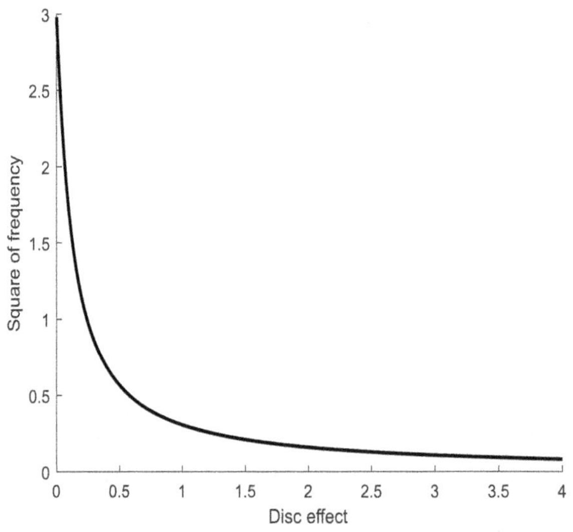

FIGURE 5.18 Variation of the critical speed function with the disc effect.

$$\bar{\omega}_{crB}^4 - \left(\frac{4}{3\mu} + 12\right)\bar{\omega}_{crB}^2 + \frac{4}{\mu} = 0 \quad \Rightarrow 3\mu\bar{\omega}_{crB}^4 - (4 + 36\mu)\bar{\omega}_{crB}^2 + 12 = 0;$$
$$\Rightarrow -4\bar{\omega}_{crB}^2 + 12 = 0 \Rightarrow \quad \bar{\omega}_{crB}^2 = 3 \tag{5.135}$$

Noting Eq. (5.131), the aforementioned equation gives

$$\left(\omega_{crB}\sqrt{\frac{ml^3}{EI}}\right)^2 = 3 \quad \Rightarrow \omega_{crB} = \sqrt{\frac{3EI}{ml^3}} = \sqrt{\frac{k_{eq}}{m}} = \omega_{nf} \quad \text{with} \quad k_{eq} = \frac{3EI}{l^3} \tag{5.136}$$

where ω_{nf} is the transverse natural frequency of the rotor system in the non-rotating condition. It gives the critical speed of disc having a point-mass for the cantilever beam case. For this case, in fact the gyroscopic effect is not present and critical speed is the same as the transverse natural frequency of the cantilever beam with a point mass at the free end. For this case, synchronous and anti-synchronous cases do not prevail.

Case 2: A disc having infinite diameter

For $\mu \to \infty$ (i.e., a disc for which all the mass is concentrated at a relatively large radius so that, $I_d \to \infty$.) no finite rotational displacement φ_x is possible since it would require an infinite resisting moment, which the shaft cannot furnish. The disc remains parallel to itself and the shaft is much stiffer than without the disc effect (i.e., $\mu = 0$). From Eq. (5.132) for $\mu \to \infty$, we get

$$\bar{\omega}_{crB}^4 - \left(\frac{4}{3\mu} + 12\right)\bar{\omega}_{crB}^2 + \frac{4}{\mu} = 0 \quad \Rightarrow \bar{\omega}_{crB}^4 - 12\bar{\omega}_{crB}^2 = 0$$

$$\text{or} \quad \bar{\omega}_{crB}^2\left(\bar{\omega}_{crB}^2 - 12\right) = 0 \quad \Rightarrow \bar{\omega}_{crB1}^2 = 0 \quad \text{and} \quad \bar{\omega}_{crB2}^2 = 12 \quad \Rightarrow \quad \omega_{crB2} = \sqrt{\frac{12EI}{ml^3}} = 2\sqrt{\frac{3EI}{ml^3}} \tag{5.137}$$

$$\text{or} \quad \omega_{crB2} = \sqrt{\frac{k_{eff}}{m}} \quad \text{with} \quad k_{eff} = \frac{12EI}{l^3} \tag{5.138}$$

It should be noted that for the present case, the solution $\bar{\omega}_{crB1}^2 = 0$ is feasible. However, $\bar{\omega}_{crB2}$, is not feasible since a small disturbance will lead to further large displacement of disc and no oscillation will be possible.

Alternatively, for the general (asynchronous) motion case from eq. (5.76) of Tiwari (2017) in the dimensional form, we have

$$v^4\left(-m\alpha_{11}\alpha_{22}I_d + m\alpha_{12}^2 I_d\right) + v^3\left(2m\alpha_{11}\alpha_{22}I_d\omega - \alpha_{12}^2 I_d\omega\right) + v^2\left(I_d\alpha_{22} + m\alpha_{11}\right) + v\left(-\alpha_{22}I_d 2\omega\right) - 1 = 0 \tag{5.139}$$

For the anti-synchronous case $v = -\omega$, and now $\omega = \omega_{crB}$ so Eq. (5.139) becomes

$$\omega_{crB}^4 mI_d\left(-\alpha_{11}\alpha_{22} + \alpha_{12}^2\right) - 2\omega_{crB}^4 mI_d\left(\alpha_{11}\alpha_{22} - \alpha_{12}^2\right) + \omega_{crB}^2\left(I_d\alpha_{22} + m\alpha_{11}\right) + \omega_{crB}^2\left(2I_d\alpha_{22}\right) - 1 = 0$$

or

$$-3\omega_{crB}^4 mI_d\left(\alpha_{11}\alpha_{22} - \alpha_{12}^2\right) + \omega_{crB}^2\left(m\alpha_{11} + 3I_d\alpha_{22}\right) - 1 = 0 \tag{5.140}$$

Defining

$$\bar{\omega}_{crB} = \omega_{crB}\sqrt{\frac{ml^3}{EI}} \quad \text{and} \quad \mu = \frac{ml^2}{I_d} \tag{5.141}$$

and for the cantilever rotor, we have

$$\alpha_{11} = \frac{l^3}{3EI}; \quad \alpha_{12} = \frac{l^2}{2EI}; \quad \alpha_{22} = \frac{l}{EI} \tag{5.142}$$

So that

$$\bar{\alpha} = \frac{\alpha_{12}^2}{\alpha_{11}\alpha_{22}} = \frac{\left(\dfrac{l^2}{2EI}\right)^2}{\left(\dfrac{l^3}{3EI}\right)\left(\dfrac{l}{EI}\right)} = \frac{\dfrac{1}{4}}{\dfrac{1}{3}} = \frac{3}{4} \tag{5.143}$$

Equation (5.140) can be written as

$$-3\bar{\omega}_{crB}^4\left(\sqrt{\frac{EI}{ml^3}}\right)^4 mI_d\left(\frac{l^3}{3EI}\frac{l}{EI} - \frac{l^4}{4E^2I^2}\right) + \bar{\omega}_{crB}^2\left(\sqrt{\frac{EI}{ml^3}}\right)^2\left(m\frac{l^3}{3EI} + 3I_d\frac{l}{EI}\right) - 1 = 0$$

$$\text{or} \quad \bar{\omega}_{crB}^4 - \bar{\omega}_{crB}^2\left(\frac{4}{3\mu} + 12\right) + \frac{4}{\mu} = 0 \tag{5.144}$$

For a general (asynchronous) motion, the frequency equation in non-dimensional form from equation (5.78) of Tiwari (2017) is given as

$$\bar{\nu}^4 - 2\bar{\omega}\bar{\nu}^3 + \frac{\mu+1}{\mu(\bar{\alpha}-1)}\bar{\nu}^2 - \frac{2\bar{\omega}}{\bar{\alpha}-1}\bar{\nu} - \frac{1}{\mu(\bar{\alpha}-1)} = 0 \tag{5.145}$$

It should be noted that the aforementioned equation, the non-dimensional parameters are different as compared to previous discussion [sections 5.3.1 {refer eq. (5.43)} and 5.3.2 {refer eq. (5.55)} of Tiwari (2017)]. So, in order to verify it, we need to convert Eq. (5.144) in non-dimensional parameters defined earlier in sections 5.3.1. and 5.3.2 of Tiwari (2017). Table 5.1 gives a comparison of non-dimensional parameters used in sections 5.3 of Tiwari (2017) (synchronous whirl condition) and 5.5 of Tiwari (2017) (asynchronous whirl condition).

For anti-synchronous case $\bar{\nu} = -\bar{\omega} = -\dfrac{\bar{\omega}_{crB}}{\sqrt{3}}$ (refer Table 5.1, since the subscripts can take both forward and backward critical speed condition), so the aforementioned equation becomes

$$\left(\frac{\bar{\omega}_{crB}}{\sqrt{3}}\right)^4 - 2\left(-\frac{\bar{\omega}_{crB}}{\sqrt{3}}\right)\left(\frac{\bar{\omega}_{crB}}{\sqrt{3}}\right)^3 + \frac{\mu+1}{\mu(\bar{\alpha}-1)}\left(\frac{\bar{\omega}_{crB}}{\sqrt{3}}\right)^2 - \frac{2}{\bar{\alpha}-1}\left(-\frac{\bar{\omega}_{crB}}{\sqrt{3}}\right)\left(\frac{\bar{\omega}_{crB}}{\sqrt{3}}\right) - \frac{1}{\mu(\bar{\alpha}-1)} = 0 \tag{5.146}$$

$$\text{or} \quad \frac{1}{9}\bar{\omega}_{crB}^4 + \frac{2}{9}\bar{\omega}_{crB}^4 + \frac{\mu+1}{3\mu(\bar{\alpha}-1)}\bar{\omega}_{crB}^2 + \frac{2}{3(\bar{\alpha}-1)}\bar{\omega}_{crB}^2 - \frac{1}{\mu(\bar{\alpha}-1)} = 0$$

TABLE 5.1
Comparison of non-dimensional parameters in Section 5.3 of Tiwari (2017) (subscript *s* for synchronous condition or critical (resonance) speed condition and it is valid for both forward and backward critical speeds) and Section 5.5 of Tiwari (2017) (subscript as for synchronous condition)

S.N.	Non-dimensional Parameter	In Section 5.3 (Tiwari, 2017) {(Synchronous Whirl Condition (Critical Speeds) Discussed for Cantilever Case Only}	In Section 5.5 (Tiwari, 2017){Asynchronous Whirl Condition (Whirl Natural Frequency) for All Kind of Boundary Conditions}	Section 5.5 (Tiwari, 2017) for Cantilever Case and Relations of Sections 5.3 and 5.5 (Tiwari, 2017) Parameters $\alpha_{11}=\frac{l^3}{3EI}$; $\alpha_{12}=\frac{l^2}{2EI}$; $\alpha_{22}=\frac{l}{EI}$
1	Dimensionless frequency (or frequency ratio)	$\bar{\nu}_s=\nu_s\sqrt{\frac{ml^3}{EI}}$	$\bar{\nu}_{as}=\nu_{as}\sqrt{\alpha_{11}m}$	$\bar{\nu}_{as}=\nu_{as}\sqrt{\frac{ml^3}{3EI}}$; $\bar{\nu}_{as}=\bar{\nu}_s\sqrt{\frac{1}{3}}$; $\nu_{as}=\nu_s\sqrt{\frac{1}{3}}$
2	Dimensionless speed (or speed ratio)	$\bar{\omega}_s=\omega_s\sqrt{\frac{ml^3}{EI}}$	$\bar{\omega}_{as}=\omega_{as}\sqrt{m\alpha_{11}}$	$\bar{\omega}_{as}=\omega_{as}\sqrt{\frac{ml^3}{3EI}}$; $\bar{\omega}_{as}=\bar{\omega}_s\sqrt{\frac{1}{3}}$; $\omega_{as}=\omega_s\sqrt{\frac{1}{3}}$
3	Disc effect	$\mu_s=\frac{ml^2}{I_d}$	$\mu_{as}=\frac{I_d\alpha_{22}}{m\alpha_{11}}$	$\mu_{as}=\frac{3ml^2}{I_d}$; $\mu_{as}=3\mu_s$
4	Coupling effect	Not used so not defined	$\bar{\alpha}_{as}=\frac{\alpha_{12}^2}{\alpha_{11}\alpha_{22}}$	$\bar{\alpha}_{as}=\frac{3}{4}$

$$\text{or}\quad \bar{\omega}_{crB}^4-\left\{\frac{1+3\mu}{\mu(1-\bar{\alpha})}\right\}\bar{\omega}_{crB}^2+\frac{3}{\mu(1-\bar{\alpha})}=0 \tag{5.147}$$

For the cantilever rotor, we have $\bar{\alpha}=\frac{3}{4}$, so we have

$$\mu_a=\frac{I_d\alpha_{22}}{m\alpha_{11}}=\frac{I_d(l/EI)}{m(l^3/3EI)}=3\frac{I_d}{ml^2}=3\mu_s \quad\text{and}\quad \bar{\nu}_a=\nu_a\sqrt{\alpha_{11}m}=\nu_a\sqrt{\frac{ml^3}{3EI}}=\frac{\nu_a}{\sqrt{3}}\sqrt{\frac{ml^3}{EI}}=\frac{\omega_{crB}}{\sqrt{3}} \tag{5.148}$$

here μ_s is disc effect and is defined in the synchronous (anti-synchronous) case and μ_a in the asynchronous case. On substituting equation (5.148) into Eq. (5.147), we get

$$\bar{\omega}_{crB}^4 - \left(\frac{1+3(3\mu_s)}{(3\mu_s)\left(1-\frac{3}{4}\right)} \right) \bar{\omega}_{crB}^2 + \frac{3}{(3\mu_s)\left(1-\frac{3}{4}\right)} = 0 \quad \Rightarrow \bar{\omega}_{crB}^4 - \left(\frac{4(1+9\mu_s)}{3\mu_s} \right) \bar{\omega}_{crB}^2 + \frac{4}{\mu_s} = 0$$

$$\text{or} \quad \bar{\omega}_{crB}^4 - \left(12 + \frac{4}{3\mu_s} \right) \bar{\omega}_{crB}^2 + \frac{4}{\mu_s} = 0 \tag{5.149}$$

which is exactly the same as Eq. (5.144).

Readers can try out similar exercise with a simply supported beam with disc location from bearings is defined by distances a and b (with two cases $a = b$ and $a = 0.3b$), the length of the shaft is l and EI is the modulus of rigidity. For the simply supported end conditions following influence coefficients (refer Appendix 2.1) can be used.

$$\alpha_{11} = \frac{a^2b^2}{3EIl}; \quad \alpha_{12} = -\frac{\left(3a^2l - 2a^3 - al^2\right)}{3EIl}; \quad \alpha_{21} = \frac{ab(b-a)}{3EIl}; \quad \alpha_{22} = -\frac{\left(3al - 3a^2 - l^2\right)}{3EIl}. \tag{5.150}$$

Exercise 5.14 Obtain the transverse critical speed for the forward synchronous motion of a cantilever shaft with a sphere (of mass m and radius r) at the free end (Figure 5.19). It is assumed that the shaft free end is connected to the centre of the sphere and is not interfering motion of the sphere at any other point of the sphere. The modulus of rigidity of the shaft is EI and its length is l. The spin speed is ω and whirl frequency is ν. Discuss the limiting cases (i) when the radius is zero and (ii) when it is infinity. Consider the gyroscopic effects. For the sphere we have $I_d = I_p = 2mr^2/5$.

Solution: For the sphere, two principal direction mass moments of inertia are same so no gyroscopic moment will be produced (refer eq. (5.52) of Tiwari (2017), which is valid for only the cantilever beam case with the synchronous motion).

$$M_{yz} = \omega^2 (I_1 - I_2)\varphi_x \tag{5.151}$$

where for the present case, we have

$$I_1 \equiv I_p = I_2 \equiv I_d$$

So, from Eq. (5.151), it can be seen that the gyroscopic moment will be zero in a synchronous whirl condition (i.e., at critical speed).

Alternatively, now let us see the frequency equation, for a general motion of *thin disc*, is given as (refer eq. (5.78) of Tiwari (2017))

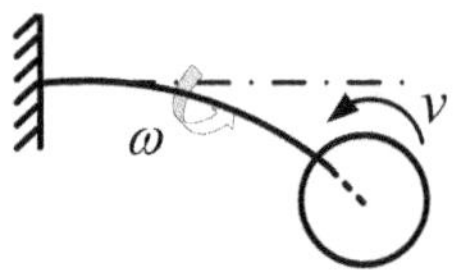

FIGURE 5.19 A cantilever shaft with a sphere at free end.

$$\bar{\nu}^4 - 2\bar{\omega}\bar{\nu}^3 + \frac{\mu+1}{\mu(\bar{\alpha}-1)}\bar{\nu}^2 - \frac{2\bar{\omega}}{\bar{\alpha}-1}\bar{\nu} - \frac{1}{\mu(\bar{\alpha}-1)} = 0 \tag{5.152}$$

So, this equation is not valid for the sphere and is valid only for thin disc case. Since it contains the expression of I_d only in the disc effect, μ, after substituting $I_p = 2I_d$ for the thin disc.

For general disc (or stick), we will have angular momentum (eqs. (5.67) and (5.68) of Tiwari (2017)), as

$$I_p\omega\cos\varphi_x + I_d\nu\varphi_x\sin\varphi_x \approx \left(I_p\omega + I_d\nu\varphi_x^2\right) \quad \text{for parallel to bearing axis} \tag{5.67}$$

and

$$I_p\omega\sin\varphi_x - I_d\nu\varphi_x\cos\varphi_x \approx I_d\varphi_x\left(I_p\omega - I_d\nu\right) \quad \text{for perpendicular to bearing axis} \tag{5.68}$$

Herein, we have retained both I_p and I_d, please note that for $I_p = I_d$ and for synchronous motion $\nu = \omega$, angular momentum in ⊥OA (eq. (5.69) of Tiwari (2017)) will be zero (whose direction was changing leading it to gyroscopic effect but now due to its magnitude is zero so no gyroscopic effect). Angular momentum ||OA anyway was not giving any moment due to no change in its magnitude and direction. So only the critical speed will have (due to disc point mass effect)

$$\omega_{cr} = \sqrt{\frac{k}{m}} = \sqrt{\frac{3EI}{ml}} \tag{5.153}$$

However, for other than synchronous case eq. (5.68) of Tiwari (2017) will not be zero.

i. For $r = 0$, both $I_p = I_d = 0$. This is a point mass for this case Eq. (5.153) will be valid.
ii. For $r \to \infty$, since $I_p = I_d$ so Eq. (5.153) still will be valid for the synchronous motion (to obtain the forward critical speed).

Exercise 5.15 Consider an industrial fan that rotates with a very high speed, and its overall body has a slow precession about its center of gravity due to flexible supports. Consider the gyroscopic effect and obtain the forward and backward critical speeds of the fan rotor system. The effective moment stiffness of the support is 1 MN-m/rad, and the effective diametral and polar mass moments of inertia of the fan are 1 kg-m^2 and 0.1 kg-m^2, respectively.

***Solution*:** The forward and backward critical speeds for pure rotational whirling motion is given as

$$\omega_{cr}^F = \sqrt{\frac{k_t}{I_d - I_p}} \quad \text{and} \quad \omega_{cr}^B = \sqrt{\frac{k_t}{I_d + I_p}} \tag{5.154}$$

where k_t is the effective moment stiffness of the support. We have following data

$k_t = 1$ MN-m/rad, $I_d = 1$ kg-m^2 and $I_p = 0.1$ kg-m^2.

On substituting in Eq. (5.154), we get

$$\omega_{cr}^F = \sqrt{\frac{1\times10^6}{(1-0.1)}} = 1{,}054.09 \text{ rad/s} \quad \text{and} \quad \omega_{cr}^B = \sqrt{\frac{1\times10^6}{(1+0.1)}} = 953.46 \text{ rad/s.}$$

Without the gyroscopic effect, the critical speed will be

$$\omega_{cr} = \sqrt{\frac{k_t}{I_d}} = \sqrt{\frac{1\times 10^6}{0.1}} = 1{,}000 \text{ rad/s}$$

The forward critical speed is more than the critical speed (without gyroscopic effect) and the backward is less. So, a splitting of the critical speed can be seen.

Exercise 5.16 Obtain the frequency equation by using the dynamic method for a general motion (i.e., asynchronous motion) of a single disc rotor system and compare the same with the frequency equation obtained using the quasi-static method. Obtain closed-form expressions of critical speeds for the forward and backward whirl conditions by the frequency equation so obtained.

***Solution*:** There are several differences between the quasi-static general formulation given in chapter 5 of Tiwari (2017) and in the dynamic approach:

1. In the quasi-static general formulation case, the disc is assumed to be thin so $I_p = 2I_d$ is assumed.
2. In the quasi-static case, a single plane motion is considered so the degree of polynomial is half (fourth degree) and that of the dynamic case (eighth degree) in which case the equations need to be written in the state-space form.

It is suggested for details to refer to Chapter 8 for the TMM and Chapter 10 for FEM of Tiwari (2017) for comparison with the quasi-static method. It is encouraged to readers to do the same as an exercise since analytical approach will be too complex to present here.

Exercise 5.17 Derive the critical speed expression of a long stick rotor for the backward synchronous whirl based on the centrifugal force concept with the help of neat diagrams and discuss the same for various disc effect parameters with the help of plots.

***Solution*:** The present problem is similar to Exercise 5.13. But the general frequency equation needs to be modified to take care of long stick ($I_p \neq 2\,I_d$) during derivation of angular momentum derivation and the corresponding frequency equation in dimensional form and then in the non-dimensional form need to be derived. But this will involve both I_p and I_d, so the non-dimensional form of the frequency equation needs to be suitably derived by choosing appropriate non-dimensional parameters involving both I_p and I_d. This aspect gets simplified in rigid disc case ($I_p = 2I_d$) and one of them (i.e., I_p) can be eliminated. See the discussion of Exercises 5.14 and 5.28 also. It is left to the reader as an exercise.

Exercise 5.18 Obtain transverse natural frequencies of a rotor system as shown in Figure 5.20. The mass of the thin disc is $m = 5$ kg, and the diametral mass moment of inertia is $I_d = 0.02$ kg-m². The lengths of the shaft are $a = 0.3$ m and $b = 0.7$ m. The diameter of the shaft is 10 mm. Bearing A has a roller support, and bearing B has a fixed support condition. Ignore the mass of the shaft; however,

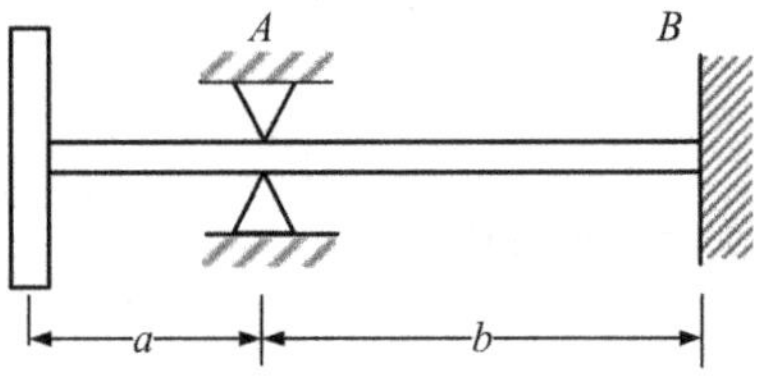

FIGURE 5.20 An overhang rotor system with an intermediate support.

consider the gyroscopic effect of the disc. Let $E = 2.1 \times 10^{11}$ N/m^2. Compare and discuss the results with the case without the gyroscopic couple.

***Solution*:** The frequency equation, for a general asynchronous motion, of a thin-disc rotor case is given as (eq. (5.78) of Tiwari (2017))

$$\bar{\nu}^4 - 2\bar{\omega}\bar{\nu}^3 + \frac{\mu+1}{\mu(\bar{\alpha}-1)}\bar{\nu}^2 - \frac{2\bar{\omega}}{\bar{\alpha}-1}\bar{\nu} - \frac{1}{\mu(\bar{\alpha}-1)} = 0 \tag{5.155}$$

where the elastic coupling parameter is represented as $\bar{\alpha}$, the disc effect as μ, the dimensionless frequency as $\bar{\nu}$ and the dimensionless speed ratio as $\bar{\omega}$. For the rotor system in Figure 5.20, the following influence coefficients are defined (refer Appendix 2.1)

$$\alpha_{yf} = \frac{a^2(4a+3b)}{12EI}; \quad \alpha_{\varphi M} = \frac{(4a+b)}{4EI}; \quad \alpha_{yM} = \alpha_{\varphi f} = \frac{a(2a+b)}{4EI}. \tag{5.156}$$

We have the following rotor parameter values given:

$$a = 0.3\text{ m}, \quad b = 0.7\text{ m}, \quad d = 0.01\text{ m}; \quad E = 2.1\times10^{11}\text{ N/m}^2; \quad I = \frac{\pi d^4}{64} = 4.9087\times10^{-10}\text{ m}^4;$$

$$\alpha_{yf} = \alpha_{11} = 2.4010\times10^{-4}\text{ m/N}; \quad \alpha_{\varphi f} = \alpha_{21} = \alpha_{yM} = \alpha_{12} = 9.4584\times10^{-4}\text{ 1/N}; \quad \alpha_{\varphi M} = \alpha_{22} = 0.0046\text{ 1/N-m};$$

$$\bar{\alpha} = \frac{\alpha_{12}^2}{\alpha_{11}\alpha_{22}} = 0.8086; \quad \mu = \frac{I_d\alpha_{22}}{m\alpha_{11}} = 0.0768.$$

Non-dimensional speed and frequency parameters are defined, respectively, as

$$\bar{\omega} = \omega\sqrt{m\alpha_{11}} \quad \text{and} \quad \bar{\nu} = \nu\sqrt{m\alpha_{11}}$$

On substituting the given rotor parameter values in Eq. (5.155), we get

$$\bar{\nu}^4 - 2\bar{\omega}\bar{\nu}^3 + \frac{0.0768+1}{(0.0768)(0.8086-1)}\bar{\nu}^2 - \frac{2\bar{\omega}}{(0.8086-1)}\bar{\nu} - \frac{1}{(0.0768)(0.8086-1)} = 0 \tag{5.157}$$

For the forward whirl case, we have $\bar{\nu} = \bar{\omega}$. On substituting in Eq. (5.157), we get

$$\bar{\omega}^4 - 2\bar{\omega}^4 + \frac{0.0768+1}{(0.0768)(0.8086-1)}\bar{\omega}^2 - \frac{2\bar{\omega}^2}{(0.8086-1)} - \frac{1}{(0.0768)(0.8086-1)} = 0$$

$$\text{or} \quad -\bar{\omega}^4 + 78.5125\bar{\omega}^2 + 68.0625 = 0$$

Solving this equation (only one positive solution is feasible), we have

$$\bar{\omega}_{crF1} = \sqrt{1.0651} = 1.0320; \quad \Rightarrow \omega_{crF1} = \frac{\bar{\omega}_{crF1}}{\sqrt{m\alpha_{11}}} = \frac{1.0320}{\sqrt{5\times2.41\times10^{-4}}} = 29.79\text{ rad/s}$$

For the backward whirl case, we have $\bar{\nu} = -\bar{\omega}$. After substituting in Eq. (5.157), we get

$$3\bar{\omega}^4 + 83.7375\bar{\omega}^2 + 68.0625 = 0$$

which gives two solutions, as

$$\bar{\omega}_{crB1}^2 = 0.8380 \quad \Rightarrow \bar{\omega}_{crB1} = 0.9154 \quad \Rightarrow \omega_{crB1} = \frac{0.9154}{\sqrt{m\alpha_{11}}} = 26.42 \text{ rad/s}$$

and

$$\bar{\omega}_{crB2}^2 = 27.0745 \quad \Rightarrow \bar{\omega}_{crB2} = 5.2033 \text{ rad/s}; \quad \Rightarrow \omega_{crB2} = \frac{5.2033}{\sqrt{m\alpha_{11}}} = 150.18 \text{ rad/s}$$

There is only one forward critical speed, which is $\omega_{crF1} = 29.79$ rad/s. There are two backward critical speeds, which are $\omega_{crB1} = 26.42$ rad/s and $\omega_{crB2} = 150.18$ rad/s. Figure 5.21 shows the Campbell diagram (plot of Eq. (5.157)), which indicates critical speeds obtained. Herein, the solid line for the forward whirl and dashed lines for the backward whirl. The line with 45° is the synchronous whirl condition ($\nu = \omega$) and the intersection with the frequency lines gives the corresponding critical speeds.

Exercise 5.19 Obtain the forward and backward transverse whirl frequencies *corresponding to thrice the rotor speed* (i.e. $\nu = 3\omega$) for a general motion of a rotor as shown in Figure 5.22. The rotor is assumed to be fixed at one end and supported at the other end. Let the mass of the disc $m = 2$ kg, the polar mass moment of inertia $I_p = 0.01$ kg-m^2, and the diametral mass moment of inertia $I_d = 0.005$ kg-m^2. The shaft is assumed to be massless, and its length and diameter are 0.2 m and 0.1 m, respectively. Assume Young's modulus of the shaft $E = 2.1 \times 10^{11}$ N/m^2.

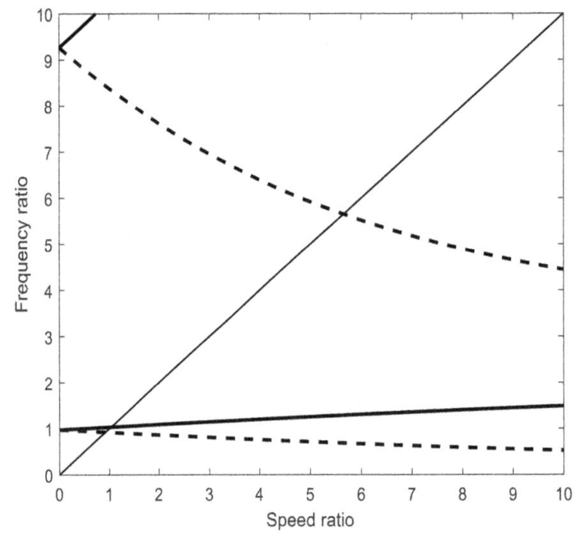

FIGURE 5.21 Campbell diagram showing critical speeds at intersection points.

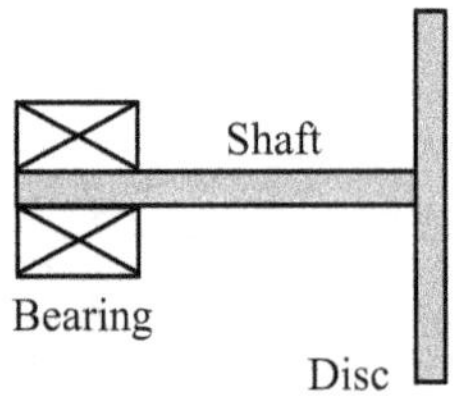

FIGURE 5.22 A cantilever rotor.

***Solution*:** For the present rotor the following data are given:

$$l = 0.2 \text{ m}; \quad d = d_s = 0.01 \text{ m}; \quad E = 2.1\times 10^{11} \text{ N/m}^2; \quad m = 2 \text{ kg};$$

$$I_p = 0.01 \text{ kg-m}^2; \quad I_d = 0.005 \text{ kg-m}^2; \quad I = \frac{\pi d^4}{64} = 4.9087\times 10^{-10} \text{ m}^4.$$

Influence coefficients for cantilever rotor are given by (refer Appendix 2.1)

$$\alpha_{11} = \frac{l^3}{3EI} = 2.5869\times 10^{-5} \text{ m/N}; \quad \alpha_{12} = \alpha_{21} = \frac{l^2}{2EI} = 1.9402\times 10^{-4} \text{ N}^{-1};$$

$$\alpha_{22} = \frac{l}{EI} = 1.9\times 10^{-3} \text{ m}^{-1}\text{N}^{-1}.$$

Now, the disc effect and coupling parameters are obtained as

$$\mu = \frac{I_d\alpha_{22}}{m\alpha_{11}} = 0.1875; \quad \bar{\alpha} = \frac{\alpha_{12}^2}{\alpha_{11}\alpha_{22}} = 0.75.$$

The frequency equation for a general motion is given as (eq. (5.78) of Tiwari (2017))

$$\bar{v}^4 - 2\bar{\omega}\bar{v}^3 + \frac{\mu+1}{\mu(\bar{\alpha}-1)}\bar{v}^2 - \frac{2\bar{\omega}}{\bar{\alpha}-1}\bar{v} - \frac{1}{\mu(\bar{\alpha}-1)} = 0 \tag{5.158}$$

On substituting the values

$$\bar{v}^4 - 2\bar{\omega}\bar{v}^3 + \frac{0.1875+1}{0.1875(0.75-1)}\bar{v}^2 - \frac{2\bar{\omega}}{(0.75-1)}\bar{v} - \frac{1}{0.1875(0.75-1)} = 0$$

$$\text{or} \quad \bar{v}^4 - 2\bar{\omega}\bar{v}^3 - 25.3333\bar{v}^2 + 8\bar{\omega}\bar{v} + 21.3333 = 0 \tag{5.159}$$

For $\bar{v} = +3\bar{\omega}$, from Eq. (5.159), we get

$$0.333\bar{v}^4 - 22.6667\bar{v}^2 + 21.3333 = 0 \tag{5.160}$$

$$\text{which gives} \quad \bar{v}^2 = 67.05,\ 0.9546 \quad \Rightarrow \bar{v} = 8.188,\ 0.977$$

Both values are positive, therefore two forward whirls are possible, as

For $\bar{v} = 8.177$

$$\bar{v}^2 = 67.05 = \left(v_{cr2}^F\sqrt{\alpha_{11}m}\right)^2 = \left(3\omega_{cr2}^F\sqrt{\alpha_{11}m}\right)^2 \quad \Rightarrow \omega_{cr2}^F = 379.5 \text{ rad/s}$$

For $\bar{v} = 0.977$

$$\bar{v}^2 = 0.9546 = \left(v_{cr1}^F\sqrt{\alpha_{11}m}\right)^2 = \left(3\omega_{cr1}^F\sqrt{\alpha_{11}m}\right)^2 \quad \Rightarrow \omega_{cr1}^F = 45.28 \text{ rad/s}$$

For backward whirl, we have $\bar{\nu} = -3\bar{\omega}$, so that

$$1.667\bar{\nu}^4 - 28\bar{\nu}^2 + 21.3333 = 0 \quad \text{or} \quad \bar{\nu}^2 = 16,\ 0.8 \quad \Rightarrow \bar{\nu} = 4,\ 0.8944$$

Both values are positive, therefore two backward whirls are possible, as

For $\bar{\nu} = 4$

$$\bar{\nu}^2 = 16 = \left(\omega_{cr2}^F \sqrt{\alpha_{11} m}\right)^2 \quad \Rightarrow \omega_{cr2}^F = 185.4 \text{ rad/s}$$

For $\bar{\nu} = 0.8944$

$$\bar{\nu}^2 = 0.8 = \left(\omega_{cr1}^F \sqrt{\alpha_{11} m}\right)^2 \quad \Rightarrow \omega_{cr1}^F = 41.25 \text{ rad/s}$$

Corresponding Campbell diagram ($\bar{\nu}$ versus $\bar{\omega}$) is shown in Figure 5.23. The solid line curves are forward whirl and the dashed line curves are backward whirl. The intersection of these curves with a line $\nu = 3\omega$ gives corresponding critical speeds as obtained previously. Two are forward critical speeds and another two are backward critical speeds corresponding to $\nu = 3\omega$. Please note that corresponding to $\nu = \omega$ (synchronous condition) we will get another set of critical speeds in a similar fashion.

Exercise 5.20 For the synchronous motion of a cantilever rotor, derive the expression of critical speed in terms of the disc mass effect, μ. Consider the *z–x* plane motion instead of *y–z* as considered in chapter 5 of Tiwari (2017). Take positive conventions with respect to displacement, force, and moment.

Solution: In *z–x* plane and *y–z* plane, the main difference comes in the positive sign convention of angular displacements. The rest of the analysis (refer section 5.3 of Tiwari (2017)) will remain the same including the final results and interpretations.

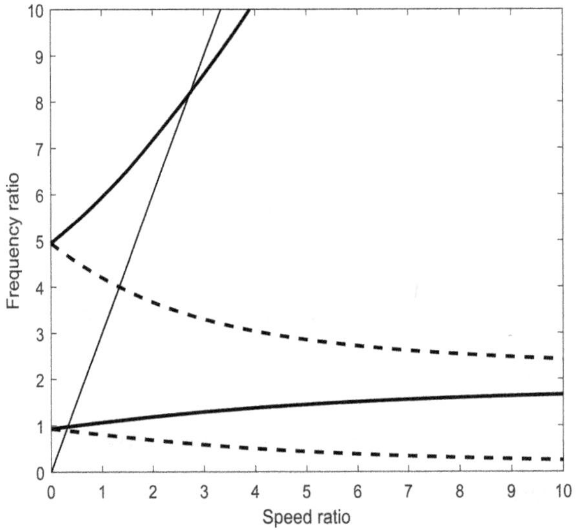

FIGURE 5.23 Campbell diagram.

$$\bar{\omega}_{cr}^2 = \left(6 - \frac{2}{\mu}\right) \pm \sqrt{\left(6 - \frac{2}{\mu}\right)^2 + \frac{12}{\mu}} \tag{5.161}$$

It is left to the reader to try out themselves.

Exercise 5.21 For the synchronous motion of a cantilever rotor, derive the expression of critical speed in terms of the disc mass effect, μ, using angular momentum principles. Compare the result achieved using the quasi-static method in chapter 5 of Tiwari (2017).

***Solution*:** For a general motion case using the angular momentum principle from (5.76) of Tiwari (2017), we have

$$\nu^4\left(-m\alpha_{11}\alpha_{22}I_d + m\alpha_{12}^2 I_d\right) + \nu^3\left(2m\,\alpha_{11}\alpha_{22}I_d\omega - 2m\alpha_{12}^2 I_d\omega\right) + \nu^2\left(\alpha_{22}I_d + m\alpha_{11}\right) + \nu\left(-\alpha_{22}I_d 2\omega\right) - 1 = 0 \tag{5.162}$$

For the synchronous case $\nu = \omega$, so the aforementioned equation becomes

$$\nu^4 mI_d\left(-\alpha_{11}\alpha_{22} + \alpha_{12}^2\right) + \nu^4 mI_d\left(2\,\alpha_{11}\alpha_{22} - 2\alpha_{12}^2\right) + \nu^2\left(\alpha_{22}I_d + m\alpha_{11}\right) + \nu^2\left(-2\alpha_{22}I_d\right) - 1 = 0$$

$$\text{or}\quad \nu^4 mI_d\left(-\alpha_{11}\alpha_{22} + \alpha_{12}^2 + 2\alpha_{11}\alpha_{22} - 2\alpha_{12}^2\right) + \nu^2\left(\alpha_{22}I_d + m\alpha_{11} - 2\alpha_{22}I_d\right) - 1 = 0$$

$$\text{or}\quad \nu^4 mI_d\left(\alpha_{11}\alpha_{22} - \alpha_{12}^2\right) + \nu^2\left(m\alpha_{11} - \alpha_{22}I_d\right) - 1 = 0 \tag{5.163}$$

For the cantilever beam case, we have the following influence coefficients (refer Appendix 2.1)

$$\alpha_{11} = \frac{l^3}{3EI};\quad \alpha_{12} = \alpha_{21} = \frac{l^2}{2EI};\quad \alpha_{22} = \frac{l}{EI} \tag{5.164}$$

On substituting Eq. (5.164) into Eq. (5.163), we get

$$\nu^4 mI_d\left\{\left(\frac{l^3}{3EI}\right)\left(\frac{l}{EI}\right) - \left(\frac{l^2}{2EI}\right)^2\right\} + \nu^2\left\{m\left(\frac{l^3}{3EI}\right) - I_d\left(\frac{l}{EI}\right)\right\} - 1 = 0$$

$$\text{or}\quad \nu^4 + \nu^2\left(\frac{12EI}{mI_d l^3}\right)\left(\frac{ml^2}{3} - I_d\right) - \frac{12E^2I^2}{mI_d l^4} = 0 \tag{5.165}$$

Now defining non-dimensional parameters, the critical speed function and the disc effect, respectively, as

$$\bar{\omega}_{cr} = \omega_{cr}\sqrt{\frac{ml^3}{EI}} \quad \text{and} \quad \mu = \frac{I_d}{ml^2} \tag{5.166}$$

where ω_{cr} is the *forward synchronous critical speed.* Equation (5.165) can be written as

$$\left(\bar{\omega}_{cr}\sqrt{\frac{EI}{ml^3}}\right)^4 + \left(\bar{\omega}_{cr}\sqrt{\frac{EI}{ml^3}}\right)^2\left\{\frac{12EI}{mI_d l^3}\left(\frac{ml^2}{3} - I_d\right)\right\} - \frac{12E^2I^2}{mI_d l^4} = 0 \tag{5.167}$$

which will simplify to

$$\bar{\omega}_{cr}^4 + \bar{\omega}_{cr}^2 \left\{ \frac{12}{I_d}\left(\frac{ml^2}{3} - I_d \right) \right\} - \frac{12ml^2}{I_d} = 0 \tag{5.168}$$

On using the disc parameter expression, we get

$$\bar{\omega}_{cr}^4 + \bar{\omega}_{cr}^2 \left(\frac{4}{\mu} - 12 \right) - \frac{12}{\mu} = 0 \tag{5.169}$$

The solution of which is

$$\bar{\omega}_{cr}^2 = \left(6 - \frac{2}{\mu} \right) \pm \sqrt{\left(6 - \frac{2}{\mu} \right)^2 + \frac{12}{\mu}} \tag{5.170}$$

which is the same as eq. (5.43) of Tiwari (2017) obtained using the quasi-static method. The plot of Eq. (5.169) is shown in Figure 5.24, which same as in section 5.3.1 based on the quasi-static method.

Exercise 5.22 Let the forward synchronous transverse critical speeds for a general motion of a rotor as shown in Figure 5.22 to be fixed at 1,700 rad/s due to some practical constraint. The rotor is assumed to be fixed at one end and free at the other end (i.e., the cantilever end conditions). Consider gyroscopic effects in the analysis. Let the mass of the disc m = 2 kg, the polar mass moment of inertia I_p = 0.01 kg-m^2, and the diametral mass moment of inertia I_d = 0.005 kg-m^2. The shaft is assumed to be massless, and its length is 0.2 m. Assume Young's modulus of the shaft $E = 2.1 \times 10^{11}$ N/m^2. Obtain the suitable diameter of the shaft. For the cantilever beam, influence coefficients are related as

$$\begin{Bmatrix} y \\ \phi_x \end{Bmatrix} = \begin{bmatrix} \alpha_{11} & \alpha_{12} \\ \alpha_{21} & \alpha_{22} \end{bmatrix} \begin{Bmatrix} f \\ M \end{Bmatrix} \quad \text{with} \quad \alpha_{11} = \frac{l^3}{3EI}; \quad \alpha_{12} = \alpha_{21} = \frac{l^2}{2EI}; \quad \alpha_{22} = \frac{l}{EI}.$$

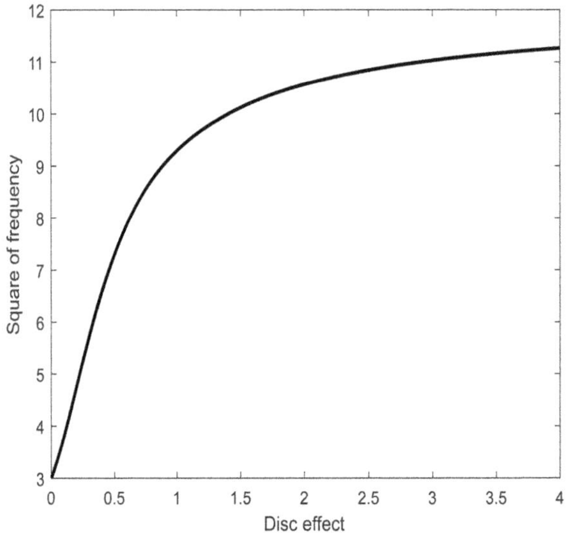

FIGURE 5.24 Plot of square of critical speed with the disc effect.

***Solution*:** In the present problem, based on the constraint of fixing the critical speed of the system, the shaft diameter is to be obtained. The following cantilever rotor system data are given

Synchronous critical speed, $\nu = +\omega_{cr}^{F} = 1{,}700\,\text{rad/s}$;
Mass of the disc $m = 2\,\text{kg}$; Polar mass moment of inertia, $I_p = 0.01$ kg-m^2;
Diametral mass moment of inertia, $I_d = 0.005$ kg-m^2;
Length of the shaft, $l = 0.2\,\text{m}$; Young's modulus $E = 2.1 \times 10^{11}$ N/m^2;
Diameter of the shaft, $d = ?$

For the asynchronous general motion for a thin single disc rotor system (translational and rotational), the frequency equation (eq. (5.78) of Tiwari (2017)) is given by,

$$\bar{\nu}^4 - 2\bar{\omega}\bar{\nu}^3 + \frac{\mu+1}{\mu(\bar{\alpha}-1)}\bar{\nu}^2 - \frac{2\bar{\omega}}{(\bar{\alpha}-1)}\bar{\nu} - \frac{1}{\mu(\bar{\alpha}-1)} = 0 \tag{5.171}$$

with non-dimensional parameters defined as

$$\bar{\nu} = \text{dimensionless frequency} = \nu\sqrt{\alpha_{11}m}; \quad \mu = \text{disc effect} = \frac{I_d\alpha_{22}}{m\alpha_{11}};$$

$$\bar{\alpha} = \text{elastic coupling} = \frac{\alpha_{12}^2}{\alpha_{11}\alpha_{22}}; \quad \bar{\omega} = \text{dimensionless speed} = \omega\sqrt{\alpha_{11}m}. \tag{5.172}$$

For the given parameters and for the cantilever rotor configuration, we have

$$\mu = \frac{I_d\alpha_{22}}{m\alpha_{11}} = \frac{I_d\left(l/EI\right)}{m\left(l^3/3EI\right)} = 0.1875 \quad \text{and} \quad \bar{\alpha} = \frac{\alpha_{12}^2}{\alpha_{11}\alpha_{22}} = \frac{\left(l^2/2EI\right)^2}{\left(l^3/3EI\right)\left(l/EI\right)} = 0.7500 \tag{5.173}$$

On substituting in Eq. (5.171), we get

$$\bar{\nu}^4 - 2\bar{\omega}\bar{\nu}^3 - 25.33\bar{\nu}^2 + 8\bar{\omega}\bar{\nu} + 21.33 = 0 \tag{5.174}$$

For the synchronous whirl, $\bar{\nu} = +\bar{\omega}$. Hence Eq. (5.174) becomes

$$\bar{\omega}^4 + 17.33\bar{\omega}^2 - 21.33 = 0 \quad \text{which gives} \quad \bar{\omega}_{cr1,2}^2 = -18.4873 \quad \text{and} \quad 1.1539$$

Negative value is not a feasible solution. Taking the positive value only, we get

$$\bar{\omega}_{cr1} = 1.07423 = \omega_{cr1}^{F}\sqrt{\alpha_{11}m}$$

On substituting the critical speed value, it gives

$$1{,}700\sqrt{\alpha_{11}m} = 1.07423 \Rightarrow \quad \alpha_{11} = \frac{1.07423^2}{1{,}700^2 \times m} = \frac{1.07423^2}{1{,}700^2 \times 2} = 1.9964 \times 10^{-7}\ \text{m/N}$$

Now, since we have

$$\alpha_{11} = \frac{l^3}{3EI} \qquad \Rightarrow \frac{0.2^3}{3 \times 2.1 \times 10^{11} \times I} = 1.9964 \times 10^{-7}\,\text{m/N}$$

which can be solved for I, as

$$I = \frac{\pi}{64} d^4 = \frac{l^3}{3E\alpha_{11}} = \frac{0.2^3}{3 \times 2.1 \times 10^{11} \times 1.9964 \times 10^{-7}} = 6.3608 \times 10^{-8}\ \mathrm{m}^4$$

It gives the diameter of the shaft as

$$d^4 = 1.2956 \times 10^{-6} \quad \text{or} \quad d = 0.0337\ \mathrm{m}$$

Alternatively, for the synchronous whirl case, the forward critical speed from eq. (5.43) of Tiwari (2017), we have

$$\bar{\omega}_{cr}^2 = \left(6 - \frac{2}{\mu}\right) + \sqrt{\left(6 - \frac{2}{\mu}\right)^2 + \frac{12}{\mu}} \tag{5.175}$$

$$\text{with} \quad \omega_{cr} = 1{,}700\ \text{rad/s}; \quad \mu = \text{disc effect} = \frac{I_d}{ml^2} = 0.0625$$

On substituting into Eq. (5.175), it gives

$$\bar{\omega}_{cr} = \omega_{cr}\sqrt{\frac{ml^3}{EI}} \quad \Rightarrow 1.8606 = 1{,}700\sqrt{\frac{2 \times 0.2^3}{2.1 \times 10^{11} \times I}}$$

$$\text{which gives} \quad I = \frac{\pi d^4}{64} = 6.3605 \times 10^{-8} \quad \Rightarrow d = 0.0337\ \text{m}.$$

Alternatively, using the dimensional form of the frequency equation can give directly I, from which d can be obtained.

Exercise 5.23 A rotor is suspended through linear springs (linear stiffness, $k = 1$ kN/m) to a rigid ceiling and through moment springs (moment stiffness, $k_t = 0.1$ kN-m/rad) to rigid supports as shown in Figure 5.25. Both the linear and moment springs resist the pure tilting motion of the rotor. The rotor consists of two thin rigid discs connected by a massless thin rigid rod. The distance between two linear springs is $l_1 = 0.8$ m and that of moment springs is $l = 1.0$ m (the moment springs are attached to the shaft close to the discs). Let the rotor spin about its shaft axis with ω = 100 rad/s. The diametral mass moment of inertia, and the polar mass moment of inertia of the whole rotor system are $I_d = 0.8$ kg-m^2 and $I_p = 0.1$ kg-m^2, respectively. For the pure tilting motion about its centre of gravity, what would be whirl frequencies and critical speeds of the rotor system? Assume asynchronous whirl conditions due to the gyroscopic couple.

Solution: In the present case, a spinning rotor is supported by a set of linear and moment springs to have its pure tilting motion about its centre of gravity. Herein, the main aim would be to obtain

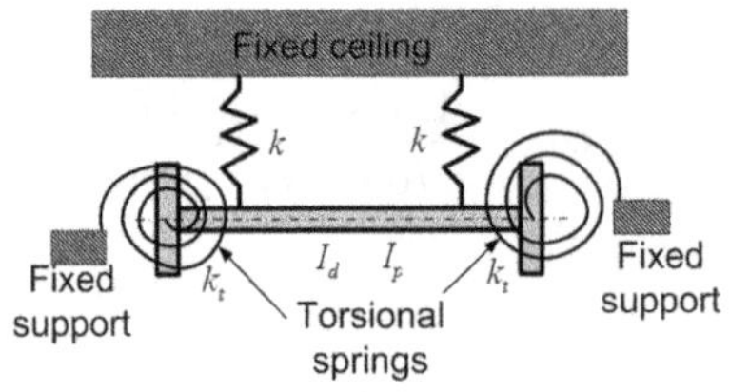

FIGURE 5.25 A rotor mounted on flexible support.

effective support moment stiffness and total inertia moments inclusive of the gyroscopic effect. Given data for the rotor system are

Linear stiffness, $k = 1$ kN/m; Moment stiffness, $k_t = 0.1$ kN-m/rad

Distance between linear springs, $l_l = 0.8$ m; Distance between moment springs, $l = 1.0$ m

Angular speed of rotor, $\omega = 100$ rad/s; Diametral mass moment of Inertia, $I_d = 0.8$ kg-m^2

Polar mass moment of Inertia, $I_p = 0.1$ kg-m^2

The total reaction couple experienced by the rotor system, as shown in Figure 5.26, is given as

$$\left(2k_t + \frac{kl_1^2}{2} \pm I_P\omega\nu\right)\varphi_x = \left(k_{eff} \pm I_P\omega\nu\right)\varphi_x \tag{5.176}$$

$$\text{with} \quad k_{eff} = 2k_t + \frac{kl_1^2}{2}$$

where "+" sign is for the forward whirl and "–" sign is for the backward whirl. The whirl natural frequency is given by

$$\nu^2 = \frac{k_{eff}}{I_d} \quad \text{or} \quad \nu^2 \mp \frac{I_p\omega}{I_d}\nu - \frac{k_{eff}}{I_d} = 0 \tag{5.177}$$

which can be solved to give

$$\nu_{1,2} = \pm\frac{I_p\omega}{2I_d} \mp \sqrt{\left(\frac{I_p\omega}{2I_d'}\right)^2 + \frac{k_{eff}}{I_d}} \quad \text{with} \quad k_{eff} = 2k_t + \frac{kl_1^2}{2} \tag{5.178}$$

For finding critical speeds, make $\nu = \omega$ and $\nu = -\omega$ for the forward and backward whirl critical speeds, respectively, and substituting in the aforementioned equations one at a time, it gives

$$\omega_{cr}{}^{F,B} = \sqrt{\frac{k_{eff}}{I_d \mp I_p}} = \sqrt{\frac{2k_t + 0.5kl_1^2}{I_d \mp I_p}} \tag{5.179}$$

On substituting parameters of the rotor system, the forward whirl critical speed is given as

$$\omega_{cr}{}^{F} = \sqrt{\frac{2k_t + 0.5kl_1^2}{I_d - I_p}} = \sqrt{\frac{2\times100 + 0.5\times1{,}000\times0.8^2}{0.8-0.1}} = \sqrt{\frac{520}{0.7}} = 27.26 \text{ rad/s} \tag{5.180}$$

And the backward whirl critical speed is given as

$$\omega_{cr}{}^{B} = \sqrt{\frac{2k_t + 0.5kl_1^2}{I_d + I_p}} = \sqrt{\frac{2\times100 + 0.5\times1{,}000\times0.8^2}{0.8+0.1}} = \sqrt{\frac{520}{0.9}} = 24.04 \text{ rad/s} \tag{5.181}$$

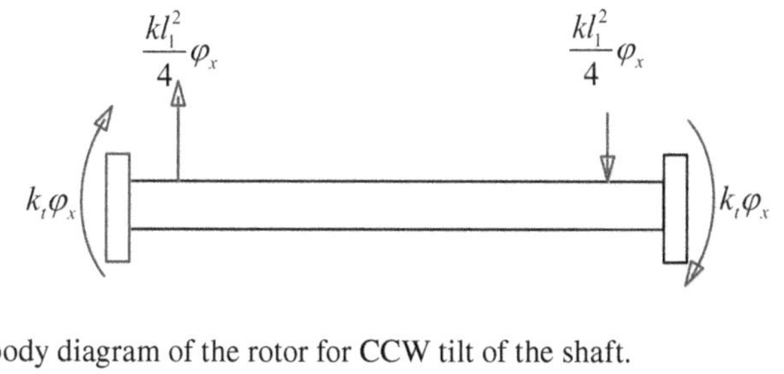

FIGURE 5.26 Free body diagram of the rotor for CCW tilt of the shaft.

Without gyroscopic effect (i.e., at zero spin speed), we will have $I_p = 0$, to get the system natural frequency (or critical speed), as

$$\omega_{nf} = \sqrt{\frac{2k_t + 0.5 \times kl_1^2}{I_d}} = \sqrt{\frac{2 \times 100 + 0.5 \times 1{,}000 \times 0.8^2}{0.8}} = \sqrt{\frac{520}{0.8}} = 25.50 \text{ rad/s} \tag{5.182}$$

It is to be noted that the whirl natural frequency at zero speed is in between the forward and backward critical speeds. So, with the gyroscopic effect, the splitting of whirl natural frequency (as well as the critical speed) takes place. Without gyroscopic effect, the whirl natural frequency remains constant with the speed of the rotor and when the speed coincides with the whirl natural frequency a single critical speed occurs. Whereas, for the case when the gyroscopic effect is present, the whirl natural frequencies (now two in numbers) change with the speed of the rotor and when speed coincides with any of the whirl natural frequencies the critical speeds occur. However, it is also important to note that the forward critical speed is very easy to excite by unbalance force but the backward critical speed will not be excited by unbalance force even when the speed is coinciding with backward critical speed. Herein, the sense of direction of excitation force is also important so we need to have sense of excitation force in the opposite direction to that of rotor speed (for example due to periodic rubbing of rotor with the stator and friction force which is always opposite to the motion may lead to the backward whirl critical speed). However, reaching this condition is a very rare case. In a laboratory setup, the backward critical speed can be demonstrated by putting a free-wheel on the rotor and driving it by an independent drive. On the freewheel some unbalance can be kept and rotated in opposite to the rotor speed, when the speed of the free-wheel coincides with the backward critical speed we will have the resonance condition.

Exercise 5.24 For the following equations of motion obtain the standard eigenvalue problem:

$$\mathbf{M}\ddot{\boldsymbol{\eta}} - \omega\mathbf{G}\dot{\boldsymbol{\eta}} + \mathbf{K}\boldsymbol{\eta} = \mathbf{0}$$

where $\mathbf{M}$ is the mass matrix, $\mathbf{G}$ is the gyroscopic matrix, $\mathbf{K}$ is the stiffness matrix, and η is the displacement vector. Discuss characteristics of eigenvalues for the stability of the system.

***Solution*:** It is the same as Exercise 5.4, but $\mathbf{C}$ is ignored in the present problem to give

$$\begin{Bmatrix} \mathbf{v} \\ \dot{\mathbf{v}} \end{Bmatrix} = \begin{bmatrix} \mathbf{1} & \mathbf{0} \\ \omega\mathbf{M}^{-1}\mathbf{G} & -\mathbf{M}^{-1}\mathbf{K} \end{bmatrix} \begin{Bmatrix} \mathbf{v} \\ \mathbf{x} \end{Bmatrix} + \begin{Bmatrix} \mathbf{0} \\ \mathbf{M}^{-1}\mathbf{f} \end{Bmatrix} \tag{5.183}$$

Then all steps will be the same as solution of Exercise 5.4 with keeping $\mathbf{C}$ matrix to zero. However, all eigenvalues will be pure imaginary. This is to the fact that the real part represents damping, which is absent here. Herein, $\lambda = \pm j\beta$ is the complex eigenvalue and its form is $\lambda = \pm j\omega_{nf}$, where ω_{nf} is the undamped natural frequency. It should be noted that we will have two pairs of complex roots $\lambda_{1,2} = \pm j\beta^+$ and $\lambda_{3,4} = \pm j\beta^-$, where in general for a given speed of the rotor $\beta^+ > \beta^-$. However, for zero speed, we will have $\beta^+ = \beta^- = \beta$. That means we will have two same pairs of complex roots at zero speed, $\lambda_{1,2} = \pm j\beta$ and $\lambda_{3,4} = \pm j\beta$. In fact, this can be used to pair a particular mode natural frequency, especially in multi-disc rotor system where multiple eigenvalues will be obtained and otherwise difficult to distinguish.

It should be noted that any two complex displacements (eigen vectors) will have either zero phase or phase of 180°. This is due to absence of the damping, in this case, all generalised coordinates will have either constant phase or 180° among themselves.

Exercise 5.25 Let the anti-synchronous transverse critical speeds for a general motion of a rotor as shown in Figure 5.27 to be fixed at 1,500 rad/s due to some practical constraint. The rotor is assumed to be fixed-supported at one end. Consider gyroscopic effects in the analysis. Take the mass of the disc $m = 2$ kg, the polar mass moment of inertia $I_p = 0.01$ kg-m^2, and the diametral mass moment of inertia $I_d = 0.005$ kg-m^2. The shaft is assumed to be massless, and its length is 0.2 m. Take Young's modulus of the shaft $E = 2.1 \times 10^{11}$ N/m^2. Obtain the suitable diameter of the shaft. For the cantilever beam, influence coefficients are related as

$$\begin{Bmatrix} y \\ \phi_x \end{Bmatrix} = \begin{bmatrix} \alpha_{11} & \alpha_{12} \\ \alpha_{21} & \alpha_{22} \end{bmatrix} \begin{Bmatrix} f \\ M \end{Bmatrix} \quad \text{with} \quad \alpha_{11} = \frac{l^3}{3EI}; \quad \alpha_{12} = \alpha_{21} = \frac{l^2}{2EI}; \quad \alpha_{22} = \frac{l}{EI}$$

Solution: In the present problem, the shaft diameter needs to be obtained based on the constraint of fixing the backward critical speed. The following rotor parameter values are given

Anti-synchronous critical speed, $\nu = -\omega_{cr}^{B} = 1{,}500$ rad/s

Mass of the disc $m = 2\,\text{kg}$; Polar mass moment of inertia, $I_p = 0.01\,\text{kg-m}^2$;

Diametral mass moment of inertia, $I_d = 0.005\,\text{kg-m}^2$

Length of the shaft, $l = 0.2\,\text{m}$; Young's modulus, $E = 2.1 \times 10^{11}$ N/m^2

Diameter of the shaft, $d = ?$

For the asynchronous general motion for thin single disc rotor system (translational and rotational motion), the frequency equation (eq. (5.78)) of Tiwari (2017) is given by, (it should be noted that eq. (5.42) or eq. (5.38) of Tiwari (2017) cannot be used since they have been derived for the synchronous condition, refer Exercise 5.13 for more discussion and clarity)

$$\bar{\nu}^4 - 2\bar{\omega}\bar{\nu}^3 + \frac{\mu + 1}{\mu(\bar{\alpha} - 1)}\bar{\nu}^2 - \frac{2\bar{\omega}}{(\bar{\alpha} - 1)}\bar{\nu} - \frac{1}{\mu(\bar{\alpha} - 1)} = 0 \tag{5.184}$$

with $\bar{\nu} = \text{dimensionless frequency} = \nu\sqrt{\alpha_{11} m}$; $\mu = \text{disc effect} = \dfrac{I_d \alpha_{22}}{m \alpha_{11}}$;

$$\bar{\alpha} = \text{elastic coupling} = \frac{\alpha_{12}^2}{\alpha_{11}\alpha_{22}}; \quad \bar{\omega} = \text{dimensionless speed} = \omega\sqrt{\alpha_{11} m}.$$

For the cantilever rotor configuration and given rotor parameters, we have

$$\mu = \frac{I_d \alpha_{22}}{m \alpha_{11}} = \frac{I_d \left({}^{l}\!/_{EI} \right)}{m \left({}^{l^3}\!/_{3EI} \right)} = \frac{3 I_d}{m l^2} = \frac{3 \times 0.005}{2 \times 0.2^2} = 0.1875 \tag{5.185}$$

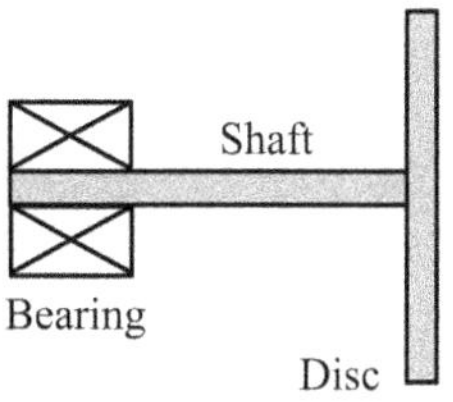

FIGURE 5.27 A cantilever rotor system.

$$\text{and} \quad \bar{\alpha} = \frac{\alpha_{12}^2}{\alpha_{11}\alpha_{22}} = \frac{\left(l^2/2EI\right)^2}{\left(l^3/3EI\right)\left(l/EI\right)} = \frac{3}{4} = 0.7500 \tag{5.186}$$

On substituting in Eq. (5.184), we get

$$\bar{\nu}^4 - 2\bar{\omega}\bar{\nu}^3 + \frac{0.1875+1}{0.1875(0.75-1)}\bar{\nu}^2 - \frac{2\bar{\omega}}{(0.75-1)}\bar{\nu} - \frac{1}{0.1875(0.75-1)} = 0$$

$$\text{or} \quad \bar{\nu}^4 - 2\bar{\omega}\bar{\nu}^3 - 25.33\bar{\nu}^2 + 8\bar{\omega}\bar{\nu} + 21.33 = 0 \tag{5.187}$$

For the anti-synchronous whirl, $\bar{\nu} = -\bar{\omega}$. Hence Eq. (5.187) becomes

$$3\bar{\omega}^4 - 33.33\bar{\omega}^2 + 21.33 = 0 \tag{5.188}$$

$$\text{which gives,} \quad \bar{\omega}_{cr1,2}^2 = 0.6818 \quad \text{and} \quad 10.4293$$

$$\text{or} \quad \bar{\omega}_{cr1}^B = 0.8257 \quad \text{and} \quad \bar{\omega}_{cr2}^B = 3.2294 \tag{5.189}$$

Case I: For $\bar{\omega}_{cr1}^B$:

$$\bar{\omega}_{cr1}^B = 0.825715 = \omega_{cr1}^B\sqrt{\alpha_{11}m} \tag{5.190}$$

On substituting the given anti-synchronous critical speed value $\omega_{cr1}^B = 1{,}500$ rad/s, it gives

$$1{,}500\sqrt{\alpha_{11}m} = 0.825715 \Rightarrow \quad \alpha_{11} = 3.03025\times10^{-7}/2 = 1.5151\times10^{-7} \text{ m/N}$$

Now, since we have

$$\alpha_{11} = \frac{l^3}{3EI} = 1.5151\times10^{-7} \text{ m/N} \tag{5.191}$$

which can be solved for I, as

$$I = \frac{\pi}{64}d^4 = \frac{l^3}{3E\alpha_{11}} = \frac{0.2^3}{3\times2.1\times10^{11}\times1.5151\times10^{-7}} = 8.3812\times10^{-8} \text{ m}^4 \tag{5.192}$$

It gives the diameter of the shaft as

$$d^4 = \frac{8.3811\times10^{-8}\times64}{\pi} = 1.7074\times10^{-6} \quad \text{or} \quad d_1 = 0.03615 \text{ m} \tag{5.193}$$

Case II: For $\bar{\omega}_{cr2}^B$:

$$\bar{\omega}_{cr2}^B = 3.22927 = \omega_{cr2}^B\sqrt{\alpha_{11}m} \tag{5.194}$$

On substituting the given anti-synchronous critical speed value $\omega_{cr2}^B = 1{,}500$ rad/s, it gives

$$1{,}500\sqrt{\alpha_{11}m} = 3.22927 \Rightarrow \quad \alpha_{11} = 4.6348\times10^{-7}/2 = 2.3174\times10^{-6} \text{ m/N}$$

Now, since we have

$$\alpha_{11} = \frac{l^3}{3EI} = 2.3174 \times 10^{-6}\,\text{m/N} \tag{5.195}$$

which can be solved for *I*, as

$$I = \frac{\pi}{64}d^4 = \frac{l^3}{3E\alpha_{11}} = \frac{0.2^3}{3\times 2.1\times 10^{11}\times 2.3174\times 10^{-6}} = 5.4797\times 10^{-9}\ \text{m}^4 \tag{5.196}$$

It gives the diameter of the shaft as

$$d^4 = \frac{5.4797\times 10^{-9}\times 64}{\pi} = 1.1163\times 10^{-7} \quad \text{or} \quad d_2 = 0.01828\ \text{m} \tag{5.197}$$

Discussions: Out of two values of diameters obtained, $d_1 = 0.03615\,\text{m}$ (Eq. (5.193)) is the higher value of diameter of the shaft will ensure that the one of anti-synchronous critical speed is $\omega_{cr}^{B} = 1{,}500$ rad/s. Corresponding to this from Eq. (5.191), we have $\alpha_{11} = 1.5151\times 10^{-7}$ m/N. The other anti-synchronous critical speed can be obtained from Eq. (5.194), as

$$\bar{\omega}_{cr2}^{B} = 3.22927 = \omega_{crL}^{B}\sqrt{\alpha_{11}m} \quad \Rightarrow \omega_{crL}^{B} = \frac{3.22927}{\sqrt{\alpha_{11}m}} = \frac{3.22927}{\sqrt{1.5151\times 10^{-7}\times 2}} = 5{,}866.36\ \text{rad/s} \tag{5.198}$$

which is higher than 1,500 rad/s. So, the lowest anti-synchronous critical speed is as per the specified value of 1,500 rad/s. So, we should choose shaft diameter of $d = 0.03615\,\text{m}$ from Eq. (5.193).

The other value of diameter from Eq. (5.197), we have $d_2 = 0.01828$ m, which is the smaller value of diameter of the shaft and will ensure that one of the anti-synchronous critical speed is 1,500 rad/s. Corresponding to this, we have $\alpha_{11} = 2.3174\times 10^{-6}$ m/N from Eq. (5.195). The other anti-synchronous critical speed can be obtained from Eq. (5.190), as

$$\bar{\omega}_{cr1}^{B} = 0.825715 = \omega_{cr1}^{B}\sqrt{\alpha_{11}m} \quad \Rightarrow \omega_{cr1}^{B} = \frac{0.825715}{\sqrt{\alpha_{11}m}} = \frac{0.825715}{\sqrt{2.3174\times 10^{-6}\times 2}} = 383.54\ \text{rad/s}$$

which is lower than 1,500 rad/s. So, we will not choose $d_2 = 0.01828$ m. Since we wanted to fix the lower anti-synchronous critical speed equal to 1,500 rad/s. So, the final choice of the diameter will be the higher one, that is $d = 0.03615$ m.

Exercise 5.26 A long symmetrical rigid rotor is supported at both ends by two identical bearings. The shaft has diameter of 0.2 m and length of 1 m, with the mass density of the shaft material 7,800 kg/m³. Consider each bearing has properties $k_{xx} = k_{yy} = k$ with other stiffness and damping properties ignored. For the pure tilting motion of the shaft, while considering the gyroscopic effects, the first forward critical speed ($\nu = \omega$) is to be fixed at 100 rad/s. Obtain the bearing stiffness, *k*.

***Solution*:** Herein, the support stiffness is to be designed for a given rotor-bearing configuration such that during its pure tilting motion, the first forward critical speed is to be fixed at a particular given frequency due to some practical constraint. The following data are given for the rotor system

Diameter of shaft, $d = 0.2$ m; Length of shaft, $l = 1$ m;

Mass density, $\rho = 7800\ \text{kg/m}^3$; First forward critical speed, $\omega_{cr1}^{F} = 100$ rad/s.

For a circular cylinder, the polar moment of inertia, I_p, and the diametral moment of inertia, I_d, are given as,

$$I_p = \frac{1}{2}mr^2 \quad \text{and} \quad I_d = \frac{1}{12}m(3r^2 + l^2) \tag{5.199}$$

with

$$m = \frac{\pi}{4} d^2 l \rho = 245.0442 \text{ kg}; \quad I_p = \frac{1}{2} m r^2 = 1.2252 \text{ kg-m}^2; \quad I_d = \frac{1}{12} m (3r^2 + l^2) = 21.0330 \text{ kg-m}^2.$$

The effective stiffness due to bearings on the rotor can be obtained by considering Figure 5.28.

The moment onto the rotor due to these bearing forces would be $0.5kl^2\varphi_x$. Hence, the effective moment stiffness will be

$$k_t = 0.5kl^2 = 0.5k \times 1^1 = 0.5k \tag{5.200}$$

If we consider, the gyroscopic effect then the effective moment stiffness becomes,

$$k_{eff} = (k_t \pm I_p \omega \nu) \tag{5.201}$$

and we know that, transverse natural frequency is given by,

$$\nu^2 = \frac{k_{eff}}{I_d} = \frac{k_t \pm I_p \omega \nu}{I_d} \quad \Rightarrow \nu^2 \pm \frac{I_p \omega}{I_d} \nu - \frac{k_t}{I_d} = 0 \tag{5.202}$$

Roots of Eq. (5.202) are

$$\nu_{1,2} = \pm \frac{I_p \omega}{2I_d} + \sqrt{\left(\frac{I_p \omega}{2I_d} \right)^2 + \frac{k_t}{I_d}} \tag{5.203}$$

which gives whirl natural frequency of the rotor system and it varies with the rotor speed. When the spin speed is equal to the whirl natural frequency, we have resonance condition and that speed is the critical speed. For obtaining the critical speed, we have

$$\upsilon = \pm \omega \quad (+ \text{ for the forward whirl and} - \text{ for the backward whirl})$$

Putting it in Eq. (5.203), we get the following expression for critical speed,

$$\omega_{cr} = \sqrt{\frac{k_t}{I_d \mp I_p}} \tag{5.204}$$

For the forward critical speed, from given data we have

$$\omega_{cr}^F = \sqrt{\frac{k_t}{I_d - I_p}} = \sqrt{\frac{0.5 \times k}{21.0326 - 1.2252}} = 100 \text{ rad/s}$$

$$\text{or} \quad 0.5 \times k = 100^2 (21.0326 - 1.2252) \Rightarrow \quad k = 3.9615 \times 10^5 \text{ N/m}$$

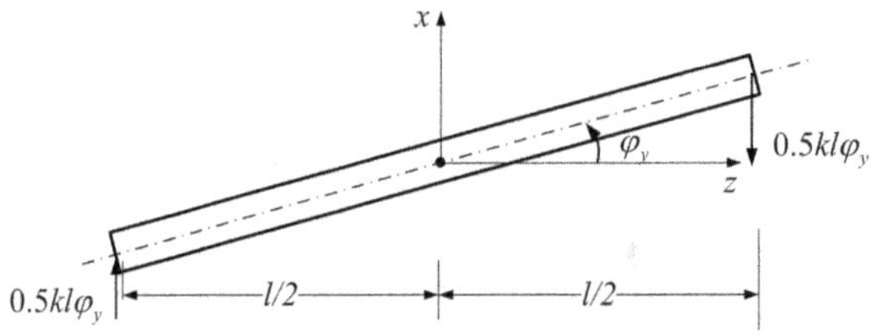

FIGURE 5.28 Free body diagram of the rotor.

Hence, the stiffness of the support should be $k_x = k_y = k = 3.9615\times10^5$ N/m. Based on the backward critical speed also on similar lines, we can obtain the corresponding support stiffness, if required.

Exercise 5.27 A massless flexible shaft of length 0.6 m and diameter 2 cm is simply supported at the ends. A thin disc of mass 10 kg, polar moment of inertia 0.2 kg-m^2, and diametral moment of inertia 0.1 kg-m^2, is at the midspan of the shaft. Consider the gyroscopic effects. Obtain the system's natural frequencies when the shaft rotates at 6,000 rpm. Obtain all critical speeds of the rotor system, and if the disc has (i) only radial eccentricity e, (ii) only pure initial tilt φ_0, and (iii) both e and φ_0, then which of the critical speeds would be encountered in each case by the rotor system? Let $E = 2.1 \times 10^{11}$ N/m^2.

Solution: The present rotor configuration is a Jeffcott rotor model. It has disc at the midspan of a simply supported flexible shaft. So as such coupling of the translatory and rotational (tilting) motion is absent. Hence, we can have translatory and rotational (tilting) motion independent of each other. The following data are given:

Length of shaft, $l = 0.6$ m; Diameter of shaft, $d = 2$ cm;
Mass of disc, $m = 10$ kg; Diametral mass moment of inertia $I_d = 0.1$ kg-m^2;
Polar mass moment of inertia, $I_p = 0.2$ kg-m^2; $I = \left(\frac{\pi}{64}\right)\times d^4 = 7.8540\times10^{-9}$ m^4;
Speed of shaft $\omega = 6{,}000$ rpm $= 628.32$ rad/s;
Modulus of elasticity $E = 2.1\times10^{11}$ N/m^2

For a simply supported rotor system with a disc at the mid-span, the influence coefficients are defined as,

$$\begin{Bmatrix} y \\ \phi_x \end{Bmatrix} = \begin{bmatrix} \alpha_{11} & \alpha_{12} \\ \alpha_{21} & \alpha_{22} \end{bmatrix} \begin{Bmatrix} f \\ M \end{Bmatrix} \tag{5.205}$$

with $\alpha_{11} = \dfrac{l^3}{48EI} = 2.7284\times10^{-6}$ m/N; $\alpha_{12} = \alpha_{21} = 0$; $\alpha_{22} = \dfrac{l}{12EI} = 3.0315\times10^{-5}$ m/N

The frequency equation (refer eq. (5.78) of Tiwari (2017)) for general motion of rotor system with a single thin disc is given by,

$$\bar{\nu}^4 - 2\bar{\omega}\bar{\nu}^3 + \frac{\mu+1}{\mu(\bar{\alpha}-1)}\bar{\nu}^2 - \frac{2\bar{\omega}\bar{\nu}}{(\bar{\alpha}-1)} - \frac{1}{\mu(\bar{\alpha}-1)} = 0 \tag{5.206}$$

$$\text{with,}\quad \bar{\nu} = \text{dimensionless frequency} = \nu\sqrt{\alpha_{11}m};\quad \mu = \text{disc effect} = \frac{I_d\alpha_{22}}{m\alpha_{11}};$$

$$\alpha = \text{elastic coupling} = \frac{\alpha_{12}^2}{\alpha_{11}\alpha_{22}};\quad \bar{\omega} = \text{dimensionless speed} = \omega\sqrt{\alpha_{11}m};$$

For the given rotor parameters, we have

$$\mu = \frac{I_d\alpha_{22}}{m\alpha_{11}} = \frac{0.1\times3.0315\times10^{-5}}{10\times2.7284\times10^{-6}} = 0.1111;\qquad \bar{\alpha} = \frac{\alpha_{12}^2}{\alpha_{11}\alpha_{22}} = 0$$

$$\bar{\omega} = \omega\sqrt{\alpha_{11}m} = 628.32\sqrt{2.72837\times10^{-6}\times10} = 3.2819;\qquad \nu = \frac{\bar{\nu}}{\sqrt{\alpha_{11}m}} = 191.4469\bar{\nu}$$

For $\bar{\alpha} = 0$, Eq. (5.206) reduces to,

$$\bar{\nu}^4 - 2\bar{\omega}\bar{\nu}^3 - \frac{\mu+1}{\mu}\bar{\nu}^2 + 2\bar{\omega}\bar{\nu} + \frac{1}{\mu} = 0 \tag{5.207}$$

The aforementioned equation can be written as,

$$(\bar{\nu}+1)(\bar{\nu}-1)\left(\bar{\nu}^2 - 2\bar{\omega}\bar{\nu} - \frac{1}{\mu}\right) = 0 \tag{5.208}$$

Roots of this equation are,

$$\bar{\nu}_1 = +1; \quad \bar{\nu}_2 = -1; \quad \bar{\nu}_{3,4} = \bar{\omega} \pm \sqrt{\bar{\omega}^2 + \frac{1}{\mu}} = 3.2819 \pm 4.4465 = 7.7284 \quad \text{and} \quad -1.1645$$

Hence, natural frequencies are,

$$\nu_1 = \frac{\bar{\nu}_1}{\sqrt{\alpha_{11}m}} = +191.45 \text{ rad/s}; \quad \nu_2 = \frac{\bar{\nu}_2}{\sqrt{\alpha_{11}m}} = -191.45 \text{ rad/s};$$

$$\nu_3 = \frac{\bar{\nu}_3}{\sqrt{\alpha_{11}m}} = 1{,}479.6 \text{ rad/s}; \quad \nu_4 = \frac{\bar{\nu}_4}{\sqrt{\alpha_{11}m}} = -222.95 \text{ rad/s}.$$

Out of these natural frequencies, the first two natural frequencies are related to the up-down motion without any tilting (wobbling), in fact both frequencies are the same. The third and fourth natural frequencies are related to the precessional motion, i.e. the wobbling without up and the down motions, the positive belongs to forward whirl and the negative belong to the backward whirl.

First two natural frequencies are independent of rotor speed as they are related to up-down motion without any tilting of the disc. Hence, the gyroscopic effect doesn't come into picture and natural frequencies remain constant. Therefore, critical speeds related to the up-down motion are the same as that of natural frequency encountered at any speed, i.e. $\omega_{cr1}^F = 191.45$ rad/s (forward whirl) and $\bar{\omega}_{cr1}^B = -191.45$ rad/s (backward whirl).

For finding the critical speed, we have to put $\bar{\nu} = \pm\bar{\omega}$ in Eq. (5.207). From Eq. (5.208), we can see that the forward critical speed related to wobbling motion is not possible. For the forward whirl $\bar{\nu} = +\bar{\omega}$, from Eq. (5.208), we have

$$\left(\bar{\omega}_{crF}^2 - 2\bar{\omega}_{crF}^2 - \frac{1}{\mu}\right) = 0 \;\Rightarrow \left(-\bar{\omega}_{crF}^2 - \frac{1}{\mu}\right) = 0 \;\Rightarrow \bar{\omega}_{crF} = \sqrt{-\frac{1}{\mu}}$$

It is not feasible. Now, for the backward whirl $\bar{\nu} = -\bar{\omega}$, from Eq. (5.208), we have

$$\left(\bar{\omega}_{crB}^2 + 2\bar{\omega}_{crB}^2 - \frac{1}{\mu}\right) = 0 \;\Rightarrow \left(3\bar{\omega}_{crB}^2 - \frac{1}{\mu}\right) = 0 \;\Rightarrow \bar{\omega}_{crB} = \sqrt{\frac{1}{3\mu}}$$

It is feasible. Only the backward critical speeds can be obtained, which is

$$\bar{\nu} = \bar{\omega}_{cr2}^B = \sqrt{\frac{1}{3\mu}} = 1.7321. \text{ Therefore, } \omega_{cr2}^B = 191.45; \quad \bar{\omega}_{cr2}^B = 331.60 \text{ rad/s}.$$

Figure 5.29 shows the Campbell diagram for the present rotor system. The solid thick lines are for the forward whirl natural frequencies and dashed lines for he backward whirl natural frequency. The horizontal line corresponding to the first forward natural frequency, which is not changing with speed since there is no coupling of the translational and rotational motions in the present case. It is well known that gyroscopic moment relates to the rotational motion only so in the present case the translational motion is not affected by the gyroscopic couple. So, the splitting of the corresponding natural frequency is also not seen so there will not be backward whirl for this case. However, higher modes have splitting and there are the forward and backward whirls. Intersection of 45° line ($\nu = \omega$) with these curves gives critical speeds. For the first mode the critical speed is same as the constant natural frequency (191.45 rad/s), for the second mode there is no intersection of the line ($\nu = \omega$) so no second forward critical speed and there is intersection with the backward whirl (dashed line), which will give backward critical speed (331.60 rad/s) as obtained earlier.

When the disc has only radial eccentricity (e), the first critical speed $\omega_{cr1}^{F} = 191.45$ rad/s (forward whirl) will be encountered, which is related to up-down motion without any tilting. Unbalance can excite only the forward whirl. When the disc has only initial transverse tilt φ_0, $\omega_{cr2}^{B} = 331.60$ rad/s critical speed will be encountered, when a massless free-wheel with unbalance (refer Figure 5.30 and Exercise 5.28) and is rotating opposite to the shaft/disc rotation, which is related to the wobbling motion. When the disc has both radial eccentricity, e, and initial tilt, φ_0, one forward critical speed will be encountered by the rotor system when the free-wheel in rotating in the same direction as the shaft/disc and one backward critical speed when the free-wheel in rotating in opposite direction as the shaft/disc.

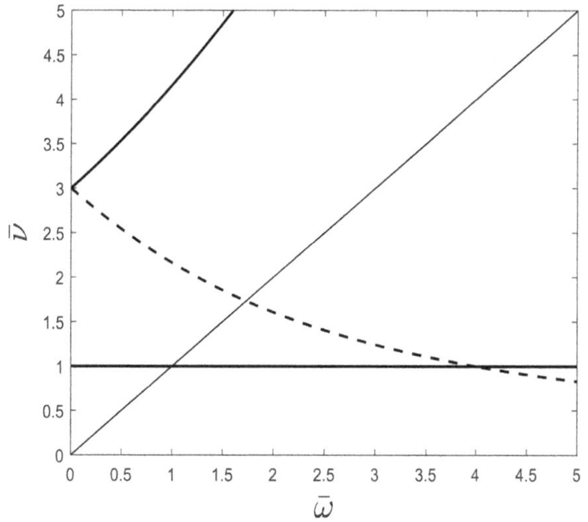

FIGURE 5.29 Campbell diagram for the rotor system.

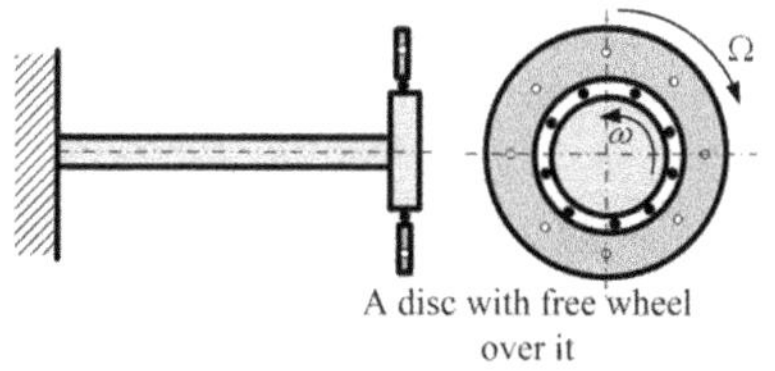

FIGURE 5.30 A cantilever rotor with free disc at the end.

Exercise 5.28 Briefly describe the following:

i. Synchronous forward whirl and asynchronous forward whirl.
ii. Whirl natural frequency and critical speed in the case of rotor systems with gyroscopic effects.
iii. A sphere is spinning about a diametral axis with ω rad/s and precessing about another diametral axis perpendicular to the spinning axis with ν rad/s. Let I_p be the polar mass moment of inertia of the sphere. What is the value of the gyroscopic moment about the third orthogonal axis?
iv. A sphere is supported at both ends of its diameter by two identical springs. The transverse stiffness of the springs at either end of the diameter is k, the mass of the sphere is m, and its polar mass moment of inertia is I_p. What is the natural frequency(ies) of the system?
v. A cantilever rotor with a massless flexible shaft has a thin heavy disc rigidly fixed at free end of the shaft (refer Figure 5.30). On the disc another massless free disc is mounted with the help of a rolling bearing, and with this arrangement the free disc can rotate (Ω) independent of the heavy disc (ω) by a separate drive arrangement mounted on the shaft itself. The free disc can be rotated independently CW or CCW at the speed of the same magnitude as the shaft speed. Also on the free disc a provision is there to keep an unbalance mass. With this arrangement, keeping the shaft speed sense of rotation the same (e.g., CCW), the free disc with an unbalance is rotated in first in the CW and then in the CCW sense of rotation. Comment on the critical speed encountered by the rotor system if speed is increased gradually from zero to beyond the critical speeds of the rotor system. Will the critical speeds observed would be the same for the CW and CCW rotation of the free disc? Explain the proper physical reasoning for this.

***Solution*:** (i) The forward whirl is defined for the case when shaft spinning and whirling sense of rotation is same. The synchronous whirl is defined when the magnitude of spin speed and whirl frequency are same as well as their sense of rotation is same. When the magnitude is same but sense of rotation is opposite then it is a condition of anti-synchronous whirl in which backward whirl prevails. Asynchronous whirl is a condition when the magnitude of rotation is not same as the whirling, the sense of whirling may or may not be same as the shaft spin. However, when the magnitude of rotation is not same as the whirling, the sense of whirling also not same as the shaft spin, then it is the backward asynchronous whirl.

(ii) In rotor with the gyroscopic effect (which is significant at high-speed rotors), the whirl natural frequency (which is a natural frequency) is dependent on the spin speed of the shaft. For example, if a perfectly balanced spinning rotor is disturbed by a tap (initial disturbance) then it will whirl at its transverse natural frequency (which is the whirl natural frequency). This is due to the fact that the rotor is under free vibration. If the speed of rotor is changed, this whirl natural frequency will also be changing due to the gyroscopic effect. If we attain a rotor spin speed for which its whirl natural frequency is also same, then that speed is called the critical speed. It should be noted that when a rotor has an unbalance then at a particular speed the frequency of whirl (which will be forced vibration) will be equal to the spin speed of the rotor. So, at the critical speed, the frequency of whirl will be equal to the whirl natural frequency.

(iii) The gyroscopic couple during the synchronous whirl (i.e. at forward critical speed) for spherical objects (having $I_p = I_d$) will be zero since it is expressed (refer eq. (5.52) of Tiwari (2017)) for the synchronous whirl, as

$$M_{yz} = \omega^2 \left(I_p - I_d\right)\varphi_x \tag{5.209}$$

Also, for the asynchronous whirl for spherical objects, the change in angular momentum from eq. (5.68) of Tiwari (2017), we have

$$I_p\omega\sin\varphi_x - I_d\nu\varphi_x\cos\varphi_x \approx \left(I_p\omega - I_d\nu\right)\varphi_x = I_d(\omega-\nu)\varphi_x \quad \text{since} \quad I_p = I_d \tag{5.210}$$

which gives gyroscopic moment, as

$$M_{yz} = I_d\varphi_x(\omega-\nu)\nu \tag{5.211}$$

which is not zero as long as the motion is asynchronous (i.e., away from critical speed). But at critical speed, the gyroscopic effect will be zero for the synchronous whirl condition (i.e. $\nu = \omega$). But for the backward whirl, the gyroscopic effect will not be zero (as can be seen in Eq. (5.211) for $\nu = -\omega$, the gyroscopic effect is not zero). The frequency equation is derived in eqs. (5.76) and (5.78) of Tiwari (2017) are based on thin disc only so they are not valid for other discs (long stick or spherical objects). To analyse such a case, we need to take the gyroscopic moment given in Eq. (5.210) and substitute in eq. (5.72) of Tiwari (2017) and proceed with the procedure given till derivation of eq. (5.78) of Tiwari (2017). But from Eq. (5.211), it can be seen that as the speed approaches the whirl natural frequency value, the gyroscopic couple for the case of spherical objects will become zero, so the gyroscopic effect will not present at the forward critical speed. One critical speed will be present but its will be same as that of non-rotated case for the forward whirl. But for backward whirl the gyroscopic effect will be seen and corresponding distinct two backward critical speeds will be seen. Readers are encouraged to attempt such formulation and analyse them in detail.

(iv) Since the sphere is a symmetrical object so there will not be any coupling between transverse linear and rotational motions. So the linear motion, the natural frequency will be

$$\omega_{nf1} = \sqrt{\frac{k_{eff}}{m}} = \sqrt{\frac{2k}{m}} \tag{5.212}$$

During a pure rotational motion as well spinning for a general motion, the effective moment stiffness experience by the sphere will be

$$k_{eff} = 0.5kD^2 \pm I_p\omega\nu \tag{5.213}$$

where a positive sign for the forward whirl and a negative sign for the backward whirl. The whirl frequency for the present case is expressed as

$$\nu_2 = \sqrt{\frac{k_{eff}}{I_d}} = \sqrt{\frac{0.5kD^2 \pm I_p\omega\nu_2}{I_p}} \quad \text{since} \quad I_p = I_d \tag{5.214}$$

$$\text{or} \quad I_p\nu_2^2 = 0.5kD^2 \pm I_p\omega\nu_2 \quad \Rightarrow \nu_2^2 \mp \nu_2\omega - \frac{0.5kD^2}{I_p} = 0 \tag{5.215}$$

$$\text{or} \quad \nu_2 = \pm 0.5\omega \pm 0.5\sqrt{(\omega)^2 + \frac{0.5kD^2}{I_p}} \quad \Rightarrow \nu_2 = \pm 0.5\omega + 0.5\sqrt{(\omega)^2 + \frac{0.5kD^2}{I_p}} \tag{5.216}$$

The negative sign in front of square root in Eq. (5.216) is ignored since it will be negative frequencies, which is not feasible. So, in Eq. (5.216), we have the forward whirl frequency for the positive sign and the backward for the negative sign.

From Eq. (5.215), for synchronous motion $\nu = \omega$ (forward or backward depending upon the negative or positive sign), it gives

$$\nu_2^2 \mp \nu_2^2 - \frac{0.5kD^2}{I_p} = 0 \quad \Rightarrow \nu_2 = \sqrt{\frac{kD^2}{4I_p}} \tag{5.217}$$

For the negative sign (backward whirl), the frequency is not feasible. It can be observed that the effect of gyroscopic couple is absent in the aforementioned expression for both forward and backward whirls when we considered pure tilting motion of a sphere. But while considering a general motion (both translational and rotational displacements) then the effect of gyroscopic couple in the forward critical speed will not be seen but it will present in the backward critical speed as we discussed in the previous question (iii).

(v) Here a free disc is rotating at the same speed that of the heavy perfectly balanced disc. As the free disc is massless, it does not contribute to the gyroscopic couple. The unbalance mass, which is attached to this free disc will induce the whirling motion. When the free disc rotates in the CCW direction, it will induce whirling motion in the CCW direction at frequency equal to that of shaft's spin speed, ω, due to unbalance. The direction of whirling is same as that of the shaft spin direction (CCW). So this is the case of forward synchronous whirl. Similarly, when the free disc rotates CW (opposite to the previous case, but shaft still rotates in CCW direction), it will induce CW whirling motion at the same frequency as that of the free-wheel spin speed. But this time, the direction of whirling will be opposite to that of shaft spin direction. Hence, this is the case of the backward synchronous whirl. Till now we considered that the free-wheel is rotating away from any of the critical speeds.

So, the critical speed induced will be different for different sense of direction of the free disc. Due to the gyroscopic couple of heavy disc, the critical speed encountered will be different for the forward synchronous motion and the backward synchronous motion. But these critical speeds encountered do depend on the spin speed of the free-wheel due to unbalance. Critical speeds will be constant for that particular speed of the heavy disc. For another speed of heavy disc, another set of natural frequencies will be there and using free-wheel we can have critical speed condition at those new frequencies. When heavy disc will be rotating at one of its critical speed (let us say forward critical speed) then anyway free-wheel will add to it when it rotates at that speed and with the same sense of direction. But backward critical speed can also be excited by rotating free-wheel in the opposite direction. Herein, it is tried to explain that when the rotor is rotating at a particular speed it will have natural frequencies (forward and backward) and it can be put into resonance condition using free-wheel by rotating in the same or opposite direction of the rotor.

Exercise 5.29 Obtain the forward and backward transverse critical speeds corresponding to twice the rotor speed ($\nu = 2 \times \omega$) for a general motion of a rotor as shown in Figure 5.22. The rotor is assumed to be fixed supported at one end and free at the other end. Let the mass of the thin disc $m = 2$ kg, the polar mass moment of inertia $I_p = 0.01$ kg-m^2, and the diametral mass moment of inertia $I_d = 0.005$ kg-m^2. The shaft is assumed to be massless, and its length and diameter are 0.2 m and 0.01 m, respectively. Take the Young's modulus of the shaft $E = 2.1 \times 10^{11}$ N/m^2.

Solution: This is a case of asynchronous motion ($\nu \neq \omega$) of the thin disc considering the gyroscopic effect. From eq. (5.75) of Tiwari (2017), we have

$$\begin{bmatrix} 1-\alpha_{11}m\nu^2 & I_d\alpha_{12}\nu(2\omega-\nu) \\ -\alpha_{12}m\nu^2 & 1+\alpha_{22}I_d\nu(2\omega-\nu) \end{bmatrix} \begin{Bmatrix} y \\ \varphi_x \end{Bmatrix} = \{0\} \tag{5.218}$$

On substituting $\nu = 2\omega$ to get the transverse critical speeds corresponding to twice the rotor speed, as

$$\begin{bmatrix} 1-\alpha_{11}m4\omega^2 & 0 \\ -\alpha_{12}m4\omega^2 & 1 \end{bmatrix} \begin{Bmatrix} y \\ \phi_x \end{Bmatrix} = \{0\} \tag{5.219}$$

For the non-trivial solution, we get

$$\begin{vmatrix} 1-4\alpha_{11}m\omega^2 & 0 \\ -4\alpha_{12}m\omega^2 & 1 \end{vmatrix} = 0 \Rightarrow \left(1-4\alpha_{11}m\omega^2\right) = 0 \quad \Rightarrow \omega_{cr} = \sqrt{\frac{1}{4\alpha_{11}m}} = \sqrt{\frac{3EI}{4l^3m}} \tag{5.220}$$

On substituting values, we get the forward transverse critical speed corresponding to twice the rotor speed, *as*

$$\omega_{cr} = \sqrt{\frac{3EI}{4l^3m}} = \sqrt{\frac{3\times 2.1\times 10^{11}\times \pi \times 0.01^4}{4\times 0.2^3\times 2\times 64}} = 69.51 \text{ rad/s.}$$

Figure 5.31 shows a Campbell diagram for the present problem, which gives a bird-eye view of all critical speeds (in non-dimensional form) at intersection points of frequency curves with line $\nu = 2\omega$, which can be obtained from Eq. (5.218) for a general motion. It can be seen that solid line represent the forward whirl natural frequency and dashed line the backward whirl. In forward whirl, only one intersection is feasible (for the present case) and that gives a forward critical speed as obtained earlier. Whereas, for the backward whirl two intersections is possible and we have two critical speeds. The backward whirl critical speeds can be obtained by substituting $\nu = -2\omega$ in Eq. (5.218) and solving as procedure described. Readers can verify the same to quench their inquisitiveness.

Exercise 5.30 Obtain transverse critical speeds of the rotor-bearing system as shown in Figure 5.32. Consider the shaft to be rigid. Both bearings have moment springs (of stiffness k_b) to resist tilting (or the transverse rotational motion) and linear springs (of stiffness k) to resist transverse translational motions. Both discs have identical inertia properties and locations from the bearings. Take $a = 0.7$ m, $b = 0.3$ m, $k_b = 1$ kN-m/rad and $k = 10$ kN/m. Each disc has $m = 5$ kg and $I_d = 0.02$ kg-m^2. Consider the gyroscopic couple effect and a pure rotational motion only.

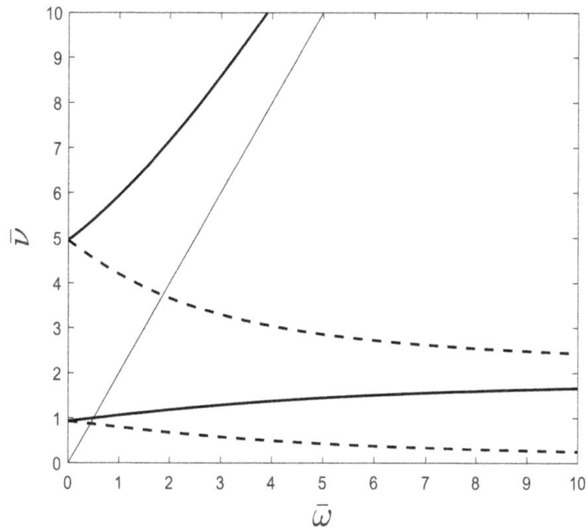

FIGURE 5.31 Campbell diagram for the rotor system.

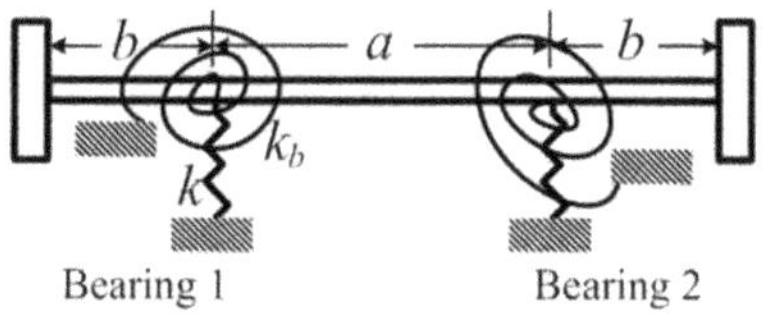

FIGURE 5.32 A rigid rotor mounted on flexible bearings.

***Solution*:** In the present case, rotor pure tilting is considered about its centre of gravity. However, asynchronous motion is considered due to the gyroscopic effect. Given rotor-bearing data are:

$a = 0.7$ m, $b = 0.3$ m, stiffness of moment spring, $k_b = 1$ kN-m/rad, stiffness of linear spring, $k = 10$ kN/m, mass of the disc, $m = 5$ kg and diametral moment of inertia of disc, $I_d = 0.02$ kg-m^2.

Refer to the free-body diagram of the rotor in Figure 5.33. Total reaction couple experienced by the rotor system is

$$M = 2\left\{k_b + k(a/2)^2 \pm I_p \nu\omega\right\}\varphi_x = k_{eff}\varphi_x \tag{5.221}$$

where "+" is for the forward whirl and "–" is for the backward whirl. In Eq. (5.221), the first two terms are from support stiffness and the third term is due to the gyroscopic moment. Referring to Figure 5.33, the effective diametral moment of inertia of the whole rotor system about the centre of mass of the system is,

$$I_{d_{eff}} = 2\left\{I_d + m(0.5a+b)^2\right\} \tag{5.222}$$

Then, the whirl natural frequency is given, as

$$\nu^2 = \frac{k_{eff}}{I_{d_{eff}}} = \frac{2\left\{k_b + k(0.5a)^2 \pm I_p \nu\omega\right\}}{I_{d_{eff}}} = \frac{k_t \pm I_{p_{eff}} \nu\omega}{I_{d_{eff}}} \tag{5.223}$$

$$\text{with} \quad k_t = 2\left\{k_b + k(0.5a)^2\right\}; \quad I_{p_{eff}} = 2I_p; \quad I_{d_{eff}} = 2\left\{I_d + m(0.5a+b)^2\right\} \tag{5.224}$$

Equation (5.223) can be written as

$$\nu^2 \mp \frac{I_{p_{eff}}\omega}{I_{d_{eff}}}\nu - \frac{k_t}{I_{d_{eff}}} = 0 \tag{5.225}$$

which on solving gives

$$\nu_{1,2} = \pm\frac{I_{p_{eff}}\omega}{2I_{d_{eff}}} \mp \sqrt{\left(\frac{I_{p_{eff}}\omega}{2I_{d_{eff}}}\right)^2 + \frac{k_t}{I_{d_{eff}}}} \tag{5.226}$$

Equation (5.226) can be used to obtain the whirl natural frequency of the system with the speed of the rotor. For finding the critical speeds, substitute $\nu = \omega$ in Eq. (5.226), the critical speeds can be obtained, as

$$\omega_{cr}^{F,B} = \sqrt{\frac{k_t}{I_{d_{eff}} \mp I_{p_{eff}}}} \tag{5.227}$$

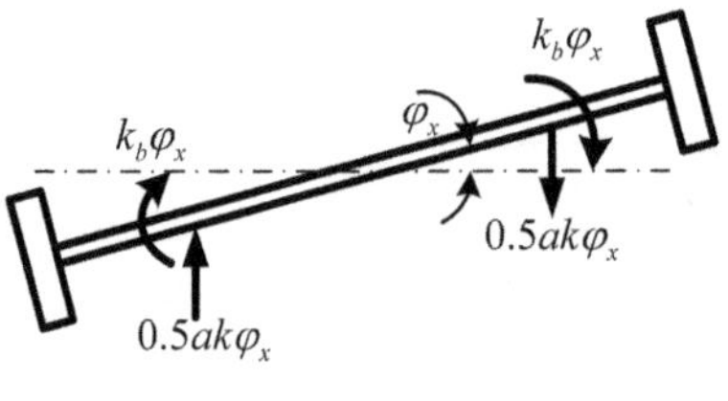

FIGURE 5.33 Free body diagram of the rotor.

where "+" is for the backward (superscript *B*) whirl and "−" is for the forward (superscript *F*) whirl. On substituting the given numerical values of the rotor-bearing system in Eq. (5.224), we get

$$I_{d_{eff}} = 2\left\{I_d + m(0.5a+b)^2\right\} = 4.265 \text{ kg-m}^2; \quad k_t = 2\left\{k_b + k(0.5a)^2\right\} = 4{,}450 \text{ N-m/rad;}$$

$$I_{P_{eff}} = 2I_p = 4I_d = 4\times 0.02 = 0.08 \text{ kg-m}^2$$

On substituting the values of k_t, I_{deff} and I_p obtained in Eq. (5.227), we get

$$\omega_{cr}{}^{F,B} = \sqrt{\frac{4{,}450}{4.265 \mp 0.08}}$$

which gives $\omega_{cr}^F = 32.61$ rad/s and $\omega_{cr}^B = 32.00$ rad/s

It should be noted that for a pure rotation motion without gyroscopic effect, the critical speed will be given as

$$\omega_{cr} = \sqrt{\frac{k_t}{I_{d_{eff}}}} = \sqrt{\frac{4{,}450}{4.265}} = 32.30 \text{ rad/s}$$

which is in between the forward and backward critical speeds, this shows the splitting behavior of the critical speed in the presence of the gyroscopic effect and it is valid for the whirl natural frequency also using Eq. (5.226).

Exercise 5.31 Obtain all critical speeds of the rotor system shown in Figures 5.34 and 5.35 by considering the gyroscopic effect. Let $a = 0.6$ m and $b = 0.4$ m. Let the shaft diameter $d = 0.015$ m and $E = 2.1 \times 10^{11}$ N/m^2. The offset disc has a mass of 2 kg and a diametral mass moment of inertia of 0.01 kg-m^2.

***Solution*:** Case I: In the present case, initially only when the end support is simple support is considered.

The frequency equation for a general motion of a *thin single disc* is given as (refer eq. (5.78) of Tiwari (2017))

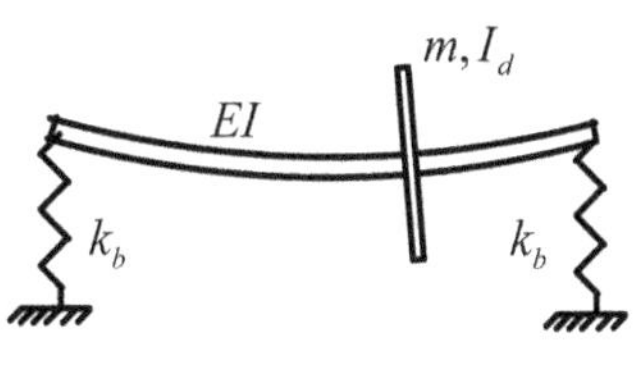

FIGURE 5.34 A flexible shaft on flexible bearings.

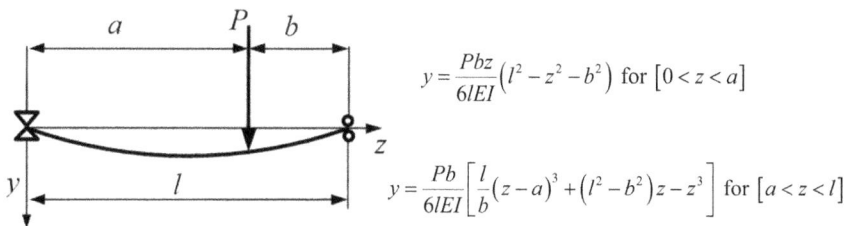

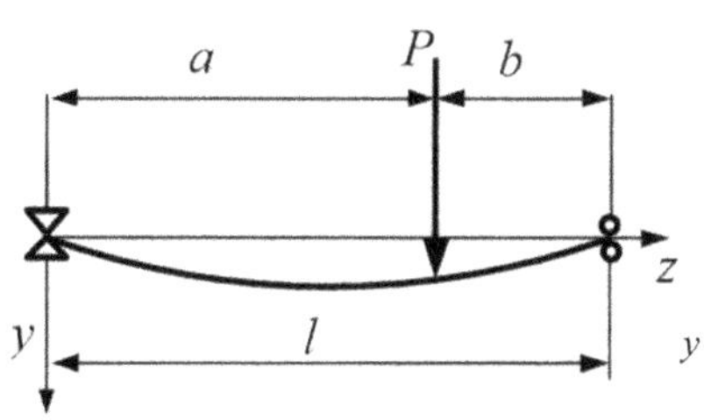

$$y = \frac{Pbz}{6lEI}\left(l^2 - z^2 - b^2\right) \text{ for } [0 < z < a]$$

$$y = \frac{Pb}{6lEI}\left[\frac{l}{b}(z-a)^3 + \left(l^2 - b^2\right)z - z^3\right] \text{ for } [a < z < l]$$

FIGURE 5.35 A simply supported shaft with a concentrated load *P* at any point.

$$\bar{\nu}^4 - 2\bar{\omega}\bar{\nu}^3 + \frac{\mu+1}{\mu(\bar{\alpha}+1)}\bar{\nu}^2 - \frac{2\bar{\omega}\bar{\nu}}{\bar{\alpha}-1} - \frac{1}{\mu(\bar{\alpha}-1)} = 0 \tag{5.228}$$

with $\bar{\nu}$ = dimensionless frequency $= \nu\sqrt{\alpha_{11}m}$; μ = disc effect $= \dfrac{I_d\alpha_{22}}{m\alpha_{11}}$

$\bar{\alpha}$ = elastic coupling $= \dfrac{\alpha_{12}^2}{\alpha_{11}\alpha_{22}}$; $\bar{\omega}$ = dimensionless speed $= \omega\sqrt{m\alpha_{11}}$

For the present problem, the following rotor data are given

$a = 0.6$ m, $b = 0.4$ m, $l = 1.0$ m, $d = 0.015$ m, $E = 2.1 \times 10^{11}$ N/m^2,
$m = 2$ kg, $I_d = 0.01$ kg-m^2.

Therefore, we have

$$I = \frac{\pi d^4}{64} = 2.485\times10^{-9}\ \text{m}^4;\quad \alpha_{11} = \frac{a^2b^2}{3EIl} = 3.679\times10^{-5}\ \frac{\text{m}}{\text{N}};$$

$$\alpha_{12} = \alpha_{12} = \frac{-\left(3a^2l - 2a^3 - aL^2\right)}{3EIl} = -3.066\times10^{-5}\ \frac{1}{\text{N}};\quad \alpha_{22} = \alpha_{22} = \frac{-\left(3al - 3a^2 - l^2\right)}{3EIl} = 1.789\times10^{-4}\ \frac{1}{\text{N}-\text{m}};$$

$$\mu = \frac{I_d\alpha_{22}}{m\alpha_{11}} = 0.0243;\quad \bar{\alpha} = \frac{\alpha_{12}^2}{\alpha_{11}\alpha_{22}} = 0.143;$$

On substituting in Eq. (5.228), we get

$$\bar{\nu}^4 - 2\bar{\omega}\bar{\nu}^3 + \frac{0.0243+1}{0.0243\times(0.143+1)}\bar{\nu}^2 - \frac{2\bar{\omega}\bar{\nu}}{0.143-1} - \frac{1}{0.0243\times(0.143-1)} = 0$$

$$\text{or}\quad \bar{\nu}^4 - 2\bar{\omega}\bar{\nu} - 49.19\bar{\nu}^2 + 2.334\bar{\omega}\bar{\nu} + 48.02 = 0 \tag{5.229}$$

For the forward whirl, we have $\bar{\omega} = \bar{\nu}$ so, from Eq. (5.229), we get

$$\bar{\omega}_{crF}^4 + 46.86\bar{\omega}_{crF}^2 - 48.02 = 0\quad \Rightarrow \bar{\omega}_{cr1,2F}^2 = -92.35\quad \text{and}\quad 1.005.$$

But, $\bar{\omega}_{crF}^2$ cannot be negative, hence,

$$\bar{\omega}_{cr1F}^2 = 1.005 = \omega_{cr1F}^2 m\alpha_{11}\quad \Rightarrow \omega_{cr1F} = \sqrt{\frac{1.005}{m\alpha_{11}}}\quad \Rightarrow \omega_{cr1F} = 116.81\ \text{rad/s} \tag{5.230}$$

For the backward whirl, we have $\bar{\omega} = -\bar{\nu}$ so, from Eq. (5.229), we get

$$\bar{\omega}_{crB}^4 - 51.524\bar{\omega}_{crB}^2 + 48.02 = 0\quad \Rightarrow \bar{\omega}_{cr1,2B}^2 = 16.19\quad \text{and}\quad 0.989.$$

Hence, we have

$$\bar{\omega}_{cr1B}^2 = 0.989 = \omega_{cr1B}^2 m\alpha_{11}\quad \Rightarrow \omega_{cr1B} = \sqrt{\frac{0.989}{m\alpha_{11}}}\quad \Rightarrow \omega_{cr1B} = 115.88\ \text{rad/s} \tag{5.231}$$

and

$$\bar{\omega}_{cr2B}^2 = 16.19 = \omega_{cr2B}^2 m\alpha_{11}\quad \Rightarrow \omega_{cr2B} = \sqrt{\frac{16.19}{m\alpha_{11}}}\quad \Rightarrow \omega_{cr2B} = 468.65\ \text{rad/s} \tag{5.232}$$

Figure 5.36 shows the Campbell diagram plotted with the help of Eq. (5.229).

Case II: A Symmetrical Flexible Shaft on Identical Bearings

For the present case (refer Exercise 4.18 for a symmetrical flexible shaft on different bearings), both the shaft and bearings are flexible as shown in Figure 5.37. All generalised coordinates are with respect to a fixed frame of reference and they are absolute displacements. The motion in single plane is considered. The analysis allows finding different instantaneous displacements of the shaft at the disc (offset from the mid–span of shaft) and at bearings. The system will behave in a similar manner to that described in the previous method (Case I), except that the flexibility of bearing will increase the overall flexibility of the support system as experienced by the rigid disc. An equivalent set of system stiffness coefficients (or influence coefficients) is first evaluated, which allows for the flexibility of the shaft in addition to that of bearings and is used in place of influence coefficients of the previous analysis (Case I). The total deflection of the disc is the vector sum of the deflection of the disc relative to the shaft ends, plus that of shaft ends in bearings. For the disc, we observe the displacement of its geometrical centre.

The deflection of the shaft ends in bearings is related to the force transmitted through bearings by the bearing stiffness coefficients as (for one of the bearing)

$$f_{by} = k_{yy} y_b \tag{5.233}$$

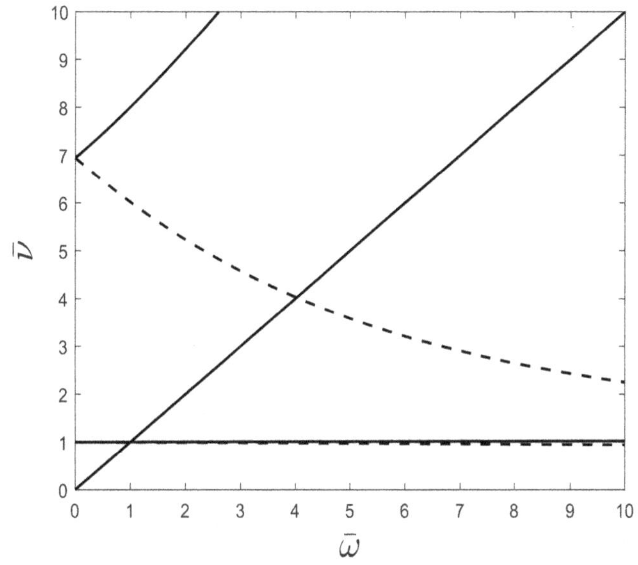

FIGURE 5.36 Campbell diagram showing one forward critical speed and two backward critical speeds at intersection points of frequency curves with the synchronous condition line ($\nu = \omega$).

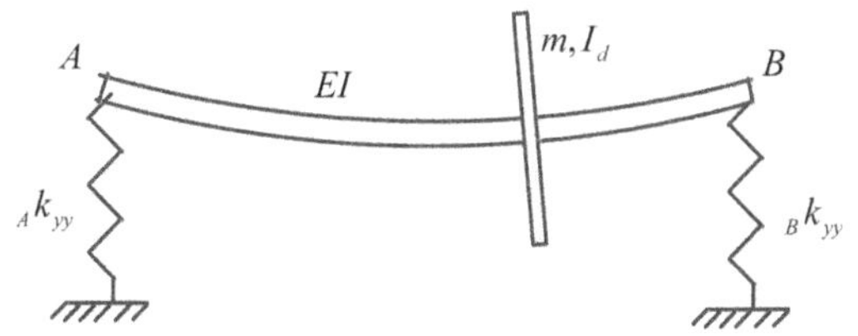

FIGURE 5.37 A flexible shaft with an off-set disc on flexible bearings.

where y_b is the instantaneous displacement of shaft ends relative to bearings in the vertical direction, and for asynchronous vibration, it takes the following form

$$y_b = Y_b e^{j\nu t} \tag{5.234}$$

where Y_b is the complex displacement in y direction and ν is the whirl natural frequency. It should be noted that bearings are modelled as a point connection with the shaft and only translational displacements are considered, since they support mainly radial (transverse) loads. Bearing forces will have the following form

$$f_{by} = F_{by} e^{j\nu t} \tag{5.235}$$

where F_{by} is the complex force in y direction. On substituting Eqs. (5.234) and (5.235) into Eq. (5.233), we get

$$F_{by} = k_{yy} Y_b \tag{5.236}$$

which is in the frequency domain and can be written in a matrix form for both bearings A (left bearing) and B (right bearing) in single plane motion in the vertical y-direction, as

$$\mathbf{F}_b = \mathbf{K}\mathbf{Y}_b \quad \text{with} \quad \mathbf{F}_b = \begin{Bmatrix} {}_A F_{by} \\ {}_B F_{by} \end{Bmatrix}; \quad \mathbf{K} = \begin{bmatrix} {}_A k_{yy} & 0 \\ 0 & {}_B k_{yy} \end{bmatrix}; \quad \mathbf{Y}_b = \begin{Bmatrix} {}_A Y_b \\ {}_B Y_b \end{Bmatrix} \tag{5.237}$$

Two bearing motions are not coupled by stiffness coefficients. The magnitude of reaction forces transmitted by bearings can also be evaluated in terms of forces applied to the shaft by the disc (refer Figure 5.38).

From Figure 5.38, the moment balance in the vertical plane will be

$$\sum M_B = 0 \Rightarrow {}_A f_{by} l = f_y (l - a) + M_{yz} \quad \text{or} \quad {}_A f_{by} = f_y \left(1 - a/l\right) + M_{yz} \left(1/l\right) \tag{5.238}$$

$$\sum M_A = 0 \Rightarrow {}_B f_{by} l = f_y a - M_{yz} \quad \text{or} \quad {}_B f_{by} = f_y \left(a/l\right) - M_{yz} \left(1/l\right) \tag{5.239}$$

Eqs. (5.238) and (5.239) can be combined in a matrix form, for single plane motion, as

$$\mathbf{f}_b = \mathbf{A}\mathbf{f}_s \quad \text{with} \quad \mathbf{f}_b = \begin{Bmatrix} {}_A f_{by} \\ {}_B f_{by} \end{Bmatrix}; \quad \mathbf{f}_s = \begin{Bmatrix} f_y \\ M_{yz} \end{Bmatrix}; \quad \mathbf{A} = \begin{bmatrix} (1 - a/l) & 1/l \\ a/l & -1/l \end{bmatrix} \tag{5.240}$$

For the asynchronous motion under free vibration, we have

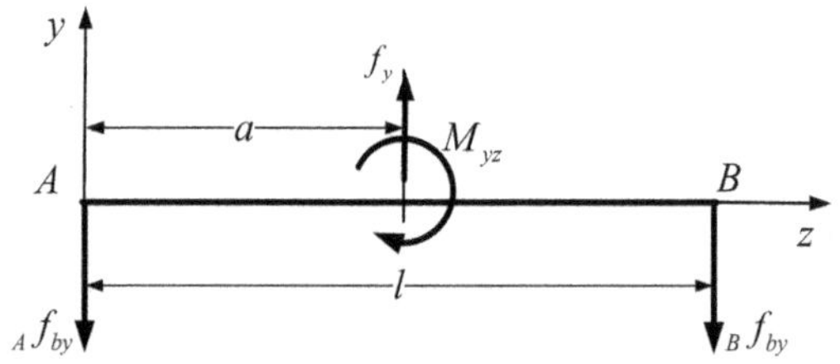

FIGURE 5.38 A free body diagram of the shaft y–z plane.

$$\mathbf{f}_b = \mathbf{F}_b e^{j\nu t} \quad \text{and} \quad \mathbf{f}_s = \mathbf{F}_s e^{j\nu t} \quad \text{with} \quad \mathbf{F}_b = \left\{ \begin{array}{c} {}_A F_{by} \\ {}_B F_{by} \end{array} \right\}; \quad \mathbf{F}_s = \left\{ \begin{array}{c} F_y \\ \bar{M}_{yz} \end{array} \right\} \tag{5.241}$$

where subscript b refers to the bearing and s refers to the shaft, and vectors $\mathbf{F}_b$ and $\mathbf{F}_s$ are complex force vector on the shaft at bearing and disc locations, respectively. On substituting Eq. (5.241) into Eq. (5.240), we get

$$\mathbf{F}_b = \mathbf{A}\mathbf{F}_s \tag{5.242}$$

which is in frequency domain now (i.e. no time dependency). In Eq. (5.240), bearing forces are related to the reaction forces and moments on the shaft by the disc. On equating Eqs. (5.237) and (5.242), we get

$$\mathbf{F}_b \equiv \mathbf{K}\mathbf{Y}_b = \mathbf{A}\mathbf{F}_s \quad \text{or} \quad \mathbf{Y}_b = \mathbf{K}^{-1}\mathbf{A}\mathbf{F}_s \tag{5.243}$$

Equation (5.243) relates the shaft end deflections to the reaction forces and moments on the shaft by the disc. The deflection at the location of the disc due to the movement of shaft ends can be obtained as follows. Consider the shaft to be rigid for time being and let us denote shaft end deflections in the vertical direction to be ${}_A y_b$ and ${}_B y_b$ at ends A and B, respectively, as shown in Figure 5.39. These displacements are assumed to be small.

The translational displacement in the y-direction can be written as (refer Figure 5.39)

$$y = {}_A y_b - \frac{({}_A y_b - {}_B y_b)}{l} a = \left(\frac{l-a}{l} \right) {}_A y_b + \left(\frac{a}{l} \right) {}_B y_b \tag{5.244}$$

And similarly, for the rotational displacements in the y–z plane, we have (refer Figure 5.39)

$$\varphi_x = \frac{({}_A y_b - {}_B y_b)}{l} = \left(\frac{1}{l} \right) {}_A y_b + \left(-\frac{1}{l} \right) {}_B y_b \tag{5.245}$$

Equations (5.244) and (5.245) can be combined in a matrix form as

$$\mathbf{u}_{s_1} = \mathbf{B}\mathbf{y}_b \quad \text{with} \quad \mathbf{u}_{s_1} = \left\{ \begin{array}{c} y \\ \varphi_x \end{array} \right\}_{s_1}; \quad \mathbf{y}_b = \left\{ \begin{array}{c} {}_A y_b \\ {}_B y_b \end{array} \right\}; \quad \mathbf{B} = \begin{bmatrix} (1 - a/l) & a/l \\ 1/l & -1/l \end{bmatrix} \tag{5.246}$$

where subscript s_1 represents that these displacements are due to a rigid body motion of the shaft. For the free vibration analysis, shaft displacements at bearing locations and at the disc centre vary sinusoidally such that

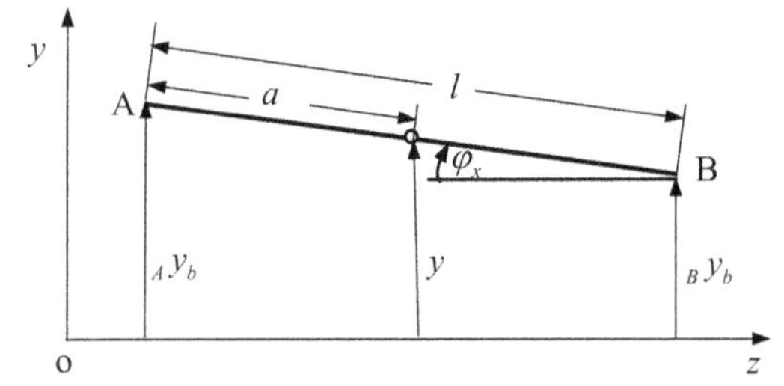

FIGURE 5.39 Rigid body movement of the shaft in y–z plane.]

$$\mathbf{u}_{s1} = \mathbf{U}_{s1} e^{j\nu t} \quad \text{and} \quad \mathbf{y}_b = \mathbf{Y}_b e^{j\nu t} \tag{5.247}$$

where ν is the whirl natural frequency in the case of free vibration. On substituting Eq. (5.247) into Eq. (5.246), we have

$$\mathbf{U}_{s1} = \mathbf{B}\mathbf{Y}_b \tag{5.248}$$

which is in the frequency domain. On substituting Eq. (5.243) into Eq. (5.248), we get

$$\mathbf{U}_{s1} \equiv \mathbf{B}\mathbf{Y}_b = \mathbf{B}\mathbf{K}^{-1}\mathbf{A}\mathbf{F}_s = \mathbf{C}\mathbf{F}_s \quad \text{with} \quad \mathbf{C} = \mathbf{B}\mathbf{K}^{-1}\mathbf{A} \tag{5.249}$$

which gives the deflection of the disc due to the shaft elastic force and moment, when the shaft is rigid. Equation (5.249) gives deflections of the disc that is caused by only the movement of shaft ends (rigid body movement) on flexible bearings. To obtain the net rotor deflection under a given load, we have to add the deflection due to the deformation of the shaft with respect to bearing locations also in Eq. (5.249). The deflection associated with the flexure of the shaft alone has already been discussed in Chapter 2 (Tiwari, 2017), which can be combined in a matrix form as

$$\mathbf{u}_{s2} = \boldsymbol{\alpha}\mathbf{f}_s \quad \text{with} \quad \mathbf{u}_{s2} = \begin{Bmatrix} y \\ \varphi_x \end{Bmatrix}_{s2}; \quad \mathbf{f}_s = \begin{Bmatrix} f_y \\ M_{yz} \end{Bmatrix}; \quad \boldsymbol{\alpha} = \begin{bmatrix} \alpha_{11} & \alpha_{12} \\ \alpha_{21} & \alpha_{22} \end{bmatrix} \tag{5.250}$$

where subscript s_2 represents that these displacements are due to the pure deformation of the shaft without any rigid body motion and α_{ij} $(i, j = 1, 2)$ is the influence coefficient. For the undamped free vibration analysis, the shaft reaction forces at the disc location and disc displacements vary sinusoidally and can be expressed as

$$\mathbf{u}_{s2} = \mathbf{U}_{s2} e^{j\nu t} \quad \text{and} \quad \mathbf{f}_s = \mathbf{F}_s e^{j\nu t} \tag{5.251}$$

On substituting Eq. (5.251) into Eq. (5.250), to get equation in the frequency domain, as

$$\mathbf{U}_{s2} = \boldsymbol{\alpha}\mathbf{F}_s \tag{5.252}$$

which is the deflection of disc due to the flexure of the shaft alone, without considering the bearing flexibility. The net deflection of the disc caused by the deflection of bearings and the flexure of the shaft is then given by (refer Eqs. (5.249) and (5.252))

$$\mathbf{U}_s = \mathbf{U}_{s1} + \mathbf{U}_{s2} = (\mathbf{C} + \boldsymbol{\alpha})\mathbf{F}_s = \mathbf{D}\mathbf{F}_s \quad \text{with} \quad \mathbf{D} = (\mathbf{C} + \boldsymbol{\alpha}) \equiv \begin{bmatrix} d_{11} & d_{12} \\ d_{21} & d_{22} \end{bmatrix} \quad \text{where } d_{12} \neq d_{21} \tag{5.253}$$

where $\mathbf{U}_s$ contains absolute displacements of the shaft at the location of the disc. Equation (5.253) describes displacements of the shaft at the disc under the action of forces and moments applied at the disc (hence the matrix **D** is effective influence coefficient matrix due to bearing **C**, and shaft $\boldsymbol{\alpha}$). On substituting matrcies **B**, **K** and **A** into expression of matrix **C**, we get

$$\mathbf{C}=\mathbf{B}\mathbf{K}^{-1}\mathbf{A}=\begin{bmatrix}\left(1-\frac{a}{l}\right) & \frac{a}{l}\\ \frac{1}{l} & -\frac{1}{l}\end{bmatrix}\begin{bmatrix}\frac{1}{_A k_{yy}} & 0\\ 0 & \frac{1}{_B k_{yy}}\end{bmatrix}\begin{bmatrix}\left(1-\frac{a}{l}\right) & \frac{1}{l}\\ \frac{a}{l} & -\frac{1}{l}\end{bmatrix}$$

$$=\begin{bmatrix}\left(1-\frac{a}{l}\right) & \frac{a}{l}\\ \frac{1}{l} & -\frac{1}{l}\end{bmatrix}\begin{bmatrix}\frac{1}{_A k_{yy}}\left(1-\frac{a}{l}\right) & \frac{1}{_A k_{yy}}\left(\frac{1}{l}\right)\\ \frac{1}{_B k_{yy}}\left(\frac{a}{l}\right) & \frac{1}{_B k_{yy}}\left(-\frac{1}{l}\right)\end{bmatrix}$$

or

$$\mathbf{C}=\begin{bmatrix}\frac{1}{_A k_{yy}}\left(1-\frac{a}{l}\right)^2+\frac{1}{_B k_{yy}}\left(\frac{a}{l}\right)^2 & \frac{1}{_A k_{yy}}\left(1-\frac{a}{l}\right)\left(\frac{1}{l}\right)+\frac{1}{_B k_{yy}}\frac{a}{l}\left(-\frac{1}{l}\right)\\ \frac{1}{_A k_{yy}}\frac{1}{l}\left(1-\frac{a}{l}\right)+\frac{1}{_B k_{yy}}\left(-\frac{1}{l}\right)\left(\frac{a}{l}\right) & \frac{1}{_A k_{yy}}\left(\frac{1}{l}\right)^2+\frac{1}{_B k_{yy}}\left(-\frac{1}{l}\right)^2\end{bmatrix}\equiv\begin{bmatrix}\beta_{11} & \beta_{12}\\ \beta_{21} & \beta_{22}\end{bmatrix} \tag{5.254}$$

with

$$\mathbf{A}=\begin{bmatrix}(1-a/l) & 1/l\\ a/l & -1/l\end{bmatrix};\quad \mathbf{B}=\begin{bmatrix}(1-a/l) & a/l\\ 1/l & -1/l\end{bmatrix};\quad \mathbf{K}^{-1}=\begin{bmatrix}1/_A k_{yy} & 0\\ 0 & 1/_B k_{yy}\end{bmatrix}$$

Since in general $\beta_{12}\neq\beta_{21}$ (and so $d_{12}\neq d_{21}$) so eq. (5.78) of Tiwari (2017) will not be valid, which has been derived for $\alpha_{12}=\alpha_{21}$. On simplifying matrix **C** for the present case $k={}_A k_{yy}={}_B k_{yy}$, we get

$$\mathbf{C}=\begin{bmatrix}\frac{1}{k}\left(1-\frac{a}{l}\right)^2+\frac{1}{k}\left(\frac{a}{l}\right)^2 & \frac{1}{k}\left(1-\frac{a}{l}\right)\left(\frac{1}{l}\right)+\frac{1}{k}\frac{a}{l}\left(-\frac{1}{l}\right)\\ \frac{1}{k}\frac{1}{l}\left(1-\frac{a}{l}\right)+\frac{1}{k}\left(-\frac{1}{l}\right)\left(\frac{a}{l}\right) & \frac{1}{k}\left(\frac{1}{l}\right)^2+\frac{1}{k}\left(-\frac{1}{l}\right)^2\end{bmatrix}$$

$$\text{or}\quad \mathbf{C}=\frac{1}{kl^2}\begin{bmatrix}(l-a)^2+a^2 & (l-a)-a\\ (l-a)-a & 1+1\end{bmatrix}=\frac{1}{kl^2}\begin{bmatrix}(l-a)^2+a^2 & (l-2a)\\ (l-2a) & 2\end{bmatrix} \tag{5.255}$$

So the effective influence coefficient matrix, which the thin disc will experience, will be

$$\mathbf{D}=\boldsymbol{\alpha}+\mathbf{C}=\begin{bmatrix}\alpha_{11} & \alpha_{12}\\ \alpha_{21} & \alpha_{22}\end{bmatrix}+\frac{1}{kl^2}\begin{bmatrix}(l-a)^2+a^2 & (l-2a)\\ (l-2a) & 2\end{bmatrix}$$

$$\text{or}\quad \mathbf{D}\equiv\begin{bmatrix}d_{11} & d_{12}\\ d_{21} & d_{22}\end{bmatrix}=\begin{bmatrix}\alpha_{11}+\frac{(l-a)^2+a^2}{kl^2} & \alpha_{12}+\frac{(l-2a)}{kl^2}\\ \alpha_{21}+\frac{(l-2a)}{kl^2} & \alpha_{22}+\frac{2}{kl^2}\end{bmatrix} \tag{5.256}$$

So, Eq. (5.253) can be used to obtain unbalance response of the system for the case of no gyroscopic effect (refer Case IV), which can be used to obtain critical speeds of the rotor-bearing system without gyroscopic effect. In the next section, the frequency equation is obtained considering the gyroscopic effect also, which can be used to obtain the critical speeds of the rotor-bearing system.

Case III: Quasi-static approach (general asynchronous motion):

For a general (asynchronous) motion case of a thin disc from eq. (5.76) of Tiwari (2017) in the dimensional form, we have

$$\nu^4\left(-m\alpha_{11}\alpha_{22}I_d + m\alpha_{12}^2 I_d\right) + \nu^3\left(2m\ \alpha_{11}\alpha_{22}I_d\omega - 2m\alpha_{12}^2 I_d\omega\right)$$
$$+\nu^2\left(I_d\alpha_{22} + m\alpha_{11}\right) + \nu\left(-\alpha_{22}I_d 2\omega\right) - 1 = 0 \tag{5.257}$$

Now since the new influence coefficient is defined considering bearing flexibility also in the form of matrix **D** (refer Eq. (5.256)). So writing the aforementioned frequency in terms of modified influence coefficients and noting that $d_{12} \neq d_{21}$, we get

$$\nu^4\left(-md_{11}d_{22}I_d + md_{12}d_{21}I_d\right) + \nu^3\left(2md_{11}d_{22}I_d\omega - 2md_{12}d_{21}I_d\omega\right)$$
$$+\nu^2\left(I_d d_{22} + md_{11}\right) + \nu\left(-d_{22}I_d 2\omega\right) - 1 = 0 \tag{5.258}$$

Forward Critical Speeds:

For the synchronous case $\nu = \omega$, and now $\omega = \omega_{crF}$, so Eq. (5.258) becomes

$$\omega_{crF}^4\left(-md_{11}d_{22}I_d + md_{12}d_{21}I_d\right) + \omega_{crF}^4\left(2md_{11}d_{22}I_d - 2md_{12}d_{21}I_d\right) + \omega_{crF}^2\left(I_d d_{22} + md_{11}\right)$$
$$+\omega_{crF}^2\left(-d_{22}I_d 2\right) - 1 = 0$$

$$\text{or}\quad \omega_{crF}^4 mI_d\left(d_{11}d_{22} - d_{12}d_{21}\right) + \omega_{crF}^2\left(md_{11} - I_d d_{22}\right) - 1 = 0 \tag{5.259}$$

Backward Critical Speeds: For the anti-synchronous case $\nu = -\omega$, and now $\omega = \omega_{crB}$, so Eq. (5.258) becomes

$$\omega_{crB}^4\left(-md_{11}d_{22}I_d + md_{12}d_{21}I_d\right) - \omega_{crB}^4\left(2md_{11}d_{22}I_d - 2md_{12}d_{21}I_d\right) + \omega_{crB}^2\left(I_d d_{22} + md_{11}\right)$$
$$-\omega_{crB}^2\left(-d_{22}I_d 2\right) - 1 = 0$$

$$\text{or}\quad 3\omega_{crB}^4 mI_d\left(d_{11}d_{22} - d_{12}d_{21}\right) - \omega_{crB}^2\left(3I_d d_{22} + md_{11}\right) + 1 = 0 \tag{5.260}$$

Equations (5.259) and (5.260) can be used to obtain forward and backward critical speeds. It should be noted that it is valid for thin disc case only.

For a general (asynchronous) motion, the frequency equation in non-dimensional form from equation (5.79) is given as

$$\bar{\nu}^4 - 2\bar{\omega}\bar{\nu}^3 + \frac{\mu+1}{\mu(\bar{\alpha}-1)}\bar{\nu}^2 - \frac{2\bar{\omega}}{\bar{\alpha}-1}\bar{\nu} - \frac{1}{\mu(\bar{\alpha}-1)} = 0 \tag{5.261}$$

It should be noted that the aforementioned equation will be equally valid for obtaining the forward and backward critical speed in the following form

$$\bar{\nu}^4 - 2\bar{\omega}\bar{\nu}^3 + \frac{\mu+1}{\mu(\bar{d}-1)}\bar{\nu}^2 - \frac{2\bar{\omega}}{\bar{d}-1}\bar{\nu} - \frac{1}{\mu(\bar{d}-1)} = 0 \tag{5.262}$$

$$\text{with } \bar{d} = \frac{d_{12}d_{21}}{d_{11}d_{22}} \tag{5.263}$$

Readers can try out a similar exercise with other boundary conditions with single thin disc rotor systems (like an overhang rotor mounted on flexible bearing).

Case IV: Dynamic approach Unbalance Response (without Gyroscopic effect):

In dynamic approach while considering the gyroscopic effect, we need to consider both orthogonal planes. But previous method (Cases I, II and III) of the quasi-static analysis used single plane motion equations only. So now how to obtain unbalance response in non-gyroscopic effect case is illustrated. Advantage of the present formulation is that equations are in analytical closed form and easy to analyse. Equation (5.253) can be written as

$$\mathbf{DF}_s = \mathbf{U}_s \quad \Rightarrow \mathbf{F}_s = \mathbf{D}^{-1}\mathbf{U}_s = \mathbf{EU}_s \quad \text{with} \quad \mathbf{E} = \mathbf{D}^{-1} \tag{5.264}$$

where the matrix **E** is similar to the stiffness matrix (it is equivalent stiffness of the shaft and bearings experienced at the disc location). Equations of motion, in the y-direction and on the y–z plane (see Figure 5.40), can be written as

$$-f_y = m\frac{d^2}{dt^2}(y + e\sin\omega t) \quad \text{or} \quad me\omega^2\sin\omega t - f_y = m\ddot{y} \tag{5.265}$$

$$\text{and} \quad -M_{yz} = I_d\ddot{\varphi}_x \tag{5.266}$$

Equations of motion (5.265) and (5.266) of the disc can be written in a matrix form as

$$\mathbf{M\ddot{u}} + \mathbf{f}_s = \mathbf{f}_{unb} \tag{5.267}$$

$$\text{with} \quad \mathbf{M} = \begin{bmatrix} m & 0 \\ 0 & I_d \end{bmatrix}; \quad \mathbf{u} = \begin{Bmatrix} y \\ \varphi_x \end{Bmatrix}; \quad \mathbf{f}_s = \begin{Bmatrix} f_y \\ M_{yz} \end{Bmatrix}; \quad \mathbf{f}_{unb} = \begin{Bmatrix} me\omega^2 \\ 0 \end{Bmatrix} e^{j\omega t} = \mathbf{F}_{unb}e^{j\omega t};$$

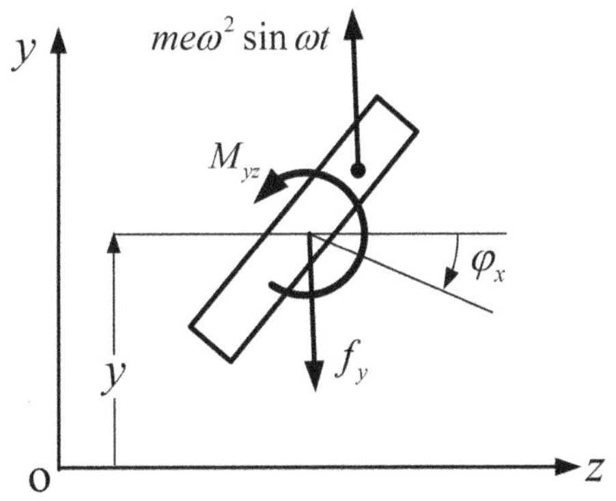

FIGURE 5.40 Free body diagram of the disc in y–z plane.

Noting $\mathbf{f}_s = \mathbf{F}_s e^{j\omega t}$ and $\mathbf{u} = \mathbf{U}_s e^{j\omega t}$, equations of motion take the following form

$$-\omega^2 \mathbf{M}\mathbf{U}_s + \mathbf{F}_s = \mathbf{F}_{unb} \tag{5.268}$$

Noting Eq. (5.264), Eq. (5.268) becomes

$$-\omega^2 \mathbf{M}\mathbf{U}_s + \mathbf{E}\mathbf{U}_s = \mathbf{F}_{unb} \tag{5.269}$$

which gives

$$\mathbf{U}_s = \mathbf{H}\mathbf{F}_{unb} \quad \text{with} \quad \mathbf{H} = \left(-\omega^2 \mathbf{M} + \mathbf{E}\right)^{-1} \tag{5.270}$$

where $\mathbf{H}^{-1} = \left(-\omega^2 \mathbf{M} + \mathbf{E}\right)$ is the equivalent dynamic stiffness matrix, as experienced by the disc, of the shaft and the bearing system. Once the response of the disc has been obtained, from Eq. (5.270) for a given unbalance force, the loading applied to the shaft by the disc can be obtained by Eq. (5.253). Then from Eq. (5.243), we can get shaft end deflections $\mathbf{Y}_b$ at each bearing, which is substituted in Eq. (5.237) to get bearing forces $\mathbf{F}_b$. Alternately, bearing forces can be used directly from Eq. (5.240). Displacements and forces have the complex form; the amplitude and the phase information can be extracted from the real and imaginary parts. Amplitudes will be the modulus of complex numbers, and phase angles of all these displacements can be evaluated by calculating the arctangent of the ratio of the imaginary to real components. The solution with both bearing and shaft flexible can also be more conveniently done by FEM (refer Chapter 12) and is left to the reader to try out.

Exercise 5.32 A rotor, as shown in Figure 5.41, consists of a long stick (cylindrical in shape with diameter D_s and length b_s; it has a mass of m_s) and a thin disc (radius r_d and it has a mass of m_d) mounted at the free end of a cantilever shaft. The shaft is assumed to be connected at centre of gravity of the cylinder-disc system (it is an idealisation of the blisk – blade over disc – in turbines, there support conditions may be different) and the shaft is in no way disturbing the motion of the blisk in the overlapped portion. The shaft flexural rigidity is EI and its length is l. Consider a pure tilting motion of the blisk and obtain critical speeds expressions for the forward and backward whirls of the rotor system.

***Solution*:** In the present problem even when it is asked a pure tilting motion, we are considering more general case when both linear and angular displacements are taking place. Synchonous condition is assumed. The system can be simplified as a cantilever beam as shown in Figure 5.42.

Due to synchronous motion of combined disc and stick (blisk), we have the resultant force, as

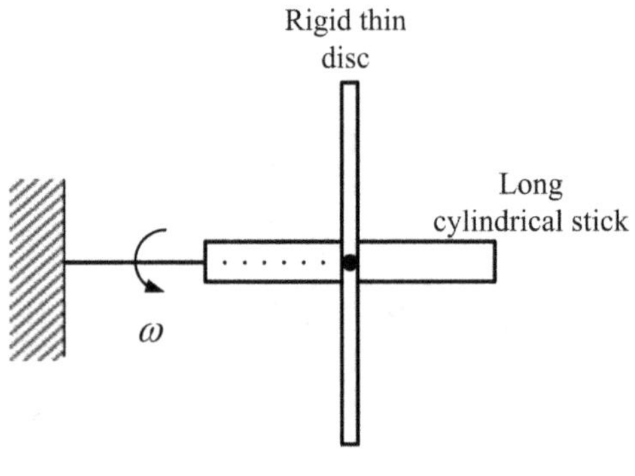

FIGURE 5.41 A cantilever shaft with a rigid thin disc and a long cylindrical stick.

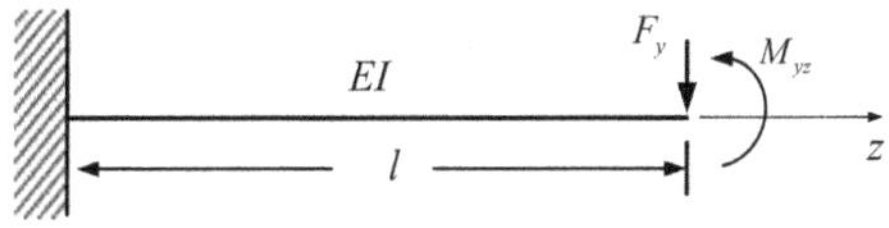

FIGURE 5.42 Cantilever rotor model.

$$F_y = m_d\omega^2\delta + m_s\omega^2\delta = m_{eq}\omega^2\delta \quad \text{with} \quad m_{eq} = m_d + m_s \tag{5.271}$$

where δ is the translatory transverse displacement and ω is the critical speed. Similarly, the resultant moment is given as

$$M_{yz} = I_{dd}\omega^2\varphi_x + I_{ds}\omega^2\varphi_x = I_{deq}\omega^2\varphi_x \quad \text{with} \quad I_{deq} = I_{dd} + I_{ds} \tag{5.272}$$

where φ_x is the rotational transverse displacement. For the cantilever end condition, forces/moments and displacements are related as (refer appendix 2.1)

$$\delta = \frac{F_y l^3}{3EI} - \frac{M_{yz}l^2}{2EI} \quad \text{and} \quad \varphi_x = \frac{F_y l^2}{2EI} - \frac{M_{yz}l}{EI} \tag{5.273}$$

On substituting Eqs. (5.271) and (5.272) into Eq. (5.273), we get

$$\delta = \frac{m_{eq}\omega^2\delta l^3}{3EI} - \frac{I_{deq}\omega^2\varphi_x l^2}{2EI} \quad \text{and} \quad \varphi_x = \frac{m_{eq}\omega^2\delta l^2}{2EI} - \frac{I_{deq}\omega^2\varphi_x l}{EI} \tag{5.274}$$

Hence, the two equations can be written as

$$\begin{bmatrix} \dfrac{m_{eq}\omega^2 l^3}{3EI} - 1 & -\dfrac{I_{deq}\omega^2 l^2}{2EI} \\ \dfrac{-m_{eq}\omega^2 \delta l^2}{2EI} & \dfrac{I_{deq}\omega^2 l}{EI} + 1 \end{bmatrix} \begin{Bmatrix} \delta \\ \varphi_x \end{Bmatrix} = \begin{Bmatrix} 0 \\ 0 \end{Bmatrix} \tag{5.275}$$

This homogeneous set of equations can have a non-trivial solution for δ and φ_x only when the determinant vanishes as

$$\begin{vmatrix} \dfrac{m_{eq}\omega^2 l^3}{3EI} - 1 & -\dfrac{I_{deq}\omega^2 l^2}{2EI} \\ \dfrac{-m_{eq}\omega^2 \delta l^2}{2EI} & \dfrac{I_{deq}\omega^2 l}{EI} + 1 \end{vmatrix} = 0 \tag{5.276}$$

which gives the frequency equation as

$$\omega^4 + \omega^2\left\{\frac{12EI}{m_{eq}I_{deq}l^3}\left(\frac{m_{eq}l^2}{3} - I_{eq}\right)\right\} - \frac{12E^2I^2}{m_{eq}I_{eq}l^4} = 0 \tag{5.277}$$

Let us define two non-dimensional parameters:

$$\bar{\omega}_{cr} = \text{non-dimensional critical speed function} = \omega\sqrt{\frac{m_{eq}l^3}{EI}} \tag{5.278}$$

$$\mu = \text{disc effect} = \frac{I_{deq}}{m_{eq}l^2} \tag{5.279}$$

Uisng Eqs. (5.278) and (5.279), Eq. (5.277) becomes

$$\bar{\omega}_{cr}^4 + \bar{\omega}_{cr}^2\left(\frac{4}{\mu} - 12\right) - \frac{12}{\mu} = 0 \tag{5.280}$$

With the solution, as

$$\bar{\omega}_{cr}^2 = \left(6 - \frac{2}{\mu}\right) \pm \sqrt{\left(6 - \frac{2}{\mu}\right)^2 + \frac{12}{\mu}} \tag{5.281}$$

The positive sign will give a positive value for $\bar{\omega}_{cr}^2$ or a real value for $\bar{\omega}_{cr}$ and the negative sign will give a complex value of the critical speed, which has no physical significance. Hence, we will have only single critical speed for a particular value of μ, so the negative sign may be ignored, to get

$$\bar{\omega}_{cr}^2 = \left(6 - \frac{2}{\mu}\right) + \sqrt{\left(6 - \frac{2}{\mu}\right)^2 + \frac{12}{\mu}} \tag{5.282}$$

Then Eq. (5.278) can be used to obtain the critical speed, ω_{cr}, of the rotor system for combined linear and angular motion of the blisk. For the completeness of the solution, for pure rotational motion ($\delta = 0$), from Eq. (5.274) the critical speed will be obtained as

$$\varphi_x = \frac{m_{eq}\omega^2\delta l^2}{2EI} - \frac{I_{deq}\omega^2\varphi_x l}{EI} \Rightarrow \varphi_x = -\frac{I_{deq}\omega^2\varphi_x l}{EI} \Rightarrow \left(1 - \frac{I_{deq}\omega^2 l}{EI}\right)\varphi_x = 0 \tag{5.283}$$

So for non-trial solution for a pure rotational motion, we have

$$\omega_{cr} = \sqrt{\frac{EI}{lI_{deq}}} \tag{5.284}$$

Exercise 5.33 Obtain transverse critical speeds of the rotor-bearing system as shown in Figure 5.43. Consider the shaft to be rigid and massless. The bearing on the left is simply supported and that on the right is supported by two springs, and each spring has the stiffness k. Let $L = 1$ m, $a = 0.3$ m, $k = 1$ kN/m, $m = 5$ kg, and $I_d = 0.02$ kg-m^2. Consider the gyroscopic couple effect.

***Solution*:** Method 1: Quasi-static method

The whole rotor (the rigid shaft and the disc) can have oscillation about left bearing. So effective diametral mass moment of inertia of the rotor system is

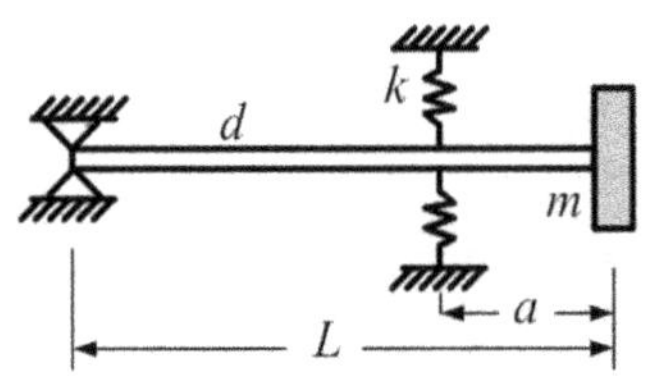

FIGURE 5.43 A rotor system flexibly supported at one end and simply supported at the other end.

$$I_{d_{eff}} = I_d + mL^2 = 0.02 + 5 \times 1^2 = 5.02 \text{ kg-m}^2 \quad (5.285)$$

The polar moment of inertia of rotor system (shaft is massless) is given as

$$I_{p_{eff}} = 2I_d = 2 \times 0.02 = 0.04 \text{ kg-m}^2 \quad (5.286)$$

The effective moment stiffness experienced by the rotor (refer Figure 5.44) is given as

$$k_{t_{eff}} = 2k(L-a)^2 = 2 \times 1 \times 10^3 \times (1-0.3)^2 = 980 \text{ N-m/rad} \quad (5.287)$$

It should be noted that k is a linear spring. For the forward critical speed for pure rotational motion, we have (refer section 5.4 of Tiwari (2017))

$$\omega_{cr}^F = \sqrt{\frac{k_{teff}}{I_{deff} - I_p}} = \sqrt{\frac{980}{5.02 - 0.04}} = 14.03 \text{ rad/s} \quad (5.288)$$

and for the backward critical speed, we have

$$\omega_{cr}^B = \sqrt{\frac{k_{teff}}{I_{deff} + I_p}} = \sqrt{\frac{980}{5.02 + 0.04}} = 13.92 \text{ rad/s} \quad (5.289)$$

Method 2: Dynamic approach

For the rotational motion of the rotor, we have

$$I_{deff}\ddot{\varphi}_y - I_p\omega\dot{\varphi}_x = -M_{zx} \quad \text{and} \quad I_{deff}\ddot{\varphi}_x + I_p\omega\dot{\varphi}_y = -M_{yz} \quad (5.290)$$

$$\text{with} \quad M_{zx} = 2k(L-a)^2\varphi_y \quad \text{and} \quad M_{yz} = 2k(L-a)^2\varphi_x \quad (5.291)$$

which gives

$$I_{deff}\ddot{\varphi}_y - I_p\omega\dot{\varphi}_x = -2k(L-a)^2\varphi_y \quad (5.292)$$

and

$$I_{deff}\ddot{\varphi}_x + I_p\omega\dot{\varphi}_y = -2k(L-a)^2\varphi_x \quad (5.293)$$

Equations (5.292) and (5.293) can be combined as

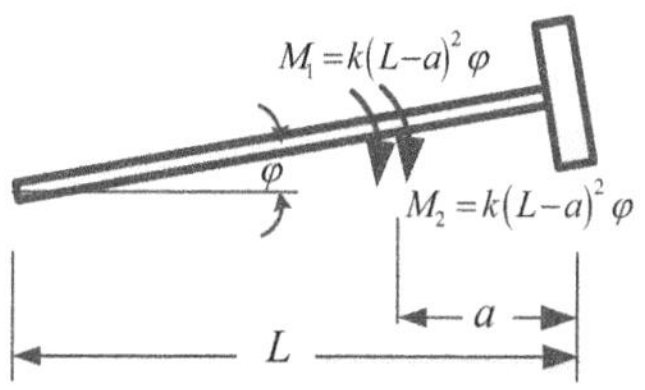

FIGURE 5.44 Free body diagram of rotor.

$$\begin{bmatrix} I_{deff} & 0 \\ 0 & I_{deff} \end{bmatrix} \begin{Bmatrix} \ddot{\varphi}_x \\ \ddot{\varphi}_y \end{Bmatrix} - \omega \begin{bmatrix} 0 & -I_p \\ I_p & 0 \end{bmatrix} \begin{Bmatrix} \dot{\varphi}_x \\ \dot{\varphi}_y \end{Bmatrix} + \begin{bmatrix} 2k(L-a)^2 & 0 \\ 0 & 2k(L-a)^2 \end{bmatrix} \begin{Bmatrix} \varphi_x \\ \varphi_y \end{Bmatrix} = \begin{Bmatrix} 0 \\ 0 \end{Bmatrix} \tag{5.294}$$

Let the solution is assumed as

$$\begin{Bmatrix} \varphi_x \\ \varphi_y \end{Bmatrix} = \begin{Bmatrix} \Phi_x \\ \Phi_y \end{Bmatrix} e^{\nu t}; \text{ so that } \begin{Bmatrix} \dot{\varphi}_x \\ \dot{\varphi}_y \end{Bmatrix} = \nu \begin{Bmatrix} \Phi_x \\ \Phi_y \end{Bmatrix} e^{\nu t} \text{ and } \begin{Bmatrix} \ddot{\varphi}_x \\ \ddot{\varphi}_y \end{Bmatrix} = \nu^2 \begin{Bmatrix} \Phi_x \\ \Phi_y \end{Bmatrix} e^{\nu t} \tag{5.295}$$

where ν is the whirl natural frequency. On substituting Eq. (5.295) into Eq. (5.294), we get

$$\nu^2 \begin{bmatrix} I_{deff} & 0 \\ 0 & I_{deff} \end{bmatrix} \begin{Bmatrix} \Phi_x \\ \Phi_y \end{Bmatrix} - \nu\omega \begin{bmatrix} 0 & -I_p \\ I_p & 0 \end{bmatrix} \begin{Bmatrix} \Phi_x \\ \Phi_y \end{Bmatrix} + \begin{bmatrix} 2k(L-a)^2 & 0 \\ 0 & 2k(L-a)^2 \end{bmatrix} \begin{Bmatrix} \Phi_x \\ \Phi_y \end{Bmatrix} = \begin{Bmatrix} 0 \\ 0 \end{Bmatrix}$$

or

$$\left(\begin{bmatrix} 2k(L-a)^2 + I_{deff}\nu^2 & I_p\nu\omega \\ -I_p\nu\omega & 2k(L-a)^2 + I_{deff}\nu^2 \end{bmatrix} \right) \begin{Bmatrix} \Phi_x \\ \Phi_y \end{Bmatrix} = \begin{Bmatrix} 0 \\ 0 \end{Bmatrix} \tag{5.296}$$

For non-trivial solution, we have

$$\begin{vmatrix} 2k(L-a)^2 + I_{deff}\nu^2 & I_p\nu\omega \\ -I_p\nu\omega & 2k(L-a)^2 + I_{deff}\nu^2 \end{vmatrix} = 0 \tag{5.297}$$

which gives

$$I_{deff}^2\nu^4 + \left\{ 4k(L-a)^2 I_{deff} + (I_p\omega)^2 \right\} \nu^2 + 4k^2(L-a)^4 = 0 \tag{5.298}$$

Equation (5.298) is the frequency equation (asynchronous motion), and can be solved to get the whirl frequency, ν, which is a function of the spin speed, ω, of the rotor. Campbell diagram (plot between ν and ω) can be drawn for given values of rotor parameters and we expect two whirl natural frequencies one belong to the forward whirl and another to the backward whirl. The intersection of these whirl frequency curves with $\nu = \omega$ line gives corresponding critical speeds.

For the forward critical speed $\nu = \omega$ (it should be noted herein that for $\nu = -\omega$ also we would get the same equation because of even power of ν is involved in frequency equation, so both forward and backward critical speeds will be expected to form the same equation, as we will see at the end of solution of this equation), so that

$$\left(I_{deff}^2 + I_p^2\right)\nu^4 + \left\{4k(L-a)^2 I_{deff}\right\}\nu^2 + 4k^2(L-a)^4 = 0 \tag{5.299}$$

which gives

$$\nu^2 = \frac{-4\left\{k(L-a)^2 I_{deff}\right\} \pm \sqrt{\left\{4k(L-a)^2 I_{deff}\right\}^2 - 4\times 4\left(I_{deff}^2 + I_p^2\right)k^2(L-a)^4}}{2\left(I_{deff}^2 + I_p^2\right)} \tag{5.300}$$

$$\text{or} \quad \nu^2 = \frac{2k(L-a)^2\left(-I_{deff} \pm I_p\right)}{\left(I_{deff}^2 + I_p^2\right)} \tag{5.301}$$

It should be noted that $I_{deff} > I_p$ for the present case since $I_{deff} = I_d + mL^2$ and $I_p = 2I_d$. Now since the first part of Eq. (5.301) (or Eq. (5.300)) is always more than the second part in magnitude, so always we will have negative roots. That will give two roots of ν and both will be imaginary, so we will get pure imaginary roots. The imaginary part will be the critical speeds of the rotor system, with the positive sign in front of square root of Eq. (5.301) as for the backward whirl (lower value) and the negative sign for the forward whirl (higher value). For the form of assumed solution and in the absence of damping it is expected that we will have pure imaginary roots of eigenvalue, ν. Hence, we have

$$\nu_F^2 = \frac{2k(L-a)^2\left(-I_{deff} - I_p\right)}{\left(I_{deff}^2 + I_p^2\right)} \quad \text{or} \quad \nu_F = j\sqrt{\frac{2k(L-a)^2\left(I_{deff} + I_p\right)}{\left(I_{deff}^2 + I_p^2\right)}} \equiv j\omega_{crF} \tag{5.302}$$

So that the forward critical speed is

$$\omega_{crF} = \sqrt{\frac{2k(L-a)^2\left(I_{deff} + I_p\right)}{\left(I_{deff}^2 + I_p^2\right)}} \tag{5.303}$$

and

$$\nu_B^2 = \frac{2k(L-a)^2\left(-I_{deff} + I_p\right)}{\left(I_{deff}^2 + I_p^2\right)} \quad \text{or} \quad \nu_B = j\sqrt{\frac{2k(L-a)^2\left(I_{deff} - I_p\right)}{\left(I_{deff}^2 + I_p^2\right)}} \equiv j\omega_{crB} \tag{5.304}$$

So that the backward critical speed is

$$\omega_{crB} = \sqrt{\frac{2k(L-a)^2\left(I_{deff} - I_p\right)}{\left(I_{deff}^2 + I_p^2\right)}} \tag{5.305}$$

On substituting values, we get

$$\omega_{crF} = 14.03 \text{ rad/s} \quad \text{and} \quad \omega_{crB} = 13.92 \text{ rad/s} \tag{5.306}$$

which is exactly the same as obtained by quasi-static methodology. In the dynamic approach, from Eq. (5.294), we have

$$\begin{bmatrix} I_{deff} & 0 \\ 0 & I_{deff} \end{bmatrix} \begin{Bmatrix} \ddot{\varphi}_x \\ \ddot{\varphi}_y \end{Bmatrix} + \omega \begin{bmatrix} 0 & I_p \\ -I_p & 0 \end{bmatrix} \begin{Bmatrix} \dot{\varphi}_x \\ \dot{\varphi}_y \end{Bmatrix}$$

$$+ \begin{bmatrix} 2k(L-a)^2 & 0 \\ 0 & 2k(L-a)^2 \end{bmatrix} \begin{Bmatrix} \varphi_x \\ \varphi_y \end{Bmatrix} = \begin{Bmatrix} 0 \\ 0 \end{Bmatrix} \tag{5.307}$$

which can be written as

$$\mathbf{M}\ddot{\boldsymbol{\varphi}} - \omega \mathbf{G}\dot{\boldsymbol{\varphi}} + \mathbf{K}\boldsymbol{\varphi} = \mathbf{0} \tag{5.308}$$

with

$$\mathbf{M} = \begin{bmatrix} I_{deff} & 0 \\ 0 & I_{deff} \end{bmatrix}; \quad G = \begin{bmatrix} 0 & -I_p \\ I_p & 0 \end{bmatrix}; \quad \mathbf{K} = \begin{bmatrix} 2k(L-a)^2 & 0 \\ 0 & 2k(L-a)^2 \end{bmatrix};$$

$$\boldsymbol{\varphi} = \begin{Bmatrix} \varphi_x \\ \varphi_y \end{Bmatrix}; \quad \mathbf{0} = \begin{Bmatrix} 0 \\ 0 \end{Bmatrix} \tag{5.309}$$

Now we will solve the same problem using the standard eigenvalue form by first converting Eq. (5.309) into the state-space form (refer chapter 12 of Tiwari (2017)). Let us define

$$\dot{\boldsymbol{\varphi}} = \mathbf{v} \tag{5.310}$$

Equation (5.308) can be rearranged as

$$\mathbf{M}\boldsymbol{\varphi} = \omega \mathbf{G}\boldsymbol{\varphi} - \mathbf{K}\boldsymbol{\varphi} \quad \Rightarrow \boldsymbol{\varphi} = \omega \mathbf{M}^{-1}\mathbf{G}\boldsymbol{\varphi} - \mathbf{M}^{-1}\mathbf{K}\boldsymbol{\varphi} \quad \Rightarrow \dot{\mathbf{v}} = -\mathbf{M}^{-1}\mathbf{K}\boldsymbol{\varphi} + \omega \mathbf{M}^{-1}\mathbf{G}\mathbf{v} \tag{5.311}$$

On combining Eqs. (5.310) and (5.311), we get

$$\begin{Bmatrix} \boldsymbol{\varphi} \\ \dot{\mathbf{v}} \end{Bmatrix} = \begin{bmatrix} \mathbf{0} & \mathbf{I} \\ -\mathbf{M}^{-1}\mathbf{K} & \omega \mathbf{M}^{-1}\mathbf{G} \end{bmatrix} \begin{Bmatrix} \boldsymbol{\varphi} \\ \mathbf{v} \end{Bmatrix} \quad \text{or} \quad \dot{\mathbf{h}} = \mathbf{A}\mathbf{h} \tag{5.312}$$

$$\text{with} \quad \mathbf{h} = \begin{Bmatrix} \boldsymbol{\varphi} \\ \mathbf{v} \end{Bmatrix}; \quad \dot{\mathbf{h}} = \begin{Bmatrix} \dot{\boldsymbol{\varphi}} \\ \dot{\mathbf{v}} \end{Bmatrix}; \quad \mathbf{A} = \begin{bmatrix} \mathbf{0} & \mathbf{I} \\ -\mathbf{M}^{-1}\mathbf{K} & \omega \mathbf{M}^{-1}\mathbf{G} \end{bmatrix} \tag{5.313}$$

For the present rotor data, we have

$$\mathbf{M} = \begin{bmatrix} 5.0200 & 0 \\ 0 & 5.0200 \end{bmatrix}; \quad G = \begin{bmatrix} 0 & -0.0400 \\ 0.0400 & 0 \end{bmatrix}; \quad \mathbf{K} = \begin{bmatrix} 980.0000 & 0 \\ 0 & 980.0000 \end{bmatrix}$$

$$\text{So that } \mathbf{A} = \begin{bmatrix} \mathbf{0} & \mathbf{I} \\ -\mathbf{M}^{-1}\mathbf{K} & \omega \mathbf{M}^{-1}\mathbf{G} \end{bmatrix} = \begin{bmatrix} 0 & 0 & 1 & 0 \\ 0 & 0 & 0 & 1 \\ -195.2191 & 0 & 0 & -0.0080\omega \\ 0 & -195.2191 & 0.0080\omega & 0 \end{bmatrix}$$

Eigenvalues of matrix **A** are obtained for various values of the spin speed one at a time. The eigenvalues come out to be pure imaginary and there are two pairs of imaginary eigenvalues. The imaginary part of the eigenvalues in the present case is the natural frequencies of the system. For example, at a spin speed of 15 rad/s, the following eigenvalues are obtained

$$-0.0000+14.0320i; \quad -0.0000-14.0320i; \quad 0.0000+13.9125i; \quad 0.0000-13.9125i$$

So the forward natural frequency is 14.0320 rad/s and the backward natural frequency is 13.9125. For the spin speed variation from 13 to 14.4 rad/s, the whirl natural frequencies are plotted in Figure 5.45, which is the Campbell diagram. The solid line is for the forward whirl and the dashed line for the backward whirl. The intersection of these curves with 45° line ($\nu = \omega$) gives two critical speeds, as

$$\omega_{crF} = 14.03 \text{ rad/s} \quad \text{and} \quad \omega_{crB} = 13.92 \text{ rad/s}$$

which are the same as those obtained from earlier methods. Figure 5.45 shows the Campbell diagram and it shows two critical speeds corresponding to the forward and backward whirls at intersection points of frequency curves with the synchronous whirl line ($\nu = \omega$).

Exercise 5.34 A motor drives a rotor having a rigid shaft and a disc at the end. The motor itself is mounted on a moment spring that resists its motion in transverse rotation direction with effective moment stiffness, $k = 2$ kN-m/rad. It is assumed that motor has no spinning motion about the rotor axis and it has rotational motion in the transverse direction about its centre of gravity. The polar mass moment of inertia of the rotor is 0.02 kg-m^2, and the diametral mass moment of inertia of the rotor and motor as a whole is 0.05 kg-m^2. Consider gyroscopic effects. Obtain critical speeds of the rotor system. If the speed of the rotor is 5,000 rpm, obtain the transverse natural frequencies of the rotor system. Compare the critical speeds of rotor without considering the gyroscopic effect.

***Solution*:** For the present case a pure rotational motion of the motor-rotor system is considered but with an asynchronous whirl (refer Figure 5.46) and refer section 5.4 of Tiwari (2017).

Following rotor-bearing data are given

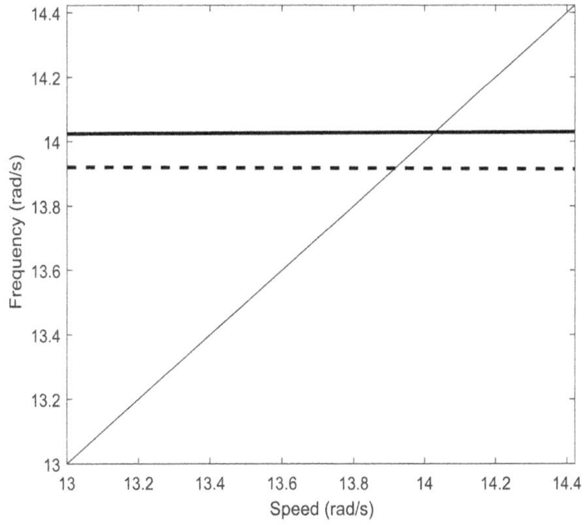

FIGURE 5.45 Campbell diagram showing two critical speeds corresponding to the forward and backward whirls at intersection points of frequency curves with the synchronous whirl line ($\nu = \omega$).

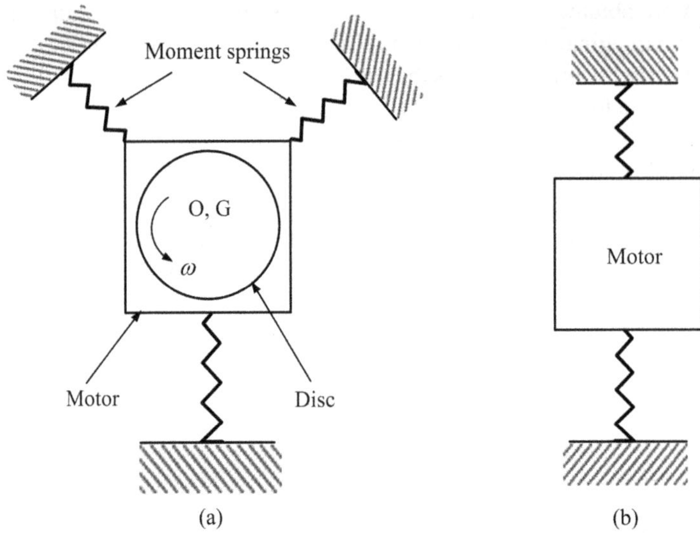

FIGURE 5.46 A motor-rotor assembly mounted on a flexible torsional support (a) end view (b) side view.

$$k_t = 2 \text{ kN-m/rad} = 2{,}000 \text{ N-m/rad}; \quad I_p = 0.02 \text{ kg-m}^2 \quad \text{and} \quad I_d = 0.05 \text{ kg-m}^2$$

If we consider the gyroscopic effect then the effective moment stiffness experienced by the motor becomes,

$$k_{eff} = (k_t \pm I_p \omega \nu) \tag{5.314}$$

where ν is the whirl frequency and ω is the spin speed of the rotor. The transverse whirl natural frequency of the rotor is given by

$$\nu^2 = \frac{k_{eff}}{I_d} = \frac{k_t \pm I_p \omega \nu}{I_d} \tag{5.315}$$

Rearranging Eq. (5.315), we get

$$\nu^2 \pm \frac{I_p \omega}{I_d} \nu - \frac{k_t}{I_d} = 0 \tag{5.316}$$

Roots of this equation are,

$$\nu_{1,2} = \pm \frac{I_p \omega}{2I_d} + \sqrt{\left(\frac{I_p \omega}{2I_d}\right)^2 + \frac{k_t}{I_d}} \tag{5.317}$$

a. For obtaining the critical speed, we have

$$\nu = \pm \omega \tag{5.318}$$

The positive sign for the synchronous whirl and the negative sign for the anti-synchronous whirl. On putting Eq. (5.318) into Eq. (5.317), we get the following expression for the critical speed,

$$\omega_{cr} = \sqrt{\frac{k_t}{I_d \mp I_p}} \tag{5.319}$$

For the forward critical speed, we have

$$\omega_{cr}^F = \sqrt{\frac{k_t}{I_d - I_p}} = \sqrt{\frac{2{,}000}{0.05 - 0.02}} = 258.20 \text{ rad/s} \tag{5.320}$$

and for the backward critical speed, we have

$$\omega_{cr}^B = \sqrt{\frac{k_t}{I_d + I_p}} = \sqrt{\frac{2{,}000}{0.05 + 0.02}} = 169.03 \text{ rad/s} \tag{5.321}$$

b. At 5,000 rpm that is at $\omega = 523.59$ rad/s, from expression (5.317), we get whirl natural frequencies, as

$$\nu_{1,2} = \pm\frac{0.02 \times 523.598}{2 \times 0.05} + \sqrt{\left(\frac{0.02 \times 523.598}{2 \times 0.05}\right)^2 + \left(\frac{2{,}000}{0.05}\right)} = \pm 104.71 + 225.75$$

which gives,

$\nu^F = 330.48$ rad/s (forward whirl frequency); $\nu^B = 121.04$ rad/s (backward whirl frequency).

c. When no gyroscopic effect is considered, both the forward and backward critical speeds remain the same. From Eq. (5.319), for $I_p = 0$, we get

$$\omega_{cr} = \sqrt{\frac{k_t}{I_d}} = \sqrt{\frac{2{,}000}{0.05}} = 200.00 \text{ rad/s} \tag{5.322}$$

which is in between the forward and backward critical speeds obtained by considering the gyroscopic effect.

Exercise 5.35 Choose a single correct answer from the multiple-choice questions.

i. In a rotor the gyroscopic moment can give rise to the instability.
   A. True B. False

   Solution: The gyroscopic moment comes in a rotor due to Coriolis component of acceleration. Its effect is seen in changing the whirl natural frequency of the system with the spin speed of the rotor. In addition to the forward critical speed, it introduces the backward critical speed also. But its effect on destablising the rotor is not observed. In fact, during the eigenvalue solution of an undamped system with the gyroscopic effect the eigenvalues are found to be pure imaginary and it is known that in unstable system the eigenvalue has positive real part. So the gyroscopic effect never leads to instability in a rotor.

ii. A cantilever rotor with a single disc and massless shaft, in a general motion, can have how many transverse natural frequencies?
   A. 1 B. 2 C. 3 D. 4 E. more than 4

Solution: In general motion with gyroscopic effect two orthogonal transverse plane motions are coupled. A disc would require two transverse translatory displacements and two rotational displacements, so total four DOFs. So we expect four transverse natural frequencies. In the absence of gyroscopic effect, two plane motions are uncoupled and because of symmetry of the shaft in each plane two natural frequencies will be there and they will be identical in both planes. So effectively only two natural frequencies will be observed for the case when the gyroscopic effect is absent.

iii. Because of the gyroscopic moment, the natural frequencies depend upon the spin speed of the rotor.
A. True B. False

Solution: The gyroscopic effect is due to Coriolis component of acceleration, which depends upon the spin speed of the rotor. While obtaining characteristic equation (or frequency equation) it contains spin speed terms along with whirl natural frequency. So with the gyroscopic effect the rotor whirl natural frequencies changes with the spin speed of the rotor.

iv. In a general motion of a unbalanced rotor, the whirl frequency and the spinning frequency (speed) are always the same.
A. true B. false

Solution: In a general motion when the gyroscopic effect is present then whirl natural frequency and spin speed are distinct. The whirl natural frequency changes with the spin speed of the rotor. These two are the same at only critical speeds.

v. Because of the gyroscopic moment, the forward whirl natural frequency of a rotor with thin disc
A. increases only B. decreases only
C. either increases or decreases D. remains constant
E. increases and decreases simultaneously.

Solution: During the forward whirl in a rotor involving a thin disc, it increases with speed. Its effect is opposite for backward whirl. Again, the effect will be reversed for a rotor with a long stick instead of a thin disc. That means for the forward whirl natural frequency decreases and backward whirl natural frequency increases.

vi. For a rotor system with gyroscopic effects and without damping, eigenvalues would be pure imaginary.
A. true B. false

Solution: The real part of an eigenvalue represents damping in the system. Since the gyroscopic effect does not impart any damping so for undamped system the eigenvalue will be pure imaginary.

vii. For a rotor system with gyroscopic effects and with damping, eigen values would be pure imaginary.
A. True B. False

Solution: The real part of an eigenvalue represents damping in the system. The gyroscopic effect does not impart any damping but for damped system the eigenvalue will be complex. The effect of the gyroscopic moment will be in imaginary part only.

viii. Because of the gyroscopic moment in the forward whirl motion of a rotor system, the effective stiffness of the shaft (or experienced by the disc) increases.

A. true B. false

Solution: For a thin disc in a rotor system due to the gyroscopic moment the disc tries to tilt less during the forward whirl, which reflects an increase in the effective stiffness. The effect is opposite for the backward whirl. For a long stick case, the forward whirl leads to decrease in effective stiffness and the backward whirl leads to increase.

ix. Because of the gyroscopic moment in the backward whirl motion of a rotor system, the effective stiffness of the shaft (or experienced by the disc) increases.

A. True B. False

Solution: Refer Exercise 5.35(viii).

x. Because of the gyroscopic moment in a rotor system with underdamping, the system will never be unstable.

A. true B. false

Solution: As the gyroscopic effect has no destabilising effect so for underdamped system the system will remain stable. The instability in rotor comes due to several effects, like the fluid-film bearing (due to cross-coupled stiffness), seals, unsymmetric shaft, asymmetry in rotor inertia, internal (or rotating) damping, and steam whirl.

xi. A Campbell diagram is a diagram of

A. the amplitude versus the rotor spin frequency
B. the phase versus the rotor spin frequency
C. both amplitude ad phase versus the rotor spin frequency
D. the whirl natural frequency versus the rotor spin frequency

Solution: In the rotor system since the whirl natural frequency is dependent on rotor spin speed due to several effects, like the gyroscopic effect and the fluid-film bearing. It is better to plot the whirl natural frequency with the rotor spin frequency to get overall variation of whirl natural frequency. This plot also can be used to get critical speeds, where the whirl natural frequency is equal to the rotor spin frequency. Such a plot is called the Campbell diagram.

xii. Because of the gyroscopic effect, in a rotor with a thin disc the backward natural whirl frequency

A. remains the same B. decreases
C. increases D. has no definite trend

Solution: During the backward whirl in a rotor involving a thin disc, it decreases with speed. Its effect is the opposite for the forward whirl. For a rotor with a long stick instead of a thin disc, again these two cases' effects will be opposite. That means for the backward whirl natural frequency increases and the backward whirl natural frequency decreases.

xiii. For a cantilever rotor with a long stick at free end (with diameter D and length b) for the pure tilting synchronous motion, the gyroscopic couple will be absent if

A. $b < D$ B. $b = D$ C. $b = \sqrt{3}D/2$ D. $b = 1.866D$

Solution: The diametral mass moment of inertia of a long stick is given as $I_d = \frac{1}{16}mD^2 + \frac{1}{12}mb^2$ and the polar moment of inertia as $I_p = \frac{1}{8}mD^2$. The gyroscopic moment for synchronous motion is given as

$$M = \omega^2\left(I_p - I_d\right)\varphi_x = \omega^2\left\{\tfrac{1}{8}mD^2 - \left(\tfrac{1}{16}mD^2 + \tfrac{1}{12}mb^2\right)\right\}\varphi_x = \omega^2\left(\tfrac{1}{8}mD^2 - \tfrac{1}{12}mb^2\right)\varphi_x \quad (5.323)$$

When $b = \sqrt{3}D/2$, the gyroscopic moment of a long stick becomes zero in Eq. (5.323).

xiv. Due to gyroscopic effects in a rotor with a thin disc
 A. frequencies of the forward and backward whirls become faster compared to the whirling frequency without gyroscopic effects
 B. frequencies of the forward and backward whirls become slower compared to the whirling frequency without gyroscopic effect
 C. frequencies of the forward and backward whirls become slower and faster, respectively, compared to the whirling frequency without gyroscopic effect
 D. frequencies of the forward and backward whirls become faster and slower, respectively, compared to the whirling frequency without gyroscopic effect.

Solution: For a thin disc rotor system, the centrifugal forces try to resist the tilt of the disc in forward whirl case and for the backward whirl it tries to tilt disc more. This results in increase in effective stiffness for the former case and decrease in the latter case. So, in the forward whirl natural frequency increases and the backward whirl natural frequency decreases. It should be noted that for the long stick rotor case, the effect will be opposite.

xv. For a general motion of a rotor with gyroscopic effects, for a special case when the elastic coupling, α, is absent, the square of the backward critical speed corresponding to the pure rotational motion would be (where the disc effect, $\mu = I_d\alpha_{22}/(m\alpha_{11})$)
 A. $1/\mu$ B. $1/\mu^2$ C. $1/(2\mu)^2$ D. $1/(3\mu)$

Solution: When the elastic coupling is absent, then the frequency equation corresponding to tilting motion becomes (eq. (5.84) of Tiwari (2017))

$$\bar{\nu}_{3,4} = \bar{\omega} \pm \sqrt{\bar{\omega}^2 + \frac{1}{\mu}} \quad (5.324)$$

For the backward critical speed, we have $\bar{\nu} = -\bar{\omega} \equiv -\bar{\omega}_{crB}$. On substituting in Eq. (5.324), we get

$$-\bar{\omega}_{crB} = \bar{\omega}_{crB} \pm \sqrt{\bar{\omega}_{crB}^2 + \frac{1}{\mu}} \quad \text{or} \quad 4\bar{\omega}_{crB}^2 = \bar{\omega}_{crB}^2 + \frac{1}{\mu} \quad \text{or} \quad \bar{\omega}_{crB}^2 = \frac{1}{3\mu} \quad (5.325)$$

xvi. For a long symmetrical rigid rotor mounted on anisotropic elastic bearings with no cross-coupling, the following mode of whirl may be assumed for analysis:
 A. Purely translation and purely conical
 B. Purely translation only
 C. Purely conical only
 D. Combination of translation and conical

Solution: For a long symmetrical rigid rotor the coupling of translational and rotational DOFs will be absent. Also since in the present case bearing cross-coupling terms are absent so the motion in two planes will be uncoupled. So, the rotor will have pure translation whirl and pure conical whirl.

xvii. While considering the gyroscopic effect in asynchronous pure rotational motion of a rotor mounted on a spring (flexible) support, if the support spring breaks suddenly, then the instantaneous whirl natural frequencies would be

A. $\nu = \pm\frac{I_d\omega}{I_p}$ B. $\nu = 0$ and $\nu = \pm\frac{I_p\omega}{I_d}$

C. $\nu = 0$ and $\nu = \pm\frac{I_d}{I_p}$ D. $\nu = \pm\frac{I_p}{I_d}$

Solution: For the present case, the frequency equation is given by (refer eq. (5.62) of Tiwari (2017))

$$\nu^2 \mp \frac{I_p\omega}{I_d}\nu - \frac{k_t}{I_d} = 0 \tag{5.326}$$

When the spring breaks then $k_t = 0$, so we get

$$\left(\nu \mp \frac{I_p\omega}{I_d}\right)\nu = 0 \quad \text{or} \quad \nu = 0 \text{ and } \nu = \pm\frac{I_p\omega}{I_d} \tag{5.327}$$

xviii. For the rotor system shown in Figure 5.47, if the gyroscopic effect is also considered, then the first forward natural frequencies as compared to those without gyroscopic effect would be

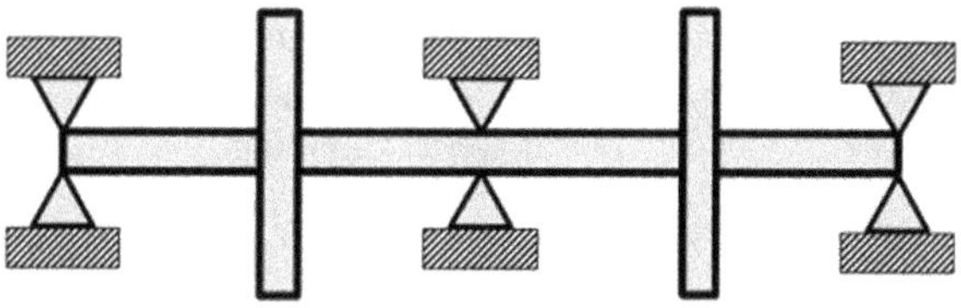

FIGURE 5.47 A multi-support rotor system.

A. remain the same B. increase
C. decrease D. increase or decrease

Solution: Since the discs are at centre of the shaft spans so in the first forward whirl, the slope at disc locations will be zero. So the disc will not have any gyroscopic effect and corresponding whirl natural frequencies will be the same as that of the rotor without gyroscopic effect.

xix. For a general motion in a single-disc rotor model with a massless elastic shaft, due to the gyroscopic effect, the total number of transverse forward critical speeds corresponding to 2× of the rotor speed is
A. 1 B. 2 C. 3 D. 4

Solution: For a general motion in a single-disc rotor model with a massless elastic shaft, due to the gyroscopic effect, the total number of transverse forward critical speeds corresponding to 2× of the rotor speed is 2 (refer figure 5.32 in Tiwari (2017)). However, corresponding to 1× of the rotor speed is 1 (eq. (5.87) of Tiwari (2017) can be used to prove this).

xx. For a long rotor with the increase in the absolute value of disc effect, the synchronous whirl frequency

A. remains the same B. increases

C. decreases D. increases then decreases

Solution: Long rotors have a similar effect as long stick. For synchronous whirl frequency is given by (eq. (5.55) of Tiwari (2017))

$$\bar{\omega}_{cr}^2 = \left(6 - \frac{2}{\mu}\right) \pm \sqrt{\left(6 - \frac{2}{\mu}\right)^2 + \frac{12}{\mu}} \tag{5.328}$$

For a long rotor μ (i.e., the disc effect) is always negative, so when the absolute value of μ is increasing that means it has a large negative value. So on taking the negative sign in Eq. (5.328), we get

$$\bar{\omega}_{cr}^2 = \left(6 + \frac{2}{\mu}\right) \pm \sqrt{\left(6 + \frac{2}{\mu}\right)^2 - \frac{12}{\mu}} \tag{5.329}$$

It should be noted that Eq. (5.328) for a positive value of μ (for the thin disc) only positive sign will give real synchronous critical speed. However, from Eq. (5.329), it can be observed that for both positive and negative sign, synchronous critical speeds will be real. That means there are two feasible synchronous critical speeds. This particular observation is made in general case of thin disc motion also. Both critical speeds decrease with increase in the negative value of μ.

xxi. Which option does not lead to instability in a rotor system?

A. gyroscopic effect B. asymmetrical shaft

C. seals D. material damping

Solution: The gyroscopic effect does not lead to instability, it is a kind of inertia term. All other factors lead to instability.

xxii. The gyroscopic couple gives

A. coupling of two orthogonal plane motions in translational displacements

B. coupling of two orthogonal plane motions in rotational displacements

C. coupling of the translational and rotational displacements in the same planes

D. coupling of the translational and rotational displacements in two orthogonal planes

Solution: The gyroscopic couple does not affect translational displacements. It couples the rotational displacements in two orthogonal transverse planes.

xxiii. A cantilever beam has a length of 0.5 m and diameter of 1 cm with a thin disc of mass 3 kg at the free end. Take Young's modulus $E = 2.1 \times 10^{11}$ N/m^2. If the disc diametral mass moment of inertia is such that its value tends to infinity, then the critical speed (rad/s) of the rotor system (while considering the gyroscopic effect) will be

A. 57.43 B. 16.6 C. 49.7 D. 99.5

Solution: The following rotor system data are given

Mass of the thin disc, $m = 3$ kg, Diameter of the shaft, $d = 0.01$ m;

Length of the shaft, $l = 0.5$ m, Young's modulus, $E = 2.1 \times 10^{11}$ N/m^2;

Second moment of area, $I = \dfrac{\pi d^4}{64} = \dfrac{\pi \times 0.01^4}{64} = 4.9087 \times 10^{-10}$ m^4;

Disc mass effect, $\mu = \dfrac{I_d}{ml^2}$, Crtical speed function, $\bar{\omega}_{cr} = \omega_{cr}\sqrt{\dfrac{ml^3}{EI}}$;

For $\mu \to \infty$, from eq. (5.46) of Tiwari (2017), we have

$$\omega_{cr} = \sqrt{\frac{12EI}{ml^3}} = \sqrt{\frac{12\times 2.1\times 10^{11}\times 4.9087\times 10^{-10}}{3\times 0.5^3}} = 57.43 \text{ rad/s} \tag{5.330}$$

Additionally, for a point mass $\mu = 0$, from eq. (5.45) of Tiwari (2017), we have

$$\omega_{cr} = \sqrt{\frac{3EI}{ml^3}} = \sqrt{\frac{3\times 2.1\times 10^{11}\times 4.9087\times 10^{-10}}{3\times 0.5^3}} = 28.72 \text{ rad/s} \tag{5.331}$$

If the radius of the disc is given then we can calculate diametral mass moment of inertia of the disc $I_d = \frac{1}{4}mr^2$, which can be used to obtain the disc mass effect, $\mu = \dfrac{I_d}{ml^2}$. Then the synchronous whirl speed (or crtical speed) is given as

$$\bar{\omega}_{cr}^2 = \left(6 - \frac{2}{\mu}\right) \pm \sqrt{\left(6 - \frac{2}{\mu}\right)^2 + \frac{12}{\mu}} \quad \text{with} \quad \bar{\omega}_{cr} = \omega_{cr}\sqrt{\frac{ml^3}{EI}} \tag{5.332}$$

which can be written as

$$\omega_{cr}^2 \frac{ml^3}{EI} = \left(6 - \frac{2}{\mu}\right) \pm \sqrt{\left(6 - \frac{2}{\mu}\right)^2 + \frac{12}{\mu}} \quad \text{or} \quad \omega_{cr} = \sqrt{\frac{EI}{ml^3}\left\{\left(6 - \frac{2}{\mu}\right) + \sqrt{\left(6 - \frac{2}{\mu}\right)^2 + \frac{12}{\mu}}\right\}} \tag{5.333}$$

xxiv. For a cantilever massless elastic shaft with a thin balanced rigid disc at the free end undergoing general asynchronous motion (both translational and rotational transverse motion with gyroscopic effect included), if the whirl frequency $\nu = 1/\sqrt{m\alpha_{11}}$ and $\mu = 1$ then the corresponding rotor speed, ω, would be

A. $1/\sqrt{m\alpha_{11}}$ B. $0.5/\sqrt{m\alpha_{11}}$

C. $2/\sqrt{m\alpha_{11}}$ D. $1/\sqrt{m\alpha_{22}}$

Solution: When we have $\nu = 1/\sqrt{m\alpha_{11}}$, we will have the dimensionless frequency (refer eq. (5.77) of Tiwari (2017))

$$\bar{\nu} = \nu\sqrt{m\alpha_{11}} = \sqrt{m\alpha_{11}}/\sqrt{m\alpha_{11}} = 1 \tag{5.334}$$

Also for a cantilever rotor, we have $\bar{\alpha} = \frac{3}{4}$ (refer page 220 of Tiwari (2017)). The frequency equation for a general motion is given as

$$\bar{\nu}^4 - 2\bar{\omega}\bar{\nu}^3 + \frac{\mu+1}{\mu(\bar{\alpha}-1)}\bar{\nu}^2 - \frac{2\bar{\omega}}{\bar{\alpha}-1}\bar{\nu} - \frac{1}{\mu(\bar{\alpha}-1)} = 0 \tag{5.335}$$

On substituting Eq. (5.334) into Eq. (5.335), we get

$$1 - 2\bar{\omega} + \frac{1+1}{1\times(\frac{3}{4}-1)} - \frac{2\bar{\omega}}{\frac{3}{4}-1} - \frac{1}{1\times(\frac{3}{4}-1)} = 0 \quad \text{or} \quad \bar{\omega} = \tfrac{1}{2} \tag{5.336}$$

which gives

$$\omega = \frac{0.5}{\sqrt{m\alpha_{11}}} \tag{5.337}$$

Alternatively, from eq. (5.87) of Tiwari (2017), we have

$$\bar{\omega} = \frac{\bar{v}^4 - 8\bar{v}^2 + 4}{2\bar{v}^3 - 8\bar{v}} \tag{5.338}$$

On substituting $\bar{v} = 1$, we get $\bar{\omega} = 0.5$.

xxv. A single-disc cantilever rotor will have four distinct natural frequencies
A. when the disc is a point mass
B. when the disc has appreciable gyroscopic effects
C. when the disc (stick) is long with negligible gyroscopic effects
D. when the disc is thin with negligible gyroscopic effects

Solution: A single-disc cantilever rotor has four DOFs. But if gyroscopic couple is absent (i.e., for disc as a point mass), then two plane motion are uncoupled and because of symmetry of the shaft on two distinct natural frequencies are observed. However, in the presence of gyroscopic effect two plane motions get coupled and four distinct whirl natural frequencies are observed.

xxvi. For a synchronous and pure tilting whirl of a long rotor (stick) for $b = 0.866\ D$ (where D is the diameter and b is the length of the stick), which effect will be absent?
A. whirl frequency B. critical speed
C. gyroscopic effect D. whirling

Solution: For a long rotor (stick) for $b = 0.866\ D$, the gyroscopic moment is zero (refer eqs. (5.52) and (5.53) in Tiwari (2017)). This is because the difference between two the mass moments inertia in two principal axis directions is zero. Refer Exercise 5.35(xiii) also.

xxvii. A cantilever shaft carries a spherical (instead of a point mass) mass as a tip mass. Which effect will be absent?
A. whirl frequency B. whirling
C. critical speed D. gyroscopic effect

Solution: For a spherical mass the difference between two mass moments inertia in two principal axis directions is zero. So, the gyroscopic effect will be zero at the first forward critical speed. Refer Exercise 5.35(xiii) also.

xxviii. A cantilever massless shaft with a tip mass (thin disc), on considering the gyroscopic effect, would have number of whirl natural frequency equal to
A. 1 B. 2 C. 3 D. 4

Solution: A single-disc cantilever rotor has four DOFs. But if gyroscopic couple is absent then two plane motion are uncoupled and because of symmetry of the shaft on two distinct natural frequencies are observed. However, in the presence of gyroscopic effect two plane motions get coupled and four distinct natural frequencies are observed. A point mass will have two DOFs in planar motion but due to symmetry of the shaft only one whirl natural frequency is observed. Since the point mass will not be having the gyroscopic effect so two plane motions will not be coupled as in the thin disc case.

xxix. The Campbell diagram is a

A. forced vibration response plot
B. free vibration response plot
C. whirl natural frequency plot
D. transient vibration response plot

Solution: The Campbell diagram is a whirl natural frequency plot with the spin speed of the shaft. Often it is useful when we have a variation of the whirl natural frequency with the spin speed. It is used to obtain the critical speed of the rotor system.

xxx. In a cantilever rotor, if $\alpha_{12} = y/M_y$ and $\alpha_{21} = \varphi_x/F_x$, then

A. $\alpha_{12}/\alpha_{21} = 1$
B. $\alpha_{12}\alpha_{21} = 1$
C. $\alpha_{12}/\alpha_{21} = -1$
D. $\alpha_{12}\alpha_{21} = -1$

Solution: Due to the cross equality, we have $\alpha_{12} = \alpha_{21}$. So, it will give $\alpha_{12}/\alpha_{21} = 1$.

xxxi. The effective stiffness of a rotor system during the backward whirl due to the gyroscopic effect has a tendency to

A. increase
B. remain same
C. decrease
D. either increase or decrease

Solution: The effective stiffness of a rotor system during the backward whirl due to the gyroscopic effect has a tendency to decrease for the case of thin disc. The long stick has the opposite effect.

xxxii. Because of the gyroscopic couple, the natural frequencies depend upon the spin speed of the shaft.

A. true
B. false

Solution: Due to the gyroscopic couple, the coupling of two transverse plane motion takes place and it affects the whirl natural frequency of the rotor system. The gyroscopic effect comes due to the Coriolis component of acceleration, which depends upon the spin speed of the rotor. Hence, the whirl natural frequency depends on the spin speed of the rotor.

xxxiii. In a general case of a perfectly balanced rotor motion, the whirl frequency and the spinning frequency (speed) are always the same.

A. true
B. false

Solution: The whirl frequency and the spinning frequency for a balanced rotor will be different when the gyroscopic effect is predominant. When these two are same, then the rotor will be rotating at its critical speed.

xxxiv. For a long stick supported at the free end of a flexible cantilever shaft, the length-to-diameter ratio of the long stick is equal to $\sqrt{3}/2$. The gyroscopic couple onto the shaft, due to the long stick, will

A. try to prevent the angular displacement of the shaft
B. try to help the angular displacement of the shaft
C. have no effect on the angular displacement of the shaft
D. none of the above

Solution: For the length-to-diameter ratio of the long stick is equal to $\sqrt{3}/2$ (refer eqs. (5.52) and (5.53) in Tiwari (2017)), there will not be any gyroscopic effect so there will have no effect on the angular displacement of the shaft due to gyroscopic couple.

xxxv. For the rotor system shown in Figure 5.48 (consider the polar moments of inertia of the end discs only and they have opposite rotation), which effect would be absent?

A. whirl frequency B. whirling

C. critical speed D. gyroscopic effect

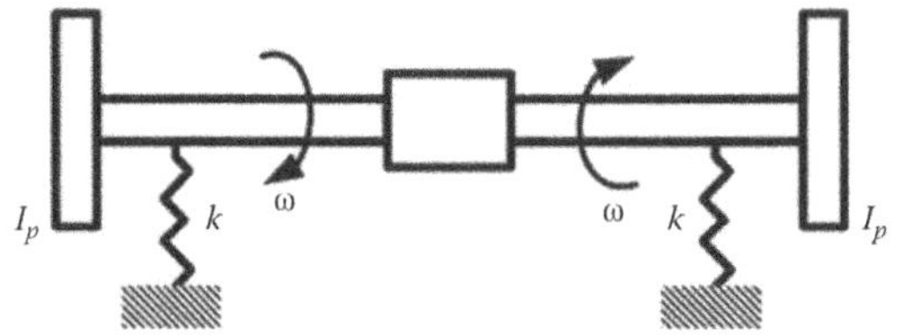

FIGURE 5.48 A rigid symmetric rotor mounted on identical flexible bearings.

Solution: Since the shaft is rigid and discs are identical but rotating in opposite direction hence the net gyroscopic moment on the rotor will be zero. This is a good way to eliminate the gyroscopic effect.

xxxvi. The whirl natural frequency of a rotor system is the frequency of oscillation of

A. a non-spinning rotor during free vibration
B. a spinning rotor during free vibration
C. a non-spinning rotor during forced vibration
D. a spinning rotor during forced vibration

Solution: The whirl natural frequency of a rotor system is the frequency of oscillation of a spinning rotor during its free vibration. If unbalance is present in rotor, then the whirl frequency is always equal to the spin speed of the shaft. When the spin speed coincides with the whirl natural frequency then the resonance takes place and it is called the critical speed.

xxxvii. The whirl frequency of a rotor system due to unbalance will be equal to

A. the spin speed of the shaft
B. the whirl natural frequency of the spinning rotor system
C. the natural frequency of the non-spinning rotor system
D. the critical speed of the rotor system

Solution: The whirl frequency of a rotor system due to unbalance will be equal to the spin speed of the shaft. This will be due to the reason that the rotor system has forced vibration in the presence of unbalance and the unbalance force has a forcing frequency equal to the spin speed of the rotor.

xxxviii. In a balanced rotor system, a free-wheel has an unbalance and is mounted on the shaft. The free-wheel is a disc, which is instead of fixed to the shaft directly. It is mounted on the shaft through a rolling bearing so that it can rotate freely, independent of shaft rotation. If free-wheel is rotated by a separate drive in the opposite sense of rotation of the shaft, on increasing the speed of free-wheel it will excite, which critical speed?

A. forward critical speed
B. backward critical speed
C. both forward and backward critical speeds
D. neither forward nor backward critical speed

Solution: Since the unbalanced force of the free-wheel has an opposite direction to that of the balanced rotor system, hence the rotor system will encounter backward critical speeds while increasing the speed of the free-wheel. It is assumed here that the gyroscopic effect is predominant in the rotor system. In the identification of bearing dyanamic parameters such excitation helps in improving the estimates of the parameters (refer chapter 14 of Tiwari (2017)).

xxxix. A Campbell diagram is a plot between

A. the whirl natural frequency versus the spin speed of a rotor
B. the critical speed versus the spin speed of a rotor
C. the whirl natural frequency versus the critical speed of a rotor
D. the amplitude of the rotor displacement versus the spin speed of a rotor

Solution: A Campbell diagram is a plot between the whirl natural frequency versus the spin speed of a rotor. It is useful when the gyroscopic effect is predominant and also for rotors mounted on fluid-film bearings, which has speed-dependent rotor dynamic properties. In such cases, the whirl natural frequency is also rotor spin speed dependent.

xl. The gyroscopic effect occurs due to

A. centrifugal force
B. Coriolis component of acceleration
C. unbalance
D. shear force

Solution: The gyroscopic effect occurs due to Coriolis component of acceleration due to simultaneous spinning and precession of rotor in two transverse orthogonal planes.

xli. In a Jeffcott rotor (with a disc at the mid-span), if the disc has no radial unbalance but it is tilted by some angle in a transverse plane, then during spinning, the critical speed excited will be (consider no gyroscopic effect with E is Young's modulus, I is the second moment of area, m is the mass of the disc, I_d is the diametral mass moment of inertia of the disc and l is the length of the shaft)

A. $\sqrt{\dfrac{12EI}{I_d l}}$ B. $\sqrt{\dfrac{48EI}{I_d l}}$ C. $\sqrt{\dfrac{12EI}{ml^3}}$ D. $\sqrt{\dfrac{48EI}{ml^3}}$

Solution: The tilt of the disc will give a moment and due to this pure tilting motion of the disc will take place. For a simple support, the influence coefficient, α_{22}, for the tilting, is given as

$$\alpha_{22} = -\frac{\left(3al - 3a^2 - l^2\right)}{3EIl} \tag{5.339}$$

For pure tilting, $a = l/2$, so we have

$$\alpha_{22} = \frac{l}{12EI} \tag{5.340}$$

So that the natural frequency in the pure tilting motion is given as

$$\omega_{nf} = \sqrt{\frac{1}{I_d \alpha_{22}}} = \sqrt{\frac{12EI}{I_d l}} \tag{5.341}$$

xlii. In a Jeffcott rotor (with a disc at the mid-span), if the disc has pure tilting motion in a transverse plane during spinning, then the critical speed will be (consider gyroscopic effect with $\mu = \frac{I_d \alpha_{22}}{m\alpha_{11}}$ is the disc effect)

A. $\frac{1}{\sqrt{3\mu}}$ B. $\frac{1}{2\sqrt{3\mu}}$ C. $\frac{1}{\sqrt{2\mu}}$ D. $\frac{1}{2\sqrt{\mu}}$

Solution: From eq. (5.82) of Tiwari (2017) for a pure tilting motion, we have

$$\bar{v}^2 - 2\bar{\omega}\bar{v} - \frac{1}{\mu} = 0 \tag{5.342}$$

For $\bar{\omega} = -\bar{v}$, it gives a feasible backward critical speed, as

$$\bar{\omega}_{cr}^2 + 2\bar{\omega}_{cr}^2 - \frac{1}{\mu} = 0 \quad \text{or} \quad \bar{\omega}_{cr} = \sqrt{\frac{1}{3\mu}} \tag{5.343}$$

For $\bar{\omega} = \bar{v}$, it gives an infeasible forward critical speed, $\bar{\omega}_{cr}^2 - 2\bar{\omega}_{cr}^2 - \frac{1}{\mu} = 0 \Rightarrow \bar{\omega}_{cr} = \sqrt{-\frac{1}{\mu}}$.

xliii. In a Jeffcott rotor (with a disc at the mid-span), if the disc has a pure tilting motion in a transverse plane during spinning, then the feasible critical speed will be

A. both forward and backward critical speeds
B. backward critical speed
C. forward critical speed
D. neither forward nor backward critical speeds

Solution: Refer solution of Exercise 5.35(xliii).

xliv. For a general motion of a single disc rotor, the gyroscopic moment is given as (where I_d is the diametral mass moment of inertia, I_p is the polar mass moment of inertia, φ_x is the angular tilt of the disc, ω is the rotor spin speed and ν is the whirl natural frequency)

A. $I_d\varphi_x(2\omega - \nu)\nu$ B. $I_p\varphi_x(2\omega - \nu)\nu$
C. $I_d\varphi_x(\omega - 2\nu)\omega$ D. $I_p\varphi_x(\omega - 2\nu)\omega$

Solution: Refer eq. (5.70) of Tiwari (2017); the gyroscopic couple is given as $I_d\varphi_x(2\omega - \nu)\nu$. It should be noted that for the synchronous whirl ($\nu = \omega$), the gyroscopic moment will be $I_d\varphi_x\omega^2$.

xlv. In a spinning rotor (the diametral mass moment of inertia of the rotor is three times its polar mass moment of inertia with $I_p = 1$ kg-m^2) in a pure tilting motion and supported on a flexible support of moment stiffness of 20 kN-m/rad) will have a forward transverse critical speed (in rad/s) as (consider gyroscopic effect

A. 100.00 B. 200.00 C. 81.65 D. 70.71

Solution: The forward critical speed for a pure tilting motion (refer eq. (5.62) of Tiwari (2017)), while considering the gyroscopic effect, is given as

$$\omega_{cr}^F = \sqrt{\frac{k_t}{I_d - I_p}} = \sqrt{\frac{20,000}{3-1}} = 100 \text{ rad/s.}$$

xlvi. In a spinning rotor (the diametral mass moment of inertia is two times its polar mass moment of inertia with $I_p = 1$ kg-m^2) in a pure tilting motion and supported on a flexible support of the moment stiffness of 20 kN-m/rad, will have a transverse critical speed (in rad/s) as (ignore the gyroscopic effect)

A. 100.00 B. 200.00 C. 81.65 D. 70.71

Solution: The critical speed of the rotor for a pure tilting motion, while ignoring the gyroscopic effect, is given as $\omega_{cr} = \sqrt{\dfrac{k_t}{I_d}} = \sqrt{\dfrac{20{,}000}{2}} = 100$ rad/s.

FINAL REMARKS

In this chapter, the most important aspect, the gyroscopic effect in rotors, is considered. Mainly, the focus is on the quasi-static method with simple rotors but for varied boundary conditions. Few examples with dynamic approaches are also covered. When the gyroscopic effect is considered, then the whirl natural frequency depends upon the spin speed of the rotor. It also split the whirl natural frequencies into the forward and backward whirls. These variations of the whirl natural frequencies with the rotor spin speed are shown in the Campbell diagram, which is used to get the critical speed of the rotor system. Apart from a general motion, pure rotational motion is also considered both by the quasi-static and dynamic methods. In a few simple cases, forced vibration due to the initial tilt of the disc is also considered. Several minute concepts are cleared in MCQs, which looks the same concept with different perspective to have much deeper understanding of them. Chapters 8 and 10 of Tiwari (2017) with gyroscopic effect by the transfer matrix method (TMM) and the finite element method (FEM), respectively, are covered for more a complex analysis.

REFERENCE

Tiwari, R., 2017, *Rotor Systems: Analysis and Identification*. Boca Raton, FL: CRC Press.

ANSWERS TO MCQs

Exercise 5.35

i. B	ii. D	iii. A	iv. B	v. A	vi. A
vii. B	viii. A	ix. B	x. A	xi. D	xii. B
xiii. C	xiv. D	xv. D	xvi. A	xvii. B	xviii. A
xix. B	xx. C	xxi. A	xxii. B	xxiii. A	xxiv. B
xxv. B	xxvi. C	xxvii. D	xxviii. D	xxix. C	xxx. A
xxxi. C	xxxii. A	xxxiii. B	xxxiv. C	xxxv. D	xxxvi. B
xxxvii. A	xxxviii. B	xxxix. A	xl. B	xli. A	xlii. A
xliii. B	xliv. A	xlv. A	xlvi. A.		

Index

For Product Safety Concerns and Information please contact our EU representative GPSR@taylorandfrancis.com
Taylor & Francis Verlag GmbH, Kaufingerstraße 24, 80331 München, Germany

www.ingramcontent.com/pod-product-compliance
Lightning Source LLC
Chambersburg PA
CBHW061349210326
41598CB00035B/5935
9781032638201